DATE DUE

MATHEMATICS
A Survey of Its Foundations

Fredric N. Misner

Ulster County Community College

MATHEMATICS

A Survey of Its Foundations

Canfield Press San Francisco

A Department of Harper & Row, Publishers, Inc.
New York   Evanston   London

Production Editor: Pat Brewer
Designer: Paula Tuerk
Copy Editor: Lionel Gambill
Photographer: Charles Marut

Library of Congress Cataloging in Publication Data

Misner, Fredric N   1936–
  Mathematics: a survey of its foundations.

  Includes index.
  1. Mathematics—1961–   I. Title.
QA39.2.M57   510   74–31370
ISBN 0–06–385480–5

76   77   10 9 8 7 6 5 4 3 2

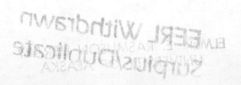

CONTENTS

PREFACE

Mathematics: A Survey of its Foundations developed from a conviction that mathematics is a human creation. In spite of the dark feelings some may have for the subject, everyone *can* understand and *can* appreciate the beauty of this creation. Having tangled with fractions, long division, and word problems, many of us come to believe that mathematics is only calculation—that higher mathematics, is, therefore, higher calculation. The mathematician becomes a lofty and unapproachable figure. Though the ranks of great mathematicians are filled with men and women of unusual genius, all of us are mathematicians to some degree. Every time we straighten out a restaurant check or play chess or bridge, we are mathematicians. We may never master calculus, but we are rational and logical creatures, which gives us a claim to that title.

This text is designed for students whose fields of concentration lie outside the physical sciences. Such students may be majoring in the humanities or social sciences, or they may be pursuing degrees in biology, business, or education. This book is appropriate for a variety of courses in which the goal is to develop an up-to-date awareness of mathematical concepts, rather than technical prowess.

The book is adaptable to courses of varying lengths. An instructor may design a short course (one semester, one or two quarters) by selecting those chapters most appropriate for the course and students. However, there is sufficient material for a two-semester course.

My experience has been that many students discover a latent interest and ability in one or more of the topics covered in a basic mathematics survey course. The chapters on algebra, probability, statistics, and computers are particularly useful as an introduction to other related courses in the curricula. To succeed with this book, the only prerequisite is an open mind. No specific mathematical expertise is assumed beyond what is already a part of everyone's thinking. No elaborate theorems are proved. Yet the process of proof is dealt with as an essential part of mathematics. The historical component of the text is important, because today's mathematics is built on the work of people who lived ten, twenty, a hundred, and several hundreds of years ago. Mathematics, like art and music, has a great heritage.

A novel aspect of the text are the readings. Selections are included that span a period of several hundred years. They range from the serious and deeply philosophical to the light and comical. A few minutes with them should convince you that they are *must* reading.

The exercises are fresh and original. Not only do they provide practice with the material developed in the section, but they also explore related ideas and prepare the student for later sections. They are carefully graded; those questions which are more challenging are preceded with an asterisk. Students are encouraged to use calculators or computer terminals where it is appropriate to do so. There are suggestions in some exercise sets for library research on interesting related topics which can result in brief reports to class. These references will provide background material to the chapter topics and introduce the student to the life and works of famous scientists and mathematicians.

An additional reading list is found at the end of the text. The chapter readings, the references in the exercise sets, and the additional readings, will all help to broaden the scope of the text and will serve to heighten interest in the material.

The publication of this book would not have been possible without near superhuman support from my friends and colleagues. I particularly want to thank Grace Shields for her superb typing. I must also express my gratitude to Professors Alvin Vaughn, Roger Yetzer, John Mikalauskas, and Ed Peifer of the Ulster Community College mathematics department and to Walter Bartlett and Steve Hilsenbeck of our computer center for reading portions of the manuscript. Special kudos go to Ann Ludwig and Pat Brewer of Canfield Press for their encouragement and help in seeing this project through.

Fredric N. Misner
Stone Ridge, New York
January, 1975

I had a feeling once about Mathematics—that I saw it all. Depth beyond Depth was revealed to me—the Byss and the Abyss. I saw—as one might see the transit of Venus or even the Lord Mayor's show—a quantity passing through infinity and changing its sign from plus to minus. I saw exactly how it happened and why the tergiversation was inevitable—but it was after dinner and I let it go.

Winston S. Churchill

1 FUNDAMENTAL CONCEPTS OF NUMBER AND COUNTING

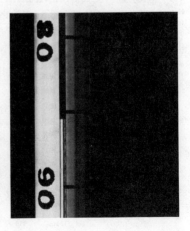

Did you ever wonder why we use the digits 0, 1, 2, 3, 4, 5, 6, 7, 8, and 9 to count? Could some shorter or longer list be used? Why does the counting system employ a zero? Questions like these motivate us to begin a survey of mathematics book at the beginning, that is, with counting.

In addition, the counting numbers provide the basis for a great part of mathematics. They were used to develop other kinds of numbers, like fractions and decimals. A discussion of counting numbers will provide a framework for our study of other numbers.

1.1 A COUNTING PROBLEM

Suppose you were to transport yourself to an early period in the history of man, say 5000 B.C. The counting system we use today did not exist in 5000 B.C. Your time machine sets you down in a small village, and the first person you meet is a young shepherd. He tells you about his flock, expressing the size of it by holding up three fingers. A few days later his grandfather dies and leaves him a large flock of sheep. The boy is now wealthy. He wants to know how wealthy he is.

The boy and his tribe probably had methods of tallying sets of objects like flocks of sheep. We can imagine that he might chisel a mark on a stone or whittle a mark on his crook, one mark for each sheep. Men could compare the marks to compare their wealth.

But you have the benefit of great mental sophistication, although we'll assume that you have forgotten all specific knowledge of numbers and counting. You have paper and pencil with you, and you decide to try to improve upon the local counting system.

How would you invent a counting method that would keep track of fairly large sets (or collections) of objects? Let us suppose that the young shepherd now had 57 sheep altogether. A neighbor has 53 sheep.

You might first make a mark on your paper, one for each sheep. That is an improvement over carving or whittling because of the time saved. You also improve their method by putting one row of marks above another and equal spacing between the marks. The resulting lists of marks might look like this:

Shepherd ///
 boy
Neighbor ///

The boy's list is longer, so he is richer.

But so many marks are hard to keep straight, and you decide to improve the lists again, this time by grouping marks. You decide to make the number of marks in a set correspond to the fingers of one hand.

Shepherd ///// ///// ///// ///// ///// ///// ///// /////
   boy   ///// ///// ///// //
Neighbor ///// ///// ///// ///// ///// ///// ///// /////
         ///// ///// ///

Now the comparison of the two lists is the same, but listing the marks in sets is easier.

Soon you realize that you have started something; all the tribesmen want you to "mark" for them. You fear you are going to run out of paper, so you think of alternatives; invent paper, or mark smaller, or replace "/////" with some other symbol! You quickly select the last option, since the invention of paper poses worse problems.

You choose to replace "/////" by "☆" because the tribesmen had a word for ☆ and because this new symbol was constructed from the same number of marks. The new comparisons are

    Shepherd boy ☆ ☆ ☆ ☆ ☆ ☆ ☆ ☆ ☆ ☆ //
    Neighbor   ☆ ☆ ☆ ☆ ☆ ☆ ☆ ☆ ☆ ///

You teach the tribesmen that "'hand' marks is one star," in their own language of course.

Then some bright young fellow catches you at your own game and says "What if I had 'hand' stars?" You realize in a flash that you could really conserve paper. You answer quickly "Well, that would be a moon, since the moon is bigger than a star" (or so it seemed then). The new lists:

    Shepherd boy   O O ☆ / /
    Neighbor   O O / / /

Everyone decides that you have gone far enough, for now it looks as if both are equally rich since "hand" objects are used in describing each set of sheep. You settle down to a lifetime of teaching the difference between O O ☆ / / and O O / / / in the public schools. You become a local hero, universally revered, except for a few who fear the new math and persist in whittling their crooks!

Exercise Set 1.1

1 In the first marking lists comparing the shepherd boy's wealth to his neighbor's, we said it was clear that the boy was richer than his neighbor. Why was it clear?

2 Suppose, for some reason, you drew the star like this: ⬡
  (a) What would have been the proper grouping of the marks?
  (b) You probably would have defined "//////" stars to be a moon. How would you have represented the shepherd boy's set of sheep then?
  (c) Would there have been any advantage to this system in representing 57 sheep? Explain.

3   If we are comparing two sets of objects by marks, say these two,

$$///////$$
$$/////$$

why is the equal spacing, and mark-above-mark advantageous?

4   Under the definitions of the symbols given in the text, what number would each of the following represent?
(a) O ☆ / /   (b) O O O / / / /   (c) O ☆ O

5   In exercise 4(c) we could have written O O ☆ since no meaning has been assigned to the order of the symbols. If we decided that smaller-valued signs to the left of larger-valued signs means to subtract (for example, ☆ O means 25 − 5, or 20, but O ☆ means 25 + 5, or 30), then how would you write each of the following:
(a) 21   (b) 26   (c) 4   (d) 99   (e) 15   (f) 39

°6   Read and report to class on "The Ability of Birds to Count" by O. Koehler in James R. Newman's *The World of Mathematics*, Volume I (New York: Simon & Schuster, 1956).

Note   Exercises and sections marked with an asterisk are optional. They can be omitted without interrupting the sequence of later material. Some of these exercises and sections are more challenging, and some are extensions of lesson material.

1.2 COUNTING AND NUMERATION SYSTEMS IN HISTORY

Anthropologists have recently discovered tribes that have no counting systems. For these people, there apparently is no need for numbers beyond "one," "two," or "three"; larger quantities are simply referred to as "many." The problem of keeping track of possessions is left to medicine men or other specialists. Some tribes, that have no very highly advanced written or oral language, count on fingers, toes, or other parts of the body. Pointing to fingers, to parts of the arm, or to locations on the head, is still the essence of the counting system practiced by Sibiler tribesmen of New Guinea.

Babylonian clay tablets, dating from 2400 B.C., show the use of the wedge-shaped symbols in Figure 1. In this system,

would be 134, since ▼► meant 100, ◄◄◄ meant 30, and ▼▼▼/▼ meant 4. It was understood that the 100, 30, and 4 were to be added. The Babylonians also used a sequence of symbols that implied multiplication.

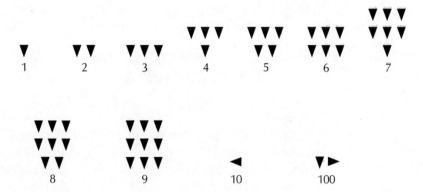

Figure 1 · Babylonian cuneiform numerals

Look at the following example.

EXAMPLE 1

◄ ▼► means "10 times 100," or "1000."
▼► ◄ means "100 plus 10," or "110."

It is not our desire to become experts at working with cuneiform symbols. But we can see the disadvantages when only one symbol, the "wedge," is used. Not only are sequential symbols cumbersome, but they also require precise spacing to make their meaning clear. Sometimes sequential symbols implied addition and sometimes they implied multiplication.

The Egyptian hieroglyphic system of 3300 B.C. employed several symbols and an additive principle for writing numbers. See Figure 2.

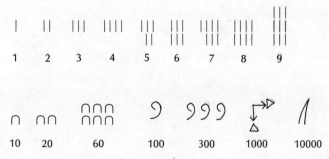

Figure 2   Egyptian hieroglyphic numerals

EXAMPLE 2

The number 23 was written ∩ ∩ ||| where ∩ ∩ meant 20 and ||| meant 3.

EXAMPLE 3

୨୨ ∩ ∩ ∩ | | | | stood for 234.

Before 600 B.C., the Greeks used a system called Herodianic signs. Over the next 200 years they changed to a system of symbols borrowed from their alphabet; see Figure 3. The Herodianic system had only a few symbols, but the later alphabetic system used 27 different symbols for representing numbers from 1 to 900.

EXAMPLE 4

The number 234 was written σλδ by adding the values of σ, λ, and δ. By comparing σλδ with the nine-symbol representation of the same number in example 3, you can see that the Greek system for writing large numbers is much more compact than the Egyptian.

EXAMPLE 5

The symbols ϕπθ stood for 589, where the values of ϕ, π, and θ are added.

All symbols for numbers, even if they are wedge-shaped, strokes, or alphabetic characters, are called *numerals*. The system of writing numbers using the basic numerals is called a *numeration system*.

The numeration system we use is called the Hindu-Arabic system. Both the Hindus and the Arabs made contributions to its development. There is evidence that the Hindus had adopted the system by the year A.D. 600, but it was not standard for the Arabs until the ninth or tenth century and was not common in Europe until about the late fifteenth century.

Our Hindu-Arabic numeration system is far superior to anything previously used to represent numbers. It uses only ten symbols (including zero) and incorporates the simple idea of positional value. "Zero" and "positional value" will be explored in Sections 1.4 and 1.5.

α	β	γ	δ	ϵ	ς	ζ	η	θ
1	2	3	4	5	6	7	8	9
ι	κ	λ	μ	ν	ξ	o	π	ϟ
10	20	30	40	50	60	70	80	90
ρ	σ	τ	υ	ϕ	χ	ψ	ω	ϡ
100	200	300	400	500	600	700	800	900

Figure 3   Greek alphabetic numerals

Exercise Set 1.2

1 Use Babylonian cuneiform numerals (Figure 1) to express each of the following numbers:

(a) 11   (b) 15   (c) 20   (d) 40   (e) 110

2 What Hindu-Arabic number does each of the following Egyptian hieroglyphic numerals represent (Figure 2)?

(a) ∩ | (b) ∩ ∩ ||| (c) 99 ∩ ∩ ||

(d) ↱→▷ 99 ∩ ∩ ∩ ∩ ∩ ||| (e) ⟋ ↱→▷ ↱→▷ 99 ∩ ∩ ∩ || (f) ↱→▷ 9 |||

3 What Hindu-Arabic number does each of the following Greek alphabetic numerals represent (Figure 3)?

(a) $\eta\epsilon\delta\beta$ (b) $\iota\epsilon$ (c) $\nu\theta\delta$

1.3 NUMBER OR NUMERAL?

You have probably noticed our use of the words "number" and "numeral." The distinction made between the words is that a number is the object itself, while a numeral is the name of the object or the symbol used to refer to the object. In similar fashion, your name is not you; you are the person and your name is a word used to refer to the person.

We often need to distinguish between an object and the name of that object. In the sentence *"Miss Jones wrote 'mud' on the board,"* we evidently mean that she wrote the name of the stuff and do not mean that she wrote with mud. But if we say *"Do not track mud into the house!"* we mean by "mud," the gooey substance itself. It would probably be impossible to track the word "mud" into the house.

We can clarify whether we mean the object itself or whether we mean its name by placing quotation marks about names of objects. The lack of quotation marks means that we are referring to the object itself.

EXAMPLES

1. Rhode Island is the smallest state.
    "Rhode Island" contains four vowels.

2. Thirteen is an unlucky number.
    "Baker's dozen" is defined as "thirteen."
    There are thirteen rolls in a baker's dozen.

3. Most cities illuminate their streets at night.
    "Illuminate" comes from the Latin, "illuminare."

4. Pennsylvania was named after William Penn.
    "Pennsylvania" means "Penn's woods."

A number is an abstract concept; we cannot touch, see, smell, or hear a number. We will not attempt to define "number" here. It is sufficient for our purposes that you know how to use numbers correctly. The answer to the question "What is 3?" is quite complex. Clearly the number 3 is not the symbol "3," for other symbols also stand for 3. Some different ways to write 3 are *three*, *III*, *tres*, and *trois*. If we ever find life on a distant planet, it is possible that we will find creatures who use numbers, but it would be extremely unlikely that they would use the same symbols we employ.

A numeration system is a way of symbolizing and organizing a set of numbers. Numerals are symbols and as such have no properties. Numbers are abstract objects and *do* have properties. For example, we can say that an even number times an even number is an even number, but we would not say that an even numeral times an even numeral is an even numeral. While 2 is an even number, there is nothing about the symbol "2" that is even. If we are talking about symbols, it might be correct to say that "two two's make a twenty-two," but if we mean the numbers themselves, we would say "two two's are four."

While we will not belabor the distinction between objects and their names, this discussion is important when we are talking about ways to represent numbers. When we wish to emphasize that we are talking about numerals, we will use quotation marks. But in some cases where the distinction between the symbol and the number itself is not important, the quotation marks will be omitted.

1.4 ZERO

In the familiar Roman numeral system (see Figure 4) there is no zero. In displaying larger and larger numbers, new symbols are introduced at four, nine, forty, ninety, four hundred, and nine hundred.

While 4000 might be written MMMM, consistency would require a new symbol to be invented for 5000; then the symbol M before that new symbol would stand for 4000.

I	II	III	IV	V	VI	VII	VIII
1	2	3	4	5	6	7	8

IX	X	XX	XXX	XL	L	LX	LXX	LXXX
9	10	20	30	40	50	60	70	80

XC	C	CC	CCC	CD	D	DC	DCC	DCCC
90	100	200	300	400	500	600	700	800

CM	M	MM	MMM	?
900	1000	2000	3000	4000

Figure 4 Roman numerals

Symbols for 5000, 10000, 50000, and so on, did exist, but were not in common usage. Probably there was little need for such large numerals.

Without a zero it is necessary to devise some complicated scheme, or to invent more and more symbols to represent larger and larger numbers.

Zero first appeared in our Hindu-Arabic counting system in about A.D. 600. Some historians feel that the Greeks invented a zero symbol and it was then transmitted by traders to the Hindus in India. Even when the Hindu-Arabic system became common in Europe in the sixteenth century, there was lingering opposition. The symbol 0 can easily be converted to 6, 8, or 9, so documents could be forged. Indeed, Roman numerals were used in accounting offices for another two hundred years.

In the sixteenth century the Spanish discovered the highly developed Mayan civilization in Central America. Research has shown that the Mayan culture reached its peak in the seventh and eighth centuries. Most notable among their achievements was the creation of a hieroglyphic written language and a numeration system. While the numeration system has been deciphered for many years now, the hieroglyphic language remains untranslated. There is evidence that the numeration system developed without any influence from Europe or the Far East. Mayan stela fragments, or stone sculpture pieces, dating from the time of Christ, show a sophisticated numeration system that includes a zero. We show the Mayan numerals in Figure 5.

We use zero as a "placeholder" in writing certain numerals. In the number 304, the 0 separates the 3 from the 4 by one place, putting the 3 in the hundreds column. Thus 304 is a different number from 34. Note that the Roman numeral equivalent to 304 is CCCIV. There is no way in Roman numerals to use the symbols III and IV to write 304 because of the absence of a placeholder.

The Hindu-Arabic counting system has been in common world-wide use for only three or four hundred years. Its development stands as one of the greatest achievements of human creativity. Its simplicity is readily apparent to us today, but is due, in no small part, to the invention of zero.

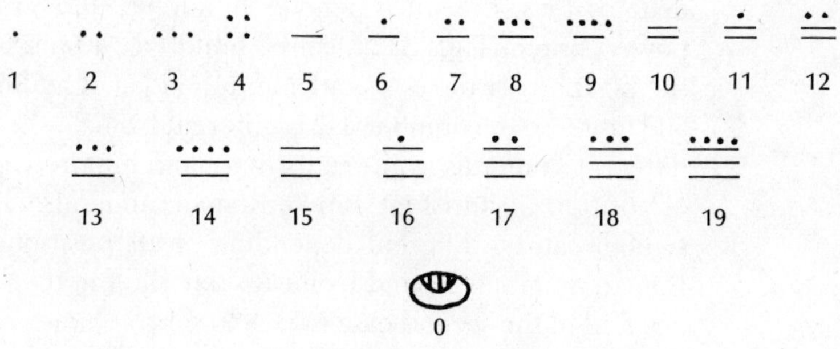

Figure 5   Mayan numerals

Exercise Set 1.3

1 In the examples that follow indicate whether the italicized words are used as numbers or as numerals.
   (a) *Seven* is English for expressing a set of *seven* objects, and *sept* is the French equivalent.
   (b) The ancient Mayans had a symbol for *zero*.
   (c) When we count, we do not start with *zero*, but *zero* is indispensable in our numeration system.
   (d) You must be *eighteen* or older to be eligible to vote.
   (e) If we say *forty, fifty, sixty,* and *seventy*, then we should say *twoty* and *threety*.

2 Write the numeral 6 in four other ways.

3 (a) Write the counting numbers from one to fifteen, using the Hindu-Arabic system.
   (b) How many different symbols did you use?
   (c) Express the first fifteen numbers in Roman numerals. How many different symbols did you use?
   (d) To continue writing in Roman numerals, you use combinations of I, V, X, L, C, D, and M to write greater and greater numbers. How would 4000 be written?

4 How does 7 differ from 70 and 700?

5 Which is more compact, 1975, or the Roman numeral equivalent?

°6 Write a brief synopsis of the article "From Numbers to Numerals and from Numerals to Computation" by Smith and Ginsburg, in *The World of Mathematics*, Volume I, by James R. Newman (New York: Simon & Schuster, 1956).

1.5 POSITIONAL NOTATION AND BASE

It is the placement of digits relative to the right-most digit of a number, along with an additive principle, that enables us to count indefinitely with only ten symbols. The right-most digit denotes the units, the second digit denotes tens, the third denotes hundreds, and so on, according to the powers of ten. Thus 342 means 3 hundreds, 4 tens, and 2 units because of the positions of the 3, 4, and 2 and the idea of adding the hundreds, tens, and units. So the number 342 is different from 243 or 432 because there are different numerals in the units, tens, and hundreds places. This principle of position is consistent, unlike Roman numerals where the meaning of a symbol can be different depending on its position; thus while X in the Roman numerals XI and IX means ten, the I in the first case means to add one and in the second case means to subtract one.

We have said that a numeral like 342 means 3 hundreds plus 4 tens plus 2 units. This can be written as

$$(3 \times 100) + (4 \times 10) + (2 \times 1)$$

or equivalently,

$$3(100) + 4(10) + 2(1)$$

If we list powers of ten as in Table 1, we can write another expression for 342: $3(10^2) + 4(10^1) + 2(10^0)$.

When we write 1,000,000 as 10^6, 10 is called the *base* and 6 is the *exponent*. The exponent tells you the number of times the base is to be used as a factor. A negative exponent implies a fraction.

EXAMPLES

$10^3 = 10 \times 10 \times 10 = 1000$
$2^4 = 2 \times 2 \times 2 \times 2 = 16$
$3^{-1} = \dfrac{1}{3^1} = \dfrac{1}{3}$

$10^{-2} = \dfrac{1}{10^2} = \dfrac{1}{100} = .01$

Any number in our system can be rewritten in expanded form using powers of ten and appropriate multiplications and additions.

EXAMPLES

$3402 = 3(10^3) + 4(10^2) + 0(10^1) + 2(10^0)$
$34.2 = 3(10^1) + 4(10^0) + 2(10^{-1})$
$34.002 = 3(10^1) + 4(10^0) + 0(10^{-1}) + 0(10^{-2}) + 2(10^{-3})$

Table 1   Powers of ten

$$10^6 = 1,000,000$$
$$10^5 = 100,000$$
$$10^4 = 10,000$$
$$10^3 = 1000$$
$$10^2 = 100$$
$$10^1 = 10$$
$$10^0 = 1$$
$$10^{-1} = .1$$
$$10^{-2} = .01$$
$$10^{-3} = .001$$
$$10^{-4} = .0001$$

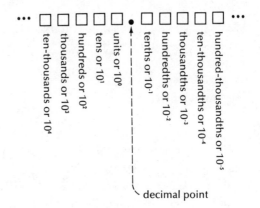

Figure 6   The decimal scheme

That we use powers of 10 and not powers of some other number is what we mean when we say our system is *decimal* or *base ten*. The word "decimal" comes from the Latin "decima" meaning ten. Our language reminds us that the base is ten when we read numbers in tens, hundreds (ten tens), thousands (ten hundreds), ten thousands, and so forth.

In Figure 6 we see the rule for all numerals. The squares only represent position and are to be filled with any of the basic ten symbols.

On the basis of this idea, which took thousands of years to evolve, we are able to count any number of objects. Basic arithmetic is inconceivable without the system, and indeed the system is inconceivable without basic arithmetic. While decimal is not the only system that can be or could have been developed, any other system must be equivalent to the decimal system if it is to do the same arithmetical job.

Exercise Set 1.4

1 "1234" is read "one thousand two hundred thirty-four" and implies the addition: $1000 + 200 + 30 + 4$. How are each of the following numerals read and what addition is implied?
   (a) 75   (b) 13   (c) 499   (d) 5613   (e) 10003
   (f) 705   (g) 342.5 (h) 13.42   (i) 29.357

2 The expanded form of 259 is $2(10^2) + 5(10^1) + 9(10^0)$. What is the expanded form of each numeral in Exercise 1?

3 If we call 259 the simplified form of $2(10^2) + 5(10^1) + 9(10^0)$, what is the simplified form of each of the following?
   (a) $7(10^3) + 5(10^2) + 2(10^1) + 4(10^0)$
   (b) $3(10^0) + 5(10^2) + 6(10^1)$
   (c) $2(10^0) + 3(10^1) + 0(10^2) + 4(10^3)$
   (d) $8(10^0) + 2(10^{-1}) + 9(10^{-2})$
   (e) $7(10^2) + 6(10^1) + 5(10^0) + 4(10^{-1}) + 3(10^{-2})$
   (f) $6(10^1) + 4(10^2) + 3(10^3) + 9(10^{-1})$

4 How are each of the following read, what addition is implied, and what is the expanded form?

(a) 23   (b) 203   (c) 2003
(d) 2.3   (e) 2.03   (f) 2.003

5 (a) Write a numeral to indicate "twelve dozen eggs."
   (b) How is this numeral read?
   (c) Expand this numeral in powers of ten.
   (d) Make the multiplications and additions indicated in the expanded form. Observe that the answer is the same numeral you wrote for part (a).
   (e) Which is easier to communicate, twelve dozen or your answer to part (b)?
   (f) What is another word for twelve dozen?
   (g) What is another word for "ten tens"?

°6 Paper is sold by the ream (which we define to be 500 sheets) or by the quire (which we define to be 25 sheets). Thus there are 20 quires to a ream. We could express 525 sheets as 1 ream 1 quire and 600 sheets as 1 ream 4 quires, and abbreviate our expressions 1 r 1 q and 1 r 4 q. Then clearly the expanded form of 3 r 5 q would be 3(500) + 5(25). So 3 r 5 q means 1500 + 125, or 1625 sheets.

   (a) Express 550 sheets in ream-quire form.
   (b) Do the same for 575, 1050, 2050, 2575, and 5825 sheets.
   (c) How many sheets are there in 5 r 3 q?
   (d) What is the expanded form of each of the following? 2 r 17 q, 9 r 19 q, 6 r 15 q.
   (e) If we say that 10 reams is 1 carton, then we can write 5000 sheets as 1 carton or simply as 1 c. So 5550 sheets can be written 1 c 1 r 2 q. How would we write 13,450 sheets?
   (f) How many sheets are there in 3 c 9 r 8 q?
   (g) While 75 sheets is 3 q, we can write 77 sheets as 3 q 2 s, where s stands for one sheet. How would we write 192 sheets?
   (h) In these carton-ream-quire-sheet symbols the maximum r quantity should be 9. Why?
   (i) What should be the maximum q quantity? The maximum s quantity?
   (j) What would be wrong with 3 c 11 r 25 q 29 s?
   (k) What is a more appropriate way to write the number of sheets indicated in (j)?

°7 While paper may be sold in cartons, reams, and quires, it is awkward in that system to express 18,759 sheets. How would this be done in c-r-q-s form? For this number of sheets the decimal representation is easier. But, would it be preferable to order 20,000 sheets or 4 cartons of paper?

°8 In surveying there are 8⅓ links to a chain, 10 chains to a furlong, and 8 furlongs to a mile. A mile is 5280 feet, a furlong is 660 feet, a chain is 66 feet, and a link is 7.92 feet. A measurement of 3 miles, 5 furlongs, 7 chains, and 4 links is thus 3(5280) + 5(660) + 7(66) + 4(7.92) = 19633.68 feet.

   (a) How many feet are there in 2 miles, 6 furlongs, 4 chains, and 3 links?
   (b) Is it simpler to say 15.84 feet or to say 2 links?
   (c) Is it easier to express *all* distances in chains and links than it is to express them in decimal?

1.6 CHANGING THE BASE; EQUIVALENT COUNTING SYSTEMS

Counting in the decimal system makes use of nine symbols set down in order in the usual way: 1, 2, 3, 4, 5, 6, 7, 8, 9. After 9 we use a tenth symbol "0" to continue: 10, 11, 12, 13, and so on. An automobile odometer (mileage indicator) is a good model of how the process works. Let us look closely at the way an odometer functions.

When a car is new, the odometer reads 0 0 0 0 0 (disregarding tenths of a mile). The wheels of the odometer turn and numerals roll into view as each mile is passed or counted; see Figure 7. By the ninth mile, we observe that each of the ten numerals on wheel 1 has been in view. This includes 0, since in order to register the first mile, the first wheel must have been at 0. Since the symbols *are* on a wheel, 0 appears again at the tenth mile on wheel 1, and 1 appears on wheel 2. The process repeats. See Figure 8.

Now suppose that a counter were built to operate the same way as an odometer, but only five numerals were used: 0, 1, 2, 3, and 4.

In this section we will mean by the phrase *set a counter*, to set all wheels at the zero position. If there are five wheels, a set counter looks like this: 0 0 0 0 0. Setting our new counter, which will have three wheels, and counting with only five numerals, we get the patterns shown in Figure 9 in the odometer windows.

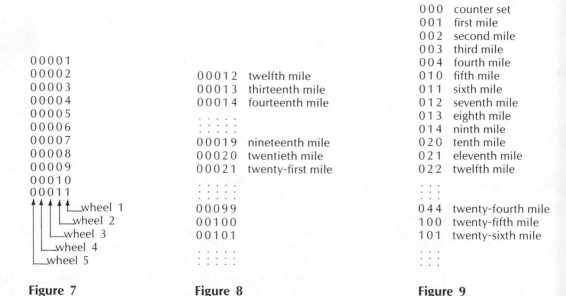

		000 counter set
		001 first mile
		002 second mile
		003 third mile
00001		004 fourth mile
00002	00012 twelfth mile	010 fifth mile
00003	00013 thirteenth mile	011 sixth mile
00004	00014 fourteenth mile	012 seventh mile
00005	: : : : :	013 eighth mile
00006	: : : : :	014 ninth mile
00007	00019 nineteenth mile	020 tenth mile
00008	00020 twentieth mile	021 eleventh mile
00009	00021 twenty-first mile	022 twelfth mile
00010	: : : : :	
00011	: : : : :	: : :
wheel 1	00099	044 twenty-fourth mile
wheel 2	00100	100 twenty-fifth mile
wheel 3	00101	101 twenty-sixth mile
wheel 4	: : : : :	: : :
wheel 5	: : : : :	: : :

Figure 7
A decimal counter

Figure 8
The decimal counter extended

Figure 9
A base-five counter

This would be called a base-five counter because only 5 different symbols are needed to represent any number of miles, and because the *number* meant by any *numeral* is understood by expanding the numeral in powers of 5.

The symbols that appear on the base-five counter are called base-five numerals. To differentiate between decimal and base-five numerals, we will use a subscript to indicate the base. Thus 44_5 means 44 (base five). A numeral like "44_5" should be read "four-four" instead of "forty-four." We will also use a subscript 10 to indicate decimal when we want to make it clear that we are referring to base ten.

Because of the odometer-like principle of our base-ten or base-five counters, it is possible to use either base to count miles. The seventh mile appears as "7" on the base-ten counter and as "12" on the base-five counter. Then $7_{10} = 12_5$. Both "7_{10}" and "12_5" are ways to express the number seven.

In Table 2 we show the equivalences of base-five and decimal numerals for the numbers from one to fifteen.

The numeral 23_5 (read "two-three") means the same as the numeral 13_{10}. Since 13_{10} means 1 ten plus 3 units, it is reasonable to say that 23_5 means 2 fives and 3 units. What does a numeral like 342_5 mean? Recall that in base ten, the position of a symbol from right to left indicates the number of units, tens, hundreds, and so on. Similarly in base five the position of a numeral from right to left will mean units, fives, twenty-fives, and so on.

So 342_5 means 2 units plus 4 fives plus 3 twenty-fives. The idea, in general, is shown in Figure 10.

Table 2 Some base-five and base-ten equivalent numerals

$1_5 = 1_{10}$	$11_5 = 6_{10}$	$21_5 = 11_{10}$
$2_5 = 2_{10}$	$12_5 = 7_{10}$	$22_5 = 12_{10}$
$3_5 = 3_{10}$	$13_5 = 8_{10}$	$23_5 = 13_{10}$
$4_5 = 4_{10}$	$14_5 = 9_{10}$	$24_5 = 14_{10}$
$10_5 = 5_{10}$	$20_5 = 10_{10}$	$30_5 = 15_{10}$

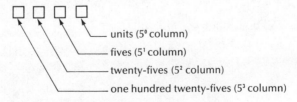

units (5^0 column)
fives (5^1 column)
twenty-fives (5^2 column)
one hundred twenty-fives (5^3 column)

Figure 10 The base-five scheme

In base ten the position of a digit from the right determines by what power of 10 we multiply the digit. Likewise, in base five the position of a digit from the right indicates what power of 5 we multiply by. For reference we list the powers of 5 in Table 3.

Table 3   Powers of 5

5^5 =	3125
5^4 =	625
5^3 =	125
5^2 =	25
5^1 =	5
5^0 =	1

EXAMPLE 1

342_5 means 2 units + 4 fives + 3 twenty-fives
or equivalently,

$$342 = 2(5^0) + 4(5^1) + 3(5^2)$$

Then from Table 3,

$$342_5 = 2(1) + 4(5) + 3(25)$$
$$= 2 + 20 + 75$$
$$= 97_{10}$$

Notice that there is no "5" in base five. But the meaning of 342_5 was in terms of powers of 5. We speak about base five from a decimal point of view. In fact, all the calculations on the right side of the equations in example 1 are in decimal. Notice that the last line says that $342_5 = 97_{10}$. By expanding 342_5 in powers of 5 and performing the calculations in decimal we have converted the base-five numeral 342 to its decimal equivalent.

EXAMPLE 2

Convert 23_5 to decimal.

$$23_5 = 2(5^1) + 3(5^0)$$

Then, since $5^1 = 5$ and $5^0 = 1$,

$$23_5 = 2(5) + 3(1)$$
$$= 10 + 3$$
$$= 13_{10}$$

Note that this agrees with Table 2.

The most important step in converting from base five to decimal is the first one, where the number is expanded in powers of 5.

If we change to base eight (where we can represent the eight digits as 0, 1, 2, 3, 4, 5, 6, 7), numerals can be changed to their decimal equivalent by expansion in powers of 8. We must also agree that $8^0 = 1$, $8^1 = 8$, $8^2 = 64$, and so on. We would also use a subscript 8 to denote base eight.

EXAMPLE 3

Convert 15_8 to decimal.

$$\begin{aligned} 15_8 &= 1(8^1) + 5(8^0) \\ &= 1(8) + 5(1) \\ &= 8 + 5 \\ &= 13_{10} \end{aligned}$$

EXAMPLE 4

Counting from one to thirteen in base eight, we have: 1, 2, 3, 4, 5, 6, 7, 10, 11, 12, 13, 14, 15. This verifies that the work in Example 3 is correct.

In base four there would be the digits 0, 1, 2, and 3. Thinking of the odometer-like principle of counting, the base-four counting numbers would be represented as 1, 2, 3, 10, 11, 12, 13, 20, 21, 22, 23, 30, 31, and so on.

EXAMPLE 5

Verify that 31_4 is the thirteenth base-four counting number. All we need to do is to convert 31_4 to decimal:

$$\begin{aligned} 31_4 &= 3(4^1) + 1(4^0) \\ &= 3(4) + 1(1) \\ &= 12 + 1 \\ &= 13_{10} \end{aligned}$$

One of the most interesting bases is base two (or binary). In base two there would be only two digits, which we can write as 0 and 1. Counting in binary proceeds as in Table 4.

If we constructed a binary odometer, there would be only two digits on each wheel. Apparently many more "binary" wheels would be needed than "decimal" wheels. Only two positions are needed to represent ten in decimal, but four positions are needed to express the same number in binary.

Table 4   Counting in binary

Binary	Decimal
000	0
001	1
010	2
011	3
100	4
101	5
110	6
111	7
1000	8
1001	9
1010	10

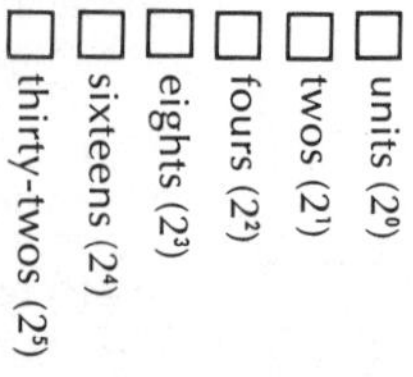

Figure 11
Binary positional notation

The binary scheme for positional notation is shown in Figure 11. Thus 111_2 means 1 four plus 1 two plus 1 unit.

EXAMPLE 6

Convert 10111_2 to decimal notation:

$$10111_2 = 1(2^4) + 0(2^3) + 1(2^2) + 1(2^1) + 1(2^0)$$
$$= 16 + 0 + 4 + 2 + 1$$
$$= 23_{10}$$

You can verify that the answer is correct by counting from one to twenty-three in binary.

The Babylonians used a base of sixty. Positional notation was made clear by separating groups of the wedge-shaped marks, although the separation was sometimes unclear. ▼ ▼▼ ▼ ▼ ▼ was used to represent one 3600 plus two 60's plus 3 (for a total of 3723 in our notation). Without spacing ▼ ▼ ▼ ▼ ▼ ▼ meant six. Yet ▼ ▼ ▼ ▼ ▼ might mean two 60's plus 3, or two 3600's plus three 60's; had there been another symbol that could have been used as a zero, the meaning would have been clear.

The Babylonians had only one symbol, the cuneiform wedge, although it was sometimes vertical, sometimes horizontal, and sometimes slanted, having a different meaning in each case. The Egyptians used a decimal system, but had 7 symbols for writing numbers up to 1,000,000. The Romans employed a decimal system using 7 symbols for numbers up to 1000, but developed a duodecimal (base twelve) system for writing fractions. Grecian numeration was decimal; originally there were 6 Herodianic signs used for numeration, and when they converted to the alphabetic numeral system, they needed 27 symbols.

The use of decimal as a base probably arose naturally because of finger counting. Base sixty seems to be related to astronomy and to the calendar, where the year was divided into 360 days. Decimal has always been the base most used by various advanced civilizations, but examples of duodecimal, vigesimal (base twenty), and centesimal (base one hundred) can be cited. The Aztec Indians of South America used base twenty, and some North American Indians used bases of three, four, five, and eight.

1.7 CHANGING FROM DECIMAL TO OTHER BASES

In Section 1.6 the key to converting from base-five (or any other base) to decimal was to expand the given numeral in powers of the base. We saw that 342_5 converts to 97_{10}. But how would we show that 97_{10} converts to 342_5?

The first step to solving this problem is to assume an answer in "counter form" without knowing at first what digits would appear in each of the counter positions. That is, assume

$$97_{10} = \underline{}\ \underline{}\ \underline{}\ \underline{}\ \underline{}\ _5$$

where the blanks (from the right) mean ones, fives, twenty-fives, one hundred twenty-fives, and six hundred twenty-fives.

The next thing to consider is what digits in the list 0, 1, 2, 3, 4 belong in each position. Certainly the following is wrong,

$$97_{10} = \underline{\ 4\ }\ \underline{}\ \underline{}\ \underline{}\ \underline{}\ _5$$

because a 4 in the fifth position from the right means 4 times 625, which is much larger than 97.

It is clear that the fifth and fourth positions should each contain zeros.

$$97_{10} = \underline{\ 0\ }\ \underline{\ 0\ }\ \underline{}\ \underline{}\ \underline{}\ _5$$

We then place a 3 in the third position since that means 3 twenty-fives. (4 is too big, since $4(25) = 100$, which is greater than 97.)

$$97_{10} = \underline{\ 0\ }\ \underline{\ 0\ }\ \underline{\ 3\ }\ \underline{}\ \underline{}\ _5$$

By placing a 3 in the third position we have attained 75 of the 97 we desire. A four in position 2 (from the right) will give us 20 more of the 97. So far we have

$$97_{10} = \underline{\ 0\ }\ \underline{\ 0\ }\ \underline{\ 3\ }\ \underline{\ 4\ }\ \underline{}\ _5$$

which means

$$
\begin{aligned}
97 &= 3(25) + 4(5) + ? \\
&= 75 + 20 + ? \\
&= 95 + ?
\end{aligned}
$$

and it should be clear that we need 2 more units.

$$97_{10} = \underline{\ 0\ }\ \underline{\ 0\ }\ \underline{\ 3\ }\ \underline{\ 4\ }\ \underline{\ 2\ }_5$$

or

$$97_{10} = 342_5$$

It is interesting that this answer is unique. We could not have started with 2 twenty-fives instead of 3, for then we would have had a balance of 47 (97 minus two twenty-fives) to attain in fives and ones. The maximum digit in any column is a 4. Of course, $97 \neq 2(25) + 4(5) + 4(1)$.

Intuitively, the rule is to put the maximum possible digit in the leftmost position. Then proceed to the right, always taking the maximum possible digit in each position until the proper sum is attained.

EXAMPLE 1

Convert 37_{10} to base five.

$$37_{10} = \underset{125's}{\underline{\ ?\ }}\ \underset{25's}{\underline{\ ?\ }}\ \underset{5's}{\underline{\ ?\ }}\ \underset{1's}{\underline{\ ?\ }}_5$$

(Below the blanks is a reminder about the meaning of the position.)

$$37_{10} = \underline{\ 0\ }\ \underline{\ ?\ }\ \underline{\ ?\ }\ \underline{\ ?\ }_5$$
$$37_{10} = \underline{\ 0\ }\ \underline{\ 1\ }\ \underline{\ ?\ }\ \underline{\ ?\ }_5$$

We can now subtract 25 from 37, getting 12, which is the balance needed in 5's and 1's.

$$37_{10} = \underline{\ 0\ }\ \underline{\ 1\ }\ \underline{\ 2\ }\ \underline{\ 2\ }_5$$

EXAMPLE 2

Convert 125_{10} to base eight.

$$125 = \underset{64's}{\underline{\ \ \ }}\ \underset{8's}{\underline{\ \ \ }}\ \underset{1's}{\underline{\ \ \ }}_8 \text{ (assuming positions)}$$

$$= \underline{\ 1\ }\ \underline{\ ?\ }\ \underline{\ ?\ }$$

$$= \underline{\ 1\ }\ \underline{\ 7\ }\ \underline{\ ?\ }$$

$$= \underline{\ 1\ }\ \underline{\ 7\ }\ \underline{\ 5\ }_8$$

$$
\begin{array}{rl}
125 & \\
-\ 64 & \text{(so far)} \\
\hline
61 & \text{(needed)} \\
-\ 56 & \\
\hline
5 & \text{(needed)} \\
-\ 5 & \\
\hline
0 & \text{balance}
\end{array}
$$

The arithmetic to the right of Example 2 is a way of keeping track of how much of the 125 we have obtained by putting successive base-eight digits in the various columns.

EXAMPLE 3

Convert 27_{10} to binary.

$$27_{10} = \underline{}_{2^5}\ \underline{}_{2^4}\ \underline{}_{2^3}\ \underline{}_{2^2}\ \underline{}_{2^1}\ \underline{}_{2^0}\,2$$

$$= \underline{0}\ \underline{1}\ \underline{}\ \underline{}\ \underline{}\ \underline{}$$

$$= \underline{0}\ \underline{1}\ \underline{1}\ \underline{}\ \underline{}\ \underline{}$$

$$= \underline{0}\ \underline{1}\ \underline{1}\ \underline{0}\ \underline{1}\ \underline{}$$

$$= \underline{0}\ \underline{1}\ \underline{1}\ \underline{0}\ \underline{1}\ \underline{1}\ _2$$

$$\begin{array}{r} 27 \\ -16 \\ \hline 11 \\ -\ 8 \\ \hline 3 \\ -\ 2 \\ \hline 1 \\ -\ 1 \\ \hline 0 \end{array}$$

Exercise Set 1.5

1 When you measure the distance from City A to City B with your automobile's odometer and find the distance to be 15 miles, then *sixteen* different numerals must have appeared in the windows of the odometer. Explain.

2 When we subtract 8 from 12, the answer is 4, but counting from 8 to 12 (8, 9, 10, 11, 12) requires five symbols. Explain.

3 How many trees are necessary to plant trees at ten-foot intervals in a row that is to be exactly 70 feet long?

4 A base-five odometer reads $\underline{0}\ \underline{2}\ \underline{3}\ \underline{1}\ \underline{4}$. How many miles has the car gone? Answer in base five only!

5 Answer question 4 in decimal.

6 Convert each base-five numeral below to decimal.
   (a) 2034   (b) 11011   °(c) 4.3
   (d) 33214   (e) 40251   (f) 444

7 One of the problems in exercise 6 is impossible. Which one and why?

8 Convert each base-ten numeral below to base five.
   (a) 47   (b) 407   (c) 4007
   (d) 470   (e) 4   °(f) 0.2

9 (a) It takes five windows of a decimal odometer to show 99,999 miles. How many windows would be necessary to express the same mileage on a base-five odometer?
   (b) If you were manufacturing odometers, which would probably be cheaper to produce, a decimal or a base-five odometer? (Assume a maximum of 100,000 decimal miles.)
   (c) A base-five odometer reads 10000. What was the reading one mile ago?

10 (a) In base eight (octal) write the symbols necessary to count three dozen objects.

(b) Convert each octal numeral below to decimal.

1. 372	2. 3072	3. 43
4. 345	5. 11111	6. 6
°7. 4.2		

(c) Convert each decimal numeral below to octal.

1. 39	2. 309	3. 1000
4. 512	5. 100,000	°6. 43.125
°7. 43.25		

11 (a) Write the symbols in binary (base two) necessary to count three dozen objects.

(b) Convert each of the following binary numerals to decimal.

1. 11 111 010	2. 11 000 111 010	3. 100 011
4. 11 100 101	5. 100 100 100 001	6. 110
°7. 100.01		

(c) Convert each of the decimal numerals below to binary.

1. 13	2. 29	3. 50
4. 128	5. 129	6. 152
7. 200	8. 183	9. 1000

°12 There is a method for converting from binary to octal.

(a) Consider the following table:

$$000_2 = 0_8 \qquad 100_2 = 4_8$$
$$001_2 = 1_8 \qquad 101_2 = 5_8$$
$$010_2 = 2_8 \qquad 110_2 = 6_8$$
$$011_2 = 3_8 \qquad 111_2 = 7_8$$

The binary 110111 is converted to octal by grouping the binary in sets of three, 110 111, and substituting the octal equivalents from the above table for each group of three. $110_2 = 6_8$ and $111_2 = 7_8$. How can we check to see that the method works?

(b) Convert each binary numeral below to octal, without first changing to decimal.

1. 10111	2. 110101	3. 100000	4. 101110111

(c) How many places are necessary to represent the decimal 100,000 in binary?

°13 Let us invent a base-three counting system using the symbols a, b, and c where the first numerals (in order) are b, c, ba, bb, bc, ca, cb, cc, baa, bab, ...

(a) Extend the system far enough to represent two dozen objects.

(b) What is the "zero" of this base-three system?

(c) Convert $bacba$ to decimal.

(d) Convert 27_{10} to this base-three system.

°14 What would be the first step in creating a base-twelve (duodecimal) system? Hint: Count from one to eighteen in decimal and then try to count from one to eighteen in duodecimal.

°15 Can we have a base-one system? Explain.

1.8 SO WHAT *IS* COUNTING?

Most of this chapter has been about the set of counting numbers. The particular numerals and base used in counting are of little importance, although our Hindu-Arabic decimal numeration system is a particularly good one.

But what exactly does it mean to count? The answer to this question is abstract. In order to discuss counting in any detail, some basic notions about sets of objects are needed. The term *set* has been used before in this text. Its meaning is assumed, except to say that a synonym might be "collection."

The objects or items in a set are called *elements* or *members*. Mathematicians indicate sets by using braces and listing elements separated by commas. The set of vowels in the alphabet can be written:

$$\{ a, e, i, o, u \}$$

Sometimes we use three dots to indicate missing elements, but this should be done only when it is clear what elements are missing. The set of all letters in the alphabet can be written:

$$\{ a, b, c, d, \ldots, z \}$$

because it is obvious which letters are missing.

Sometimes capital letters are used to stand for an entire set. For example, it is customary to use N to stand for the set of counting numbers. The N is probably borrowed from another name for the counting numbers, the *natural numbers*. We write:

$$N = \{ 1, 2, 3, 4, \ldots \}$$

The three dots used in listing the elements of N indicate that there is no last element. The set N is infinite, although an exact meaning of that term will be delayed until Chapter 11 on Infinity.

The set $\{ e, e, l, r \}$ is the same as $\{ e, l, r \}$. For this reason the same element should not be listed more than once. If it is necessary to repeat elements, we can always use subscripts to distinguish between them.

EXAMPLE 1

If letters are drawn in a word game and two A's and one T turn up, the set of three letters can be written:

$$\{ A_1, A_2, T \}$$

The two A's are not the same and we have distinguished between them by using subscripts.

Sometimes it is necessary to compare two sets by matching their elements. In Section 1.1 we compared the shepherd boy's set of marks with his neighbor's. Given two sets, we can match the elements and tell whether or not the sets have the same number of elements.

EXAMPLE 2

Let $A = \{a, b, c, d\}$ and let $B = \{1, 2, 3, 4, 5\}$. We can match the elements of A with those of B as follows:

$$
\begin{array}{cccc}
a & b & c & d \\
\updownarrow & \updownarrow & \updownarrow & \updownarrow \\
1 & 2 & 3 & 4 \quad 5
\end{array}
$$

The arrows in Example 2 indicate a matching or comparison of the letters with the numerals. Once matched this way, it is evident that there are more numerals than letters.

EXAMPLE 3

Let $A = \{1, 2, 3\}$ and $B = \{r, s, t\}$. One matching possible between these sets is:

$$
\begin{array}{ccc}
1 & 2 & 3 \\
\updownarrow & \times & \\
r & s & t
\end{array}
$$

In Example 3, 1 is matched with r, 2 with t, and 3 with s. It should be clear that because of this matching, sets A and B have the same number of elements. Of course the matching of 1 with r, 2 with s, and 3 with t would lead to the same conclusion. Such matchings are demonstrations that a *one-to-one correspondence* exists between the two sets.

DEFINITION 1   One-to-One Correspondence

Let A and B be any two sets. If each element of A can be paired with one and only one element of B, and each element of B can be paired with one and only one element of A, then sets A and B are said to be in *one-to-one correspondence*.

The word "paired" in the definition means that if the element x in set A happens to be matched with the element y in set B, then y is also matched with x.

EXAMPLE 4

Let $A = \{1, 2, 3, 4, 5\}$ and $B = \{a, e, i, o, u\}$. Since we can indicate a matching between sets A and B as

$$
\begin{array}{ccccc}
1 & 2 & 3 & 4 & 5 \\
\updownarrow & \updownarrow & \updownarrow & \updownarrow & \updownarrow \\
a & e & i & o & u
\end{array}
$$

The sets A and B are in one-to-one correspondence. We could also show the matching this way:

$$
\begin{array}{ccccc}
1 & 2 & 3 & 4 & 5 \\
\updownarrow & \updownarrow & \updownarrow & \updownarrow & \updownarrow \\
o & i & a & u & e
\end{array}
$$

There are many such pairings, all of which demonstrate the one-to-one correspondence.

EXAMPLE 5

Let $R = \{1, 2\}$ and $S = \{a, b\}$; the matching:

$$
\begin{array}{cc}
1 & 2 \\
\downarrow & \uparrow \\
a & b
\end{array}
$$

is not evidence of a one-to-one correspondence, since 1 is associated with a but a is not associated with 1. That is, 1 and a are not paired.

Even some infinite sets can be placed in one-to-one correspondence.

EXAMPLE 6

Let $N = \{1, 2, 3, 4, \ldots\}$ and let $E = \{2, 4, 6, 8, 10, \ldots\}$. The following is a matching that shows that N can be placed in one-to-one correspondence with E.

$$
\begin{array}{cccccc}
1 & 2 & 3 & 4 & 5 & \ldots \\
\updownarrow & \updownarrow & \updownarrow & \updownarrow & \updownarrow & \\
2 & 4 & 6 & 8 & 10 & \ldots
\end{array}
$$

The rule for the required pairing is that any number in N is associated with twice itself in E and any number in E is associated with half itself in N.

If two sets, A and B, can be placed in one-to-one correspondence, we call them *equivalent* sets, and write:

$$A \leftrightarrow B$$

In Example 6, we have shown that $N \leftrightarrow E$. Notice that equivalent does not mean "the same." Certainly set N is not the same as set E.

Now that the notion of a one-to-one correspondence has been introduced, we can return to our discussion of counting. When we count objects, we sound-off natural (or counting) numbers in order starting with one, establishing as we go along a one-to-one correspondence between the numbers mentioned and the objects being counted. The last natural number used is taken as the answer to the question "How many objects are there?"

EXAMPLE 7

In counting the names of the days of the week { Sunday, Monday, Tuesday, ..., Saturday } we say (or think) "one, two, three, four, five, six, seven," so there are 7 days in the week. The one-to-one correspondence is between the set of names and the set of the first seven counting numbers.

The set of natural numbers is insufficient for counting every set that exists for two reasons. First, some sets are infinite and there is no natural number we can associate with infinite sets. Second, some sets have no elements in them (for example, the set of two-headed dogs). Such sets are termed *empty* sets. There is no natural number that answers the question "How many?" for empty sets.

We use *cardinal numbers* to answer the question "How many?" for all sets, even empty and infinite sets. Zero is a cardinal number. The cardinal number of, or the cardinality of, the empty set is 0. The cardinality of infinite sets is considered in a later chapter.

If we consider only finite sets, the set of cardinal numbers is essentially the same as the set of the names of the natural numbers, with the additional number 0. Convenient notation for the set of cardinal numbers would be N^{+0}.

$$N = \{ 1, 2, 3, 4, 5, 6, 7, \ldots \} \qquad \text{natural numbers}$$
$$N^{+0} = \{ 0, 1, 2, 3, 4, 5, 6, 7, \ldots \} \qquad \text{cardinal numbers for finite sets}$$

If A is a certain set, then the number of elements in A, or the cardinal number associated with A, or the cardinality of set A, is written:

$$n(A)$$

EXAMPLE 8

Let $B = \{ a, b, c, d \}$, $C = \{ a, b, c, d, \ldots, z \}$, $D = \{ a, b, c, d, \ldots, j \}$, $E = \{ a, e, i, o, u \}$, and $F = \{ a, b, a \}$. Then $n(B) = 4$, $n(C) = 26$, $n(D) = 10$, and $n(E) = 5$. Note that F should be written as $\{ a, b \}$, so $n(F) = 2$.

While we know *how* to count, the abstract idea of associating natural numbers with sets explains *what* we are doing when we count. Young children usually first learn to count by memorizing the counting numbers blindly, as though they were nonsense words. In a second step, when children become more mature, they learn the association between numbers and sets of objects. Eventually they count objects correctly, though they may not comprehend exactly what they are doing.

Exercise Set 1.6

1 For the sets $A = \{1, 2, 3\}$ and $B = \{a, b, c\}$ we could indicate a one-to-one correspondence this way:

$$1 \quad 2 \quad 3$$
$$\updownarrow \quad \updownarrow \quad \updownarrow$$
$$a \quad b \quad c$$

We could also indicate the correspondence this way:

$$1 \quad 2 \quad 3$$
$$\updownarrow \quad \updownarrow \quad \updownarrow$$
$$a \quad c \quad b$$

How many possible one-to-one correspondences could be demonstrated? Write them all down, trying to organize all the possibilities so as not to miss any.

2 You see two women walking on campus and say to yourself "There are Emily and Martha, but I don't remember who is Emily and who is Martha." You also remember that their last names are Smith and Jones, but you forget which first name goes with which last name. What are the possibilities for the pairings of first name with last name?

°3 How many one-to-one correspondences can be formed between the sets:
   (a) $\{ \star, !, ¢, {}^* \}$ and $\{ 2, 4, 6, 8 \}$?
   (b) $\{ A, B, C, D, F \}$ and $\{$ Bob, Alice, Ted, Carol, Ferdinand $\}$?
   (c) $\{ r, s, t \}$ and $\{ \square, \triangle, \bigcirc, \square \}$?

4 Among three boys (Bob, Donald, and Steve) and three girls (Alice, Betty, and Carol) the following facts are true: Bob loves Betty, Donald loves Carol, and Steve loves Alice.
   (a) Do these facts necessarily set up a one-to-one correspondence between the set of boys and the set of girls?
   (b) The following further facts might also be true: Betty loves Bob, Carol loves Donald, and Alice loves Steve. Do we now have a one-to-one correspondence?
   (c) Assume that the facts for part (a) are true and that these additional facts are true: Betty loves Bob, Carol loves Steve, and Alice loves Donald. Is this a one-to-one correspondence?

Comment   While one-to-one correspondences do not require that there be any meaning or rule for the association, the most important correspondences that exist probably follow some rule. In Exercise 4, whether or not one person loves another is the rule for associating boy with girl and girl with boy.

5 Let S be the set of states in the union. What is $n(S)$?

6 $A = \{$ a, e, h, l, m, p, t $\}$ $\qquad$ $B = \{$ p, a, m, p, h, l, e, t $\}$
   $C = \{$ p, a, m, h, l, e, t $\}$ $\qquad$ $D = \{$ a, b, c, d, e, f, g, h $\}$
   $E = \{$ 1, 2, 3, 4, 5, 6, 7 $\}$
   (a) What is $n(A)$? $\qquad\qquad$ (b) What is $n(B)$?
   (c) What is $n(C)$? $\qquad\qquad$ (d) What is $n(D)$?
   (e) What is $n(E)$?
   (f) Which of these sets are the same?
   (g) Which of these sets have the same number of elements?

7 (a) What letters are used to spell the word "red"? The word "reed"? The word "deer"?
   (b) How many different letters are there in each of the three words?
   (c) To agree with this section, would we prefer to express the set of letters in the word "reed" as $\{$ e, d, r $\}$ or as $\{$ e, e, d, r $\}$?

8 Give three examples of sets with cardinality 0.

SUMMARY

We have explored the development of counting systems by inventing a hypothetical tally system and by reviewing some history. Our present-day system of Hindu-Arabic numeration was in existence in A.D. 500 but took several hundred years to become standard in Europe. The concept of "zero" as a numeral was probably one of the last developments. Both the simplicity and elegance of today's numeration system depend heavily on the "zero" concept.

From the investigation of positional notation and base we were led to consider base ten and to compare it with other bases. We could have inherited a counting and calculation system that did not employ ten digits (0, 1, 2, 3, 4, 5, 6, 7, 8, and 9). Base five uses only five symbols and binary employs only two. In order to focus attention on the inner workings of the decimal system, we converted from decimal to other bases and vice versa.

We concluded with a more abstract discussion of counting, and found it necessary to introduce a few ideas about sets. Counting was described in terms of a one-to-one correspondence between the natural numbers and the objects being counted.

Before attempting the Review Test you should be sure you are familiar with the following techniques and ideas:

1. Expanding numerals in powers of ten.

2. Reading numerals.

3. Use of quotation marks in indicating numerals.

4. Converting from decimal to other bases.

5. Converting from other bases to decimal.

6. Counting in other bases.

7. One-to-one correspondences.

8. The notation $n(A)$ for cardinality.

REVIEW TEST

1 Express the decimal numeral 375 in words.

2 What is the simplified form for the expanded numeral $3(10^3) + 5(10^1) + 9(10^0) + 2(10^{-1})$?

3 What base is implied in $2(6^2) + 3(6^1) + 5(6^0)$?

4 (a) In the sentence "John is 19 years old," is the 19 a number or a numeral?
   (b) In the sentence "47 contains a seven in the units place," is the 47 a number or a numeral?

5 A base-five odometer reads "134." How many miles have been counted? Answer in decimal.

6 Convert each of the following to decimal:
   (a) 132_4 (b) 27_8 (c) 10111_2 (d) 88_9

7 Convert each of the following decimal numerals to the base indicated:
   (a) 13 to base five   (b) 27 to base seven   (c) 40 to base two

8 Which of the following pairings are one-to-one correspondences?
   (a) 1 2 3   (b) r s t (c) 1 2 3 4
       ↕ ↕ ↕   ↕ ↓↗   ↕ ↕ ↕
       a b c a b a b c

9 $A = \{a, b, c, d, e\}$ $B = \{a, a, b, c, c\}$
   (a) What is $n(A)$?   (b) What is $n(B)$?

Levi Leonard Conant (1857–1916) was a professor of
mathematics at Worcester Polytechnic Institute.
Though he wrote several specialized mathematics
texts, he was concerned with general education as
well. This article is taken from *The Number Concept:
Its Origin and Development*, published in 1896. His
intensive research into tribal counting systems shows
the importance he gave to the most fundamental of
mathematical ideas.

COUNTING
LEVI LEONARD CONANT

Among the barbarous tribes whose languages have been studied,
. . . none have ever been discovered which did not show some familiarity
with the number concept. The knowledge thus indicated has often
proved to be most limited; not extending beyond the numbers 1 and 2, or
1, 2, and 3. Examples of this poverty of number knowledge are found
among the forest tribes of Brazil, the native races of Australia and
elsewhere. . . . At first thought it seems quite inconceivable that any
human being should be destitute of the power of counting beyond 2. But
such is the case; . . . The Chiquitos of Bolivia had no real numerals
whatever, but expressed their idea for "one" by the word *etama*,
meaning alone. The Tacanas of the same country have no numerals
except those borrowed from Spanish, or from Aymara or Peno, languages
with which they have long been in contact. A few other South American
languages are almost equally destitute of numeral words. But even here,
rudimentary as the number sense undoubtedly is, it is not wholly
lacking; and some indirect expression, or some form of circumlocution,
shows a conception of the difference between *one* and *two* or at least,
between *one* and *many*.

❖ ❖ ❖

We know of no language in which the suggestion of number does not
appear, and we must admit that the words which give expression to the
number sense would be among the early words to be formed in any

From Conant, *The Number Concept: Its Origin and Development*, (New York: Macmillan
and Co., 1896) pp. 1–4, 6–8, 17–20. (Footnotes have been deleted.)

language. They express ideas which are, at first, wholly concrete, which are of the greatest possible simplicity, and which seem in many ways to be clearly understood, even by the higher orders of the brute creation. The origin of number would in itself, then appear to lie beyond the proper limits of inquiry; and the primitive conception of number to be fundamental with human thought.

In connection with the assertion that the idea of number seems to be understood by the higher orders of animals, the following brief quotation from a paper by Sir John Lubbock may not be out of place: "Leroy . . . mentions a case in which a man was anxious to shoot a crow. 'To deceive this suspicious bird, the plan was hit upon of sending two men to the watch house, one of whom passed on, while the other remained; but the crow counted and kept her distance. The next day three went, and again she perceived that only two retired. In fine, it was found necessary to send five or six men to the watch house to put her out in her calculation. The crow, thinking that this number of men had passed by, lost no time in returning.' From this he inferred that crows could count up to four. Lichtenberg mentions a nightingale which was said to count up to three. Every day he gave it three mealworms, one at a time. When it had finished one it returned for another, but after the third it knew that the feast was over . . . ."

✿   ✿   ✿

Practical methods of numeration are many in number and diverse in kind. But the one primitive method of counting which seems to have been almost universal throughout all time is the finger method. It is a matter of common experience and observation that every child, when he begins to count, turns instinctively to his fingers; and, with these convenient aids as counters, tallies off the little number he has in mind. This method is at once so natural and obvious that there can be no doubt that it has always been employed by savage tribes, since the first appearance of the human race in remote antiquity. All research among uncivilized peoples has tended to confirm this view, were confirmation needed of anything so patent. Occasionally some exception to this rule is found; or some variation, such as is presented by the forest tribes of Brazil, who, instead of counting on the fingers themselves, count on the joints of their fingers. . . .

The variety in practical methods of numeration observed among savage races, and among civilized peoples as well, is so great that any detailed account of them would be almost impossible. In one region we find sticks or splints used; in another, pebbles or shells; in another, simple scratches, or notches cut in a stick, Robinson Crusoe fashion; in another, kernels or little heaps of grain; in another, knots on a string; and so on, in diversity of method almost endless. Such are the devices which have been, and still are, to be found in the daily habit of great numbers of . . . tribes; while to pass at a single step to the other extremity of intellectual development, the German student keeps his beer score by chalk marks on the table or on the wall. But back of all these devices, and forming a common origin to which all may be referred, is the universal finger method; the method with which all begin, and which all find too convenient ever to relinquish entirely, even though their civilization be of the highest type. Any such mode of counting, whether involving the use of the fingers or not, is to be regarded simply as an extraneous aid in the expression or comprehension of an idea which the mind cannot grasp, or cannot retain, without assistance. The German student scores his reckoning with chalk marks because he might otherwise forget; while the Andaman Islander counts on his fingers because he has no other method of counting,—or, in other words, of grasping the idea of number.

✿　✿　✿

In the Muralug Island, in the western part of Torres Strait, a somewhat remarkable method of counting formerly existed, which grew out of, and is to be regarded as an extension of, the digital method. Beginning with the little finger of the left hand, the natives counted up to 5 in the usual manner, and then, instead of passing to the other hand, or repeating the count on the same fingers, they expressed the numbers from 6 to 10 by touching and naming successively the left wrist, left elbow, left shoulder, left breast, and sternum. Then the numbers from 11 to 19 were indicated by the use, in inverse order, of the corresponding portions of the right side, arm, and hand, the little finger of the right hand signifying 19. The words used were in each case the actual names of the parts touched; the same word, for example, standing for 6 and 14; but they were never used in the numerical sense unless accompanied by the proper gesture, and bear no resemblance to the common numerals, which are but few in number.

✿　✿　✿

The place occupied, in the intellectual development of man, by finger counting and by the many other artificial methods of reckoning, —pebbles, shells, knots, the abacus, etc.—seems to be this: The abstract processes of addition, subtraction, multiplication, division, and even counting itself, present to the mind a certain degree of difficulty. To assist in overcoming that difficulty, these artificial aids are called in; and, among savages of a low degree of development, like the Australians, they make counting possible. A little higher in the intellectual scale, among the American Indians, for example, they are employed merely as an artificial aid to what could be done by mental effort alone. Finally, among semi-civilized and civilized peoples, the same processes are retained, and form a part of the daily life of almost every person who has to do with counting, reckoning, or keeping tally in any manner whatever. They are no longer necessary, but they are so convenient and so useful that civilization can never dispense with them.

❀   ❀   ❀

In the elaborate calculating machines of the present, such as are used by life insurance actuaries and others having difficult computations to make, we have the extreme of development in the direction of artificial aid to reckoning. But instead of appearing merely as an extraneous aid to a defective intelligence, it now presents itself as a machine so complex that a high degree of intellectual power is required for the mere grasp of its construction and method of working.

2 SETS—THE ABC'S OF MATHEMATICS

Avoiding the use of sets and set language is possible in certain areas of mathematics. However, the use of sets makes many mathematical ideas easy to explain. We saw in Chapter 1 that even the simple idea of counting is best explained in terms of sets.

We shall not take a highly formal point of view in our study of sets, although that would be possible. Some of the most current philosophical and mathematical problems have to do with sets themselves, while others deal with sets of axioms or specific sets of numbers.

In mathematics everything seems related to sets. The word *set* is perhaps the most basic in all of mathematics. Any attempt to define the word would undoubtedly require even more difficult terms and language.

In this chapter our goal will be to introduce set notation, set language, and set relationships. The power of set usage is great; we shall use the ideas of this chapter to simplify many explanations in later sections of the book.

2.1 BASIC SET NOTIONS

We do not have to restrict ourselves to studying sets of numbers. While numbers frequently make very good examples, the structure of sets is independent of particular examples. We insist only that our examples of sets be *well-defined*, that is, we must be able to tell whether any object belongs or does not belong to the set in question.

EXAMPLE 1

"The set of white houses" is a well-defined set. A house is either white or it is not, and hence belongs or does not belong to the set mentioned. (While we wish to be accurate, we have no wish to be picayune; thus we will not argue over exact meanings of common words like "white" or "house.")

EXAMPLE 2

"The set of books in our college library" is well-defined. Any object is either a book or it is not a book. If an object is not a book, then it clearly does not belong to the set mentioned. If an object is a book, then it is either in or not in our college library.

EXAMPLE 3

"The set of beautiful paintings" is not well-defined. While we can tell whether an object is a painting or not, we cannot tell whether or not it is beautiful. "Beauty is in the eye of the beholder," so there will be no clear-cut way to tell whether a painting is beautiful.

It must be possible to determine whether an object is an element of, or belongs to, a given set. If there is argument or ambiguity or opinion involved in making the decision about membership, we will simply term the set "not well-defined" and disregard the set. Of course, sets that are not well-defined are often interesting and can sometimes completely involve a person's imagination. Consider the sets: "good Republicans," "beautiful people," "moral attributes," "law-abiding citizens," "best governments." Because these are not well-defined sets, we cannot deal with them mathematically. Many of the most interesting aspects of life deal with sets that must be excluded from our study.

Let S be a set. If an object, say x, is in set S, then x is called an element of S and we will write:

$$x \in S$$

If x is not an element of S, we will write:

$$x \notin S$$

There are two common ways to denote sets. The first is to simply list the elements within braces, separating each element by a comma. This is called the *roster method* of denoting a set.

EXAMPLE 4

$\{ a, b, c, d, \ldots, t \}$ is a set of letters in the alphabet up to and including t.

EXAMPLE 5

$\{ *, !, \# \}$ is a set of three symbols commonly found on a typewriter.

EXAMPLE 6

$\{ 2, 4, 6, 8, \ldots \}$ is the set of even counting numbers.

The second way to denote a set is to state a rule for forming the set without listing its members.

EXAMPLE 7

1. The set of all past Presidents of the United States. 2. The set of all vowels in the alphabet. 3. The set of all blue flowers. 4. The set of all even counting numbers. 5. The set of all natural numbers less than 10.

We can sometimes shorten these verbal descriptions of sets by using *set-builder* notation (sometimes called set-generator notation) as follows: We may write the set of all Presidents of the United States as

$$\{ x \mid x \text{ is a President of the United States} \}$$

The braces indicate a set, and can be read "the set of all"; then follows a typical member of the set represented by any symbol; then follows the sign "|" which can be read "such that;" and then follows the rule for membership in the set:

The set of all x such that x is a President of the United States.

$\{\ x\ |\ x$ is a President of the United States $\}$

EXAMPLE 8

"The set of all vowels in the alphabet that come before t" can be written $\{\ \theta\ |\ \theta$ is a letter of the alphabet and θ is a vowel and θ comes before $t\ \}$. Of course, this set can be written more easily by the roster method: $\{\ a, e, i, o\ \}$. Sometimes, one of the two methods for denoting sets will be easier than the other.

EXAMPLE 9

"The set of all thoughts" or $\{\ x\ |\ x$ is a thought$\}$ certainly is easier to represent by the rule method than by the roster method, since we could never list all thoughts.

Exercise Set 2.1

1 Determine whether each set is well-defined:
  (a) The set of great statesmen
  (b) The set of Princeton graduates
  (c) The set of men who have landed on the moon
  (d) The set of men who have landed on Mars
  (e) The set of short students
  (f) The set of tallest skyscrapers
  (g) The set of the four tallest skyscrapers
  (h) The set of young children in Wyoming

2 Decide whether each statement is true or false:
  (a) $w \in \{\ a, b, c, \ldots\ z\ \}$
  (b) $w \in \{\ a, b, c, \ldots, t\}$
  (c) $a \in \{\ x\ |\ x$ is a vowel $\}$
  (d) $\{\ a\ \} \in \{\ a, b, c\ \}$
  (e) $\{\ p\ \} \in \{\ \{\ p\ \}, \{\ q\ \}, \{\ r\ \}\ \}$
  °(f) $10 \notin \{\ x\ |\ x$ is an even number and x is a multiple of 5 $\}$
  (g) $(1, 8) \in \{\ (1, 1), (1, 2), (1, 3), (1, 4), \ldots\ \}$

3 Express each set using set-builder notation:
  (a) $\{\ 1, 2, 3\ \}$ (b) $\{\ 3, 6, 9, 12, 15, \ldots\ \}$
  (c) $\{\ 5, 10, 15, 20, \ldots, 75\ \}$ (d) $\{\ $red, orange, yellow, blue, green, violet $\}$
  (e) $\{\ $"1," "2," "3," "4," "5," $\ldots\ \}$ (f) $\{\ $maple, spruce, fir, pine, $\ldots$, aspen $\}$
  (g) $\{\ 1, 4, 9, 16, 25, 36, \ldots\ \}$

4 Express each set below using the roster method:
   (a) $\{\, \alpha \mid \alpha \in N$ and α is less than or equal to 7 $\}$
  °(b) $\{\, \beta \mid \beta \in N$ and β is a multiple of 6 $\}$
   (c) $\{\, t \mid t \in N$ and t is even and $t^2 = 16 \}$
   (d) $\{\, t \mid t \in N$ and t is even and $t^2 = 7\}$
   (e) $\{\, x \mid x$ is the name of a Great Lake $\}$
   (f) $\{\, x \mid x$ is in the alphabet and x is not a vowel$\}$
   (g) $\{\, x \mid x$ is a positive fraction $\}$
   (h) $\{\, x \mid x$ is the name of a President following J. F. Kennedy and preceding L. B. Johnson $\}$
  °(i) $\{\, (x, y) \mid x \in \{\, 1, 2 \}$ and $y \in \{\, a, b, c \} \}$
   (j) $\{\, (x, y) \mid x$ is heads or tails and y is heads or tails $\}$
   (k) $\{\, x \mid x$ is in N and x satisfies the equation $2x = 14 \}$
   (l) $\{\, x \mid x \in N$ and $2x = 5\}$
  (m)$\{\, x \mid x \in N^{+0}$ and $2x + 5 = 5\}$
   (n) $\{\, x \mid x \in N$ and $2x + 5 = 5\}$

2.2 THE UNIVERSAL SET AND VENN DIAGRAMS

The *universal set* or the *universe* is a set containing all the elements that might come under consideration during the course of a discussion. The symbol used for the universal set is U.

EXAMPLE 1

Let U be the set of all flowers in Mrs. Perkins's garden. Consider subsets of those flowers that are white and those that are scented.

For now, it is sufficient to think of *subset* as simply part of the universal set. Later the term will be defined more precisely.

The universal set can be sketched as a rectangle and the subsets can be drawn as circles within the rectangle. Such figures, which serve as visual aids, are called *Venn diagrams*. The Venn diagram for Example 1 is shown in Figure 1.

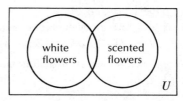

Figure 1   Venn diagram where U is the set of flowers in Mrs. Perkins's garden

In the preceding Venn diagram you will note that the circles overlap. This is to show that some of the elements may be in more than one subset. Venn diagrams allow for all the possibilities of membership.

In Figure 1 the circle labeled "white flowers" acts as a cell containing any flower that is white. The scented flowers are all contained in the other circle. Since some white flowers may also be scented, the two circles are shown overlapping. Any flower that was white and scented would belong to the overlapping portion of the two circles.

Some of the regions in a Venn diagram may not contain any elements; that is, a region may be empty. If Mrs. Perkins's garden contained no white flowers, then the region labeled "white flowers" would be empty. Likewise, the set labeled "scented flowers" may or may not be empty. In fact, if Mrs. Perkins's garden contained only ragweed, the universal set would be empty. The nonexistence of elements in certain sets does not prevent us from leaving room for them.

EXAMPLE 2

The universal set can be taken to be the set of all U.S. citizens. We can then discuss subsets of the universal set, say those U.S. citizens of voting age, those who are unemployed, or those who reside in Florida.

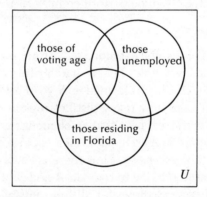

Figure 2   Venn diagram for Example 2, where U is the set of all U.S. citizens

How do we know that a particular Venn diagram accounts for all the possibilities of membership or nonmembership in the various subsets? Let us first investigate the situation where we have a universal set U and two subsets A and B.

An element of U is either in A or not in A. At the same time, it is either in B or not in B. There are four possibilities, indicated by the four separate (distinct) regions in Figure 3.

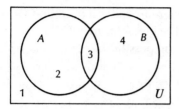

	A	B	Region
	in	in	3
	in	not in	2
	not in	in	4
	not in	not in	1

Figure 3   Venn diagram, any two subsets

Though you may be convinced that we have accounted for all the possibilities of membership in two subsets of a given universe, we will state a rule that will assure us that we are correct. The rule applies to many different counting situations. We will use it many times throughout this text.

THE FUNDAMENTAL PRINCIPLE OF COUNTING

If an act (or task) can be performed in N ways, and then a second act (or task) can be performed in M ways, the succession of acts (act 1 followed by act 2) can be performed in $N \times M$ ways.

This can be extended to 3 or more acts simply by repeated application of the principle. That is, if a third act can be performed in P ways, then all three can be performed in $N \times M \times P$ ways.

To see how many distinct regions there must be for two subsets of a universal set, note that an element has two choices for set A (it is either *in A* or *not in A*). The same element also has two choices for set B (it is either *in B* or *not in B*). Thus there are 2×2, or 4 ways for an element to be located relative to two sets A and B.

Figure 3 allows for all four possibilities. Since we were working with *any* two sets, Figure 3 can be called the most general Venn diagram for two subsets of a given universe. It should be clear that Figure 4 does not allow for all the possibilities:

Figure 4

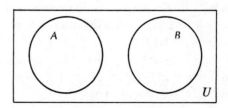

There is no region for an element that happens to belong to both *A* and *B*.

EXAMPLE 3

Let $U = \{a, b, c, d, \ldots, m\}$, $A = \{a, b, c, d, e\}$, and $B = \{d, e, f, g, h, i, j\}$. Distribute the elements of *U*, *A*, and *B* in an appropriate Venn diagram.

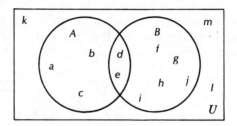

Figure 5

Suppose now, that there are three subsets, *A*, *B*, and *C*, in a given universe. What is the most general Venn diagram? Any element of the universe is in *A* or not in *A* (two choices), it is in *B* or not in *B* (two choices), and it is in *C* or not in *C* (two choices). Thus our Venn diagram must allow for $2 \times 2 \times 2$ or 8 distinct regions. Table 1 shows all the possibilities.

The Venn diagram in Figure 6 with regions labeled 1, 2, 3, 4, 5, 6, 7, and 8, is the most general diagram for three sets of a given universal set. It allows for all eight possibilities. You should check to see that the region labeled "1" is that region where an element would belong if it were in *A*, in *B*, and in *C* (line 1, Table 1). Also that region 2 would contain elements in *A*, in *B*, and not in *C*, and so on.

Table 1

Showing membership possibilities for three subsets

A	B	C	Region in Figure 6
in	in	in	1
in	in	not in	2
in	not in	in	3
in	not in	not in	4
not in	in	in	5
not in	in	not in	6
not in	not in	in	7
not in	not in	not in	8

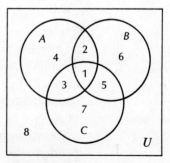

Figure 6

Venn diagram, any three subsets

Venn diagrams provide a way to organize data in certain survey problems. Consider the following example:

EXAMPLE 4

On a certain plane, there are 61 persons. Twenty-eight of these drink coffee, 25 drink tea, and 36 drink milk. Also 16 drink coffee and milk, 17 drink milk and tea, 13 drink coffee and tea, and 12 drink all three. How many passengers drink coffee and tea but not milk? How many drink neither tea nor coffee nor milk?

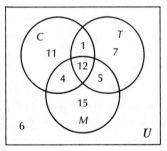

Figure 7   Diagram for Example 4

We can make this seemingly difficult problem easy if we record the given information on a Venn diagram that allows for all the possibilities. The numbers in the distinct regions in Figure 7 refer to the number of passengers in each category.

We start the problem with the fact that 12 persons drink all three beverages. We place a "12" in the intersection of all three circles. Then, since 13 drink coffee and tea, we locate the overlapping of the *C* and *T* circles and make sure the *total* is 13. Thus the two parts of this overlap region contain 12 persons and 1 person, respectively. Continue this way until all of the given information is distributed. That 28 persons drink coffee means that in the *C* circle the total must be 28, which is the case in Figure 7.

Once the number of persons in each part of the *C*, *T*, and *M* circles is ascertained, we add the numbers 11, 1, 7, 4, 12, 5, and 15 and get 55. Since there are 61 persons altogether, there must be 6 outside the circles. That is, 6 drink neither tea nor coffee nor milk. (The answer to our second question is 6.)

Study carefully the region containing "1." Do you see that this corresponds to the set of persons drinking coffee and tea but not milk? (The answer to the first question in Example 4 is 1.)

Exercise Set 2.2

1 Let $U = \{a, b, c, d, e, f, g, h, i, j\}$ and $B = \{a, e, i\}$ and $C = \{a, b, c, d, f\}$.
   (a) Record this information in a Venn diagram.
   (b) How many letters are in C, but not in B?
   (c) How many letters are in U but in neither B nor C?
   (d) What letters are in B and in C or else in neither B nor C?

2 (a) If an object can be in any of four subsets of a universal set, how many distinct regions would the Venn diagram contain?
   (b) Sketch such a diagram. Hint: Use some closed figure other than a circle for one of the sets.

3 Why is the following diagram not the most general diagram showing three subsets of a given universe?

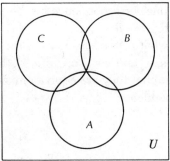

Figure 8

4 Sketch a Venn diagram for the following information:

$$U = \{1, 2, 3, 4, 5, 6, 7, 8, 9, 10\}, A = \{1, 3, 5, 7, 9\}, B = \{2, 4, 6, 8, 10\}$$

   (a) What two regions are empty?
   (b) Now introduce a third set $C = \{1, 4, 9\}$ and sketch another Venn diagram. How many regions are empty?
   (c) What elements are in A and C but not in B?
   (d) What elements are in A and B but not in C?

5 Registration cards for 100 students, drawn at random, show that there are 15 students taking mathematics, chemistry and physics, 5 students taking math and physics alone, 10 students taking chemistry and physics alone, 15 taking math and chemistry alone, 15 taking just math, 10 taking just chemistry, and 20 taking just physics.
   (a) How many cards show that a student was not taking any of these courses?
   (b) If there had been 120 cards, how would the answer to question (a) have changed?
   (c) If there had been 80 cards, how would you have responded to (a)?

6 If there are five routes from New York to Chicago, and four routes from Chicago to Denver, and two routes from Denver to Los Angeles, how many different routes are there from New York to Los Angeles by way of Chicago and Denver?

7 How many different 7-digit telephone numbers can be dialed? If the first digit cannot be a 0 or a 1, how many 7-digit numbers can be dialed?

°8 Sketch a Venn diagram showing five subsets of a given universe allowing for all possibilities of membership and nonmembership. Use closed curves for each subset but do not try to use circles.

2.3 RELATIONSHIPS BETWEEN SETS

Now that we know how to denote sets and how to represent them by Venn diagrams, we turn to a study of the comparison between two sets.

In Chapter 1 one-to-one correspondences introduced the notion of equivalence of sets. Recall that two sets A and B are equivalent if a one-to-one correspondence can be established between them; in that case we write $A \leftrightarrow B$.

But what if two sets contain the same elements? For example, if $R = \{ e, i, o, u, a \}$ and $S = \{ u, o, e, a, i \}$, then R and S contain the same elements. It is true that they are equivalent since we can demonstrate a one-to-one correspondence.

$$R: \quad e \quad i \quad o \quad u \quad a$$
$$\quad \updownarrow \quad \updownarrow \quad \updownarrow \quad \updownarrow \quad \updownarrow$$
$$S: \quad u \quad o \quad e \quad a \quad i$$

However, we know that two sets may contain different elements and still be equivalent. A stronger statement can be made about the sets R and S above. When two sets contain the same elements, we will say that they are *equal*. The following definition gives the criteria for equality between sets.

DEFINITION 1   Equal Sets

Let A and B be two sets. If every element of A is also an element of B and if every element of B is also an element of A, then A is *equal* to B and we write $A = B$.

EXAMPLE 1
Let $T = \{ a, b, c, d \}$ and $S = \{ d, c, a, b \}$.
Clearly $T = S$.

EXAMPLE 2
Let $A = \{ b, c, d, e \}$ and $B = \{ b, c, d, e, a \}$.
In this case A is not equal to B. While every element of A is an element of B, it is not true that every element of B is an element of A.

EXAMPLE 3

Let $A = \{ a, b, c \}$ and $B = \{ a, a, b, c \}$.
Note that every element of A is an element of B, and every element of B is an element of A. Therefore, $A = B$.

In Example 3 the cardinality of (or the number of elements in) set A is 3 and the cardinality of set B seems to be 4. However, we have said that elements should not be repeated in sets, so B can be more properly written $\{ a, b, c \}$. For this reason the equality of the sets in Example 3 is clear.

Equal sets are the same sets. If a Venn diagram is sketched showing that $A = B$, then A and B would have to be the same circle, and $A = B$ simply tells us that there are two labels for the same set. See Figure 9.

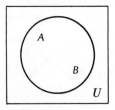

Figure 9   Equal sets are the same sets.

Now we know that if two sets have the same number of elements, they are equivalent, and they may be equal. The sets $A = \{ a, b, c \}$ and $B = \{ a, b, c, d \}$ are neither equivalent nor equal, but there is a relationship between them. Every element of A is an element of B. We will call A a *subset* of B.

DEFINITION 2   Subset

Let A and B be two sets. If every element of A is an element of B, then A is a *subset* of B. This is written $A \subseteq B$.

EXAMPLE 4

Let $R = \{ 1, 2, 3, 4, 5 \}$ and $S = \{ 1, 2, 3 \}$.
Then $S \subseteq R$ since each element of S is an element of R.

In Example 4 not all of the elements of R are elements of S, so R is not a subset of S. This is written:

$$R \nsubseteq S$$

where the slash through the subset symbol negates its meaning.

EXAMPLE 5

Let $Q = \{\, i, c, e \,\}$ and $R = \{\, e, i, c \,\}$.
Here $Q = R$, but Q is a subset of R also, since every element in Q is also found in R. It is also true that $R \subseteq Q$.

Example 5 suggests that equality of sets could have been defined in terms of subsets. Set A is equal to set B if $A \subseteq B$ and $B \subseteq A$.

One of the interesting subsets of a given set contains no elements. The set that contains no elements is called the *null set* or the *empty set* and is designated by either of these two symbols:

$$\emptyset \text{ or} \{\ \ \}$$

Why is $\emptyset$ a subset of any set? Think of any set you wish and call it A. Then $\emptyset \subseteq A$ since any element found in $\emptyset$ (there are none) is also found in A.

EXAMPLE 6

Let $A = \{\, a, b, c, d, e, f \,\}$. Then the empty set, $\emptyset$, is a subset of A, since there are no elements in the empty set that are not in A.

The intuitive idea of subset does not seem to allow a set to be a subset of itself, but Definition 2 does imply that every set is a subset of itself. It is conventional to allow A to be a subset of A, but mathematicians say that A is not a *proper* subset of A.

DEFINITION 3   Proper Subset

Set A is a proper subset of set B if first, A is a subset of B, and second, B contains at least one element not found in A. The symbol for proper subset is $\subset$.

EXAMPLE 7

Let $A = \{\, 1, 2, 3 \,\}$ and $B = \{\, 1, 2, 3, 4 \,\}$.
Then $A \subset B$ since $A \subseteq B$ and B contains at least one element not in A.

After some thought it should be apparent that all the subsets of a given set are proper subsets except the set itself.

If a set is given, how many subsets are there? Suppose we take $S = \{\, a, b, c \,\}$. Then element a is in any subset or it is not in that subset; element b

is in any subset or it is not in that subset; element c is in any subset or it is not. Thus there are $2 \times 2 \times 2$ or 8 subsets of S. The following are the subsets of S:

$$\{\ \}, \{a\}, \{b\}, \{c\}, \{a, b\}, \{a, c\}, \{b, c\}, \{a, b, c\}$$

It should be clear that if $T = \{a, b, c, d\}$, there are $2 \times 2 \times 2 \times 2$ or 16 subsets. And, if we had 5 elements in a set, there would be $2 \times 2 \times 2 \times 2 \times 2$ or 32 subsets. In general, if a set contains n elements, there are 2^n subsets.

$$\underbrace{2 \times 2 \times 2 \times \ldots \times 2}_{n \text{ twos}}$$

(The repeated multiplication of 2's follows from the Fundamental Principle of Counting.)

Exercise Set 2.3

1 Show that if $L = \{a, b, c, \ldots, t\}$ and $M = \{a, b, c, \ldots, w\}$, then $L \neq M$. Try to make your answer exactly fit the definition of equality.

2 If $R = \{a, b, c\}$ and $S = \{1, 2, 3\}$, then $R \leftrightarrow S$. Clearly $R \neq S$. Are equivalent sets ever equal sets? Are equal sets always equivalent?

3 True or false: If $A = \{x \mid x \in N \text{ and } x \text{ is less than } 7\}$ and $B = \{1, 2, 3, 4, 5, 6\}$, then $A = B$.

4 Let A be the set of letters used to spell the word "steak" and let B be the set of letters used to spell the word "skate." Does $A = B$?

5 Let $U = \{1, 2, 3, 4, 5, 6, 7, 8\}$, $A = \{2, 4, 6, 8\}$, $B = \{1, 3, 5, 7\}$, $C = \{8, 2, 4, 6\}$, and $D = \{a, b, c, d\}$.
Decide whether or not each of the following is true:
(a) $A \subseteq U$ (b) $B \subseteq B$ (c) $C \subset C$ (d) $A = B$
(e) $A = C$ (f) $A \leftrightarrow D$ (g) $A \leftrightarrow C$

6 Make up two examples of sets that are empty. What is the cardinality of the empty set?

7 (a) Write all the subsets of $\{1, 2, 3, 4\}$.
(b) Which subset is not a proper subset?

8 (a) Is $\emptyset$ a subset of $N = \{1, 2, 3, \ldots\}$?
(b) Is $\emptyset$ a proper subset of N?
°(c) Is $\emptyset$ a subset of $\emptyset$?
°(d) Is $\emptyset$ a proper subset of $\emptyset$?

9 (a) How many subsets does $\{a, b\}$ have?
(b) How many subsets does a set of four elements have? Five elements? Six elements? One hundred elements?

10 Recall the meaning of the symbols N and N^{+0} (Chapter 1).
(a) Is $N \subseteq N^{+0}$?   (b) Is $N \subset N^{+0}$?   (c) Is $N^{+0} \subseteq N$?

2.4 UNION, INTERSECTION, AND COMPLEMENT

When two sets are given, the elements can be combined to form a new set called the *union* of the given sets.

DEFINITION 4   Union

By the *union* of two sets A and B, we mean a third set whose elements are found in A or in B or in both. The union of A and B is written $A \cup B$.

EXAMPLE 1

Let $A = \{ a, b, c \}$ and $B = \{ c, d, e \}$. Then $A \cup B = \{ a, b, c, d, e \}$.

Observe that the union of two sets may contain fewer elements than in the sum of the elements of A and the elements of B. Using cardinality notation and referring to Example 1, we see that $n(A) = 3$, $n(B) = 3$, but $n(A \cup B) = 5$.

EXAMPLE 2

Let $R = \{ 1, 2, 3, 4, 5, 6, 7 \}$ and $S = \{ 1, 3, 5, 7 \}$.
$R \cup S = \{ 1, 2, 3, 4, 5, 6, 7 \}$. Note that $R \cup S = R$.

We can depict the union of two sets by shading those regions of a Venn diagram that satisfy the definition, as in Figure 10. Points or elements in the region labeled "1" do not belong to the union since they are not in A and not in B. Points in regions labeled "2," "3," and "4" do belong to the union set, since an element in any of these three regions is in A or in B. (Points in region "3" lie in both A and B; Definition 4 specifically tells us to include such points in the union.) Thus $A \cup B$ refers to the shaded portion in Figure 10.

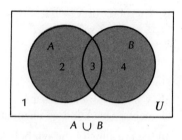

$A \cup B$

Figure 10   The union of two sets

We can also speak of the set of elements two given sets have in common. This set is called the *intersection* of the given sets.

DEFINITION 5   Intersection

Let A and B be two sets. The *intersection* of A and B is a third set whose elements are found in A *and* in B. The intersection of sets A and B is denoted $A \cap B$.

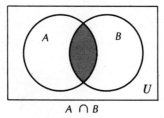

$A \cap B$

Figure 11   The intersection of two sets

Intuitively, we may think of the intersection as the "overlap" set. The Venn diagram depicting the intersection of sets A and B is shown in Figure 11.

EXAMPLE 3

Let $A = \{ 1, 2, 3, 4 \}$ and $B = \{ 3, 4, 5, 6, 7 \}$.
Then $A \cap B = \{ 3, 4 \}$.

EXAMPLE 4

Let $R = \{ x \mid x \in N$ and x is less than 10 $\}$ and let $S = \{ y \mid y \in N$ and y is even and y is less than 12 $\}$.
Then $R \cap S = \{ 2, 4, 6, 8 \}$.

EXAMPLE 5

The intersection of the set of people living in Florida with the set of U.S. citizens who are unemployed, is the set of unemployed U.S. citizens living in Florida.

EXAMPLE 6

$\phi \cap U = \phi$ That is, the intersection of the empty set with the universal set is the empty set.

It is important to keep the distinction between union and intersection clear. In the definition of union the key word used is *or*. In the definition of intersection the key word is *and*. We can abbreviate our thinking this way:

$$A \cup B \text{ means "in } A \text{ or in } B, \text{ or in both"}$$
$$A \cap B \text{ means "in } A \text{ and in } B\text{"}$$

When a subset of a universe is given, the elements of the universe not in the subset are called the *complement* of the subset.

DEFINITION 6   Complement

Let *U* be a universal set with subset *A*. Then the *complement* of set *A*, written *A′* ("A prime"), is a set whose elements are in *U* but not in *A*.

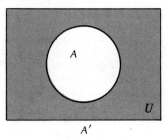

Figure 12   The complement of set *A*

EXAMPLE 7

Let $U = \{1, 2, 3, 4, 5, 6, 7, 8, 9, 10\}$ and $R = \{1, 2, 3, 4, 5, 6\}$. Then $R \subseteq U$ and so $R' = \{7, 8, 9, 10\}$.

The Venn diagram of the complement of set *A* is shown in Figure 12. Notice that *A′* is outside of *A* but in *U*.

Unions, intersections, and complements are operations on sets and can be combined. For example, given sets *A*, *B*, and *C* we can find $A \cup B'$ or $A \cap (C \cup B)$ or $(B \cap C)'$. There are many possibilities. Before turning to some examples, we will clarify the use of parentheses and the order in which compound operations are performed.

1. $R \cup S$ is the same as $(R \cup S)$. Here the parentheses are used only to clarify the grouping.
2. $R \cup S'$ is termed the union of *R* and *S′*. Before forming the union, we need to know *S′*.
3. $(R \cup S)'$ is the complement of $R \cup S$. Before finding the complement, we find the union.
4. Be careful not to assume that $(R \cup S)'$ is the same as $R' \cup S'$; the first expression is a "complement of the union of two sets" and the second is a "union of the complements of two sets."

EXAMPLE 8

Let $U = \{1, 2, 3, 4, 5, 6\}$, $R = \{1, 3, 5\}$, and $S = \{3, 4, 5, 6\}$.

(a) Find $R \cup S'$. First, we find S', which is the set of elements in U but not in S. So, $S' = \{1, 2\}$. Then, since R is $\{1, 3, 5\}$, the required union, $R \cup S'$, is $\{1, 2, 3, 5\}$.

(b) Find $(R \cup S)'$. First, we find $R \cup S$, the set of elements in R or in S, or in both. $R \cup S = \{1, 3, 5, 4, 6\}$. Then we find the complement of $R \cup S$, which is the set of elements in U but not in $R \cup S$. $(R \cup S)' = \{2\}$.

(c) Find $R' \cup S'$. First, we find that R' is $\{2, 4, 6\}$ and that S' is $\{1, 2\}$. Then we form the required union. $R' \cup S' = \{1, 2, 4, 6\}$.

In general, parenthesized operations are done first. Thus in $A \cup (B \cap C)$ we need to know $B \cap C$ before we form the union with A. Also, in $A \cup (B \cap C)'$ we would find the intersection, then its complement, and then the union with set A.

EXAMPLE 9

Let $U = \{1, 2, 3, 4, 5, 6, 7, 8\}$, $A = \{1, 3, 5\}$, $B = \{1, 2, 3, 4\}$, and $C = \{1, 2, 4, 8\}$.

(a) Find $A \cup (B \cap C)$.

$$B \cap C = \{1, 2, 4\}$$
$$A = \{1, 3, 5\}$$

so

$$A \cup (B \cap C) = \{1, 2, 3, 4, 5\}$$

(b) Find $A \cup (B \cap C)'$.

$$B \cap C = \{1, 2, 4\}$$

so

$$(B \cap C)' = \{3, 5, 6, 7, 8\}$$
$$A = \{1, 3, 5\}$$

Then $A \cup (B \cap C)' = \{1, 3, 5, 6, 7, 8\}$

Figures 10, 11, and 12 give the basic Venn diagrams corresponding to union, intersection, and complement. For combinations of these operations we can also find appropriate Venn diagrams.

EXAMPLE 10

Shade those regions of a Venn diagram corresponding to $A \cup B'$, assuming the universe U. First, we select the most general Venn diagram (Figure 13) for two subsets of U.

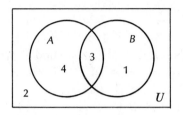

Figure 13

The four distinct regions are labeled 1, 2, 3, 4, in any order. Then we find the regions associated with the components of $A \cup B'$. Then $A \cup B'$ is the union of regions 4, 3, and 2, 4. So $A \cup B'$ corresponds to regions 2, 3, and 4, and we shade the Venn diagram as in Figure 14.

	Corresponds to regions
A	4, 3
B'	2, 4

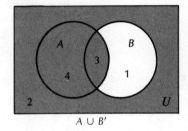

Figure 14 $A \cup B'$

EXAMPLE 11

Shade those regions of a Venn diagram corresponding to $(A \cup B) \cap C'$.

Clearly, we need a Venn diagram with three overlapping circles and eight regions, which we label arbitrarily as in Figure 15.

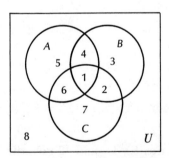

Figure 15

		Corresponds to regions
	A	1, 4, 5, 6
	B	1, 2, 3, 4
Therefore,	$A \cup B$	1, 2, 3, 4, 5, 6
	C	1, 2, 6, 7
So,	C'	3, 4, 5, 8
Hence,	$(A \cup B) \cap C'$	3, 4, 5

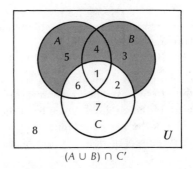

Figure 16 $(A \cup B) \cap C'$

After a while you may be able to quickly locate the regions to shade for compound statements, without making lists. But, the list method will be very useful in later work with proofs concerning sets.

Exercise Set 2.4

1 $U = \{1, 2, 3, 4, 5, 6, 7, 8, 9, 10\}$ $R = \{1, 2, 3, 4, 5\}$
   $S = \{3, 4, 5, 6, 7\}$ $T = \{1, 3, 5, 7, 9\}$
   Determine the elements in each set:
   (a) $R \cup S$ (b) $R \cap S$
   (c) $R \cap U$ (d) $T \cup U$
   (e) R' (f) S'
   (g) $R' \cup S'$ (h) $R' \cap S'$
   (i) $(R \cup S)'$ (j) $(R \cap S)'$
   (k) $T \cup (R \cap S)$ (l) $(T \cup R) \cap (T \cup S)$
   (m) $(T \cap R)'$ (n) $T' \cup R'$
   (o) $(T \cup R)'$ (p) $T' \cap R'$

2 Using the sets given in Exercise 1, find each of the following:
   (a) $R \cup (S \cup T)'$ (b) $(R \cup S) \cup T$
   (c) $R \cup (S \cup T)$ (d) $R \cap (S \cup T)$
   (e) $R \cap (S \cup T)'$ (f) $R \cup (S \cap T)$
   (g) $(R \cup S) \cap T$ (h) $(R \cup S) \cap (R \cup T)$
   (i) $R' \cup S' \cup T'$ (j) $(R \cup S \cup T)'$

3 In an appropriate Venn diagram, shade those regions corresponding to each of the following:
   (a) $W \cap T'$ (b) $A' \cup B'$
   (c) $(A \cup B)'$ (d) $A \cap (B \cup C)$
   (e) $(A \cup B) \cap C$ (f) $A \cap B \cap C$, which is the same as
                                             $(A \cap B) \cap C$ or $A \cap (B \cap C)$
   (g) $(A \cup B \cup C)'$ (h) $(A \cap B) \cap C'$

°4 For the sets given in Exercise 1, decide whether each of the following is true:
   (a) $(R \cup S) \subseteq T$ (b) $(R \cap S) \subseteq T$ (c) $R \leftrightarrow S \leftrightarrow T$
   (d) $R = S = T$ (e) $R \subseteq R$ (f) $R' \subseteq U$
   (g) $R \cap R' = \emptyset$ (h) $(R \cap R') \cup U = U$ (i) $(R \cap R') \cap U = \emptyset$

°5 $\xleftarrow{\hspace{0.3cm}\overset{A\qquad B}{\hspace{2cm}}\hspace{0.3cm}}$ l Shown at the left is a line l and a line segment AB on line l.

(a) Is $AB \subseteq l$?   (b) What is $AB \cap l$?   (c) What is $AB \cup l$?

°6 A circle can be thought of as a set of points in a plane at an equal distance (radius) from a fixed point (center). Any other point in the plane is either in the interior of the circle (distance from the center is less than the radius) or in the exterior of the circle (distance from center is greater than the radius). Thus we have three sets of points in the plane P: the circle C, its interior C_I, and its exterior C_E.

(a) What is $C \cup C_I$? (A description will do.)
(b) What is $C \cup C_E$?
(c) What is $C \cup C_I \cup C_E$?
(d) What is $C \cap C_I$?
(e) What is $C_I \cup C_E$?
(f) What is $C_I \cap C_E$?
(g) Taking U as P, what is $(C_I \cup C)'$?

*7 (a) Describe each of the eight regions in the Venn diagram below in terms of union, intersection, and complement by completing the table. There may be more than one correct answer.

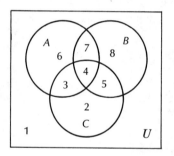

Figure 17

A	B	C	Region	Set language description
in	in	in	4	$A \cap B \cap C$
in	in	out	7	$(A \cap B) \cap C'$
in	out	in	3	$A \cap (B' \cap C)$
in	out	out	6	
out	in	in	5	
out	in	out	8	
out	out	in	2	
out	out	out	1	

(b) Have you discovered a method for doing this? Hint: The answers to (a) can be only in terms of $\cap$ and $'$.

(c) What regions are described by each of the following?
1. $(A \cup B) \cap C$ 2. $C \cap B'$
3. $(A \cap B \cap C)'$ 4. $(A \cup B \cup C)'$
5. $(A \cap B) \cap (B \cap C)$ 6. $(A \cap B) \cup (A \cap C) \cup (B \cap C)$
7. $(B \cap C)'$ 8. $B' \cap C'$

2.5 PROVING STATEMENTS INVOLVING SETS

Some of the preceding exercises were designed to suggest that different combinations of union, intersection, and complement may yield the same set. You might have noticed in Question 1, Exercise Set 2.4, $R' \cup S'$ had the same elements as $(R \cap S)'$, and $(T \cup R)'$ had the same elements as $T' \cap R'$.

These specific examples suggest but do not prove that the sets are equal. We need a method of proof that is general and not dependent upon an example. Examples are often misleading. In the following example several conclusions are reached that are true for the given sets but not true for all sets.

EXAMPLE 1

Let $U = \{1, 2, 3, 4, 5, 6, 7\}$, $A = \{1, 2, 3, 4\}$, and $B = \{5, 6, 7\}$. Then

$$A' = \{5, 6, 7\}$$
$$B' = \{1, 2, 3, 4\}$$
$$A \cup B = \{1, 2, 3, 4, 5, 6, 7\}$$
$$(A \cup B)' = \{\quad\} = \emptyset$$
$$A \cap B = \emptyset$$
$$(A \cap B)' = \{1, 2, 3, 4, 5, 6, 7\}$$
$$A' \cap B' = \emptyset$$
$$A' \cup B' = \{1, 2, 3, 4, 5, 6, 7\}$$

So we can conclude, from this example, that $A' \cup B' = A \cup B$ or that $A \cap B = A' \cap B'$. We also see that $(A \cup B)' = A' \cap B'$ and that $(A \cup B)' = A' \cap B'$. But are all of these equalities *theorems* about sets, that is, are these statements true for all sets? Can we prove the statements in general? We will see that the last two equalities mentioned are true in general, but that the first two equalities are not. To show that $A' \cup B' = A \cup B$ and $A \cap B = A' \cap B'$ are *not* true in general, we need only to change the sets A and B.

EXAMPLE 2

Let $U = \{1, 2, 3, 4, 5\}$, $A = \{1, 2, 4\}$, and $B = \{2, 4\}$. Then

$$A' = \{3, 5\}$$
$$B' = \{1, 3, 5\}$$
$$A' \cup B' = \{1, 3, 5\}$$
$$A \cup B = \{1, 2, 4\}$$
$$A' \cap B' = \{3, 5\}$$
$$A \cap B = \{2, 4\}$$

In this example $A' \cup B' \neq A \cup B$ and $A \cap B \neq A' \cap B'$.

We now turn to a technique of proof that can be used for any statement of equality between compound sets.

Theorem 1 $A \cup (B \cap C) = (A \cup B) \cap (A \cup C)$

First, we set up a Venn diagram and label the regions 1 through 8 as in Figure 18. Remember that we have determined that the figure below is the most general Venn diagram for any three sets.

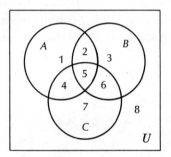

Figure 18

Proof

Step 1   Find the regions corresponding to $A \cup (B \cap C)$.

	Corresponds to regions
A	1, 2, 4, 5
$B \cap C$	5, 6
So, $A \cup (B \cap C)$	1, 2, 4, 5, 6

Step 2   Find the regions corresponding to $(A \cup B) \cap (A \cup C)$.

	Corresponds to regions
$A \cup B$	1, 2, 3, 4, 5, 6
$A \cup C$	1, 2, 4, 5, 6, 7
So, $(A \cup B) \cap (A \cup C)$	1, 2, 4, 5, 6

Step 3   Since the regions corresponding to $A \cup (B \cap C)$ are 1, 2, 4, 5, 6 (by step 1) and the regions corresponding to $(A \cup B) \cap (A \cup C)$ are also 1, 2, 4, 5, 6 (by step 2), we conclude that the theorem is true.

The method of proof is to show that each side of the statement of equality corresponds to the same set of regions in a labeled Venn diagram. Regardless of how the regions are labeled we will get the same regions in terms of their relative position to each of the others. If we were to shade those regions of a Venn diagram corresponding to each side of Theorem 1, we would get a figure like Figure 19.

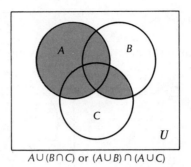

Figure 19 $A \cup (B \cap C)$ or $(A \cup B) \cap (A \cup C)$

Theorem 2 $A \cap (B \cup C) = (A \cap B) \cup (A \cap C)$
(The proof is left as an exercise for the student.)

Theorem 3 $(A \cup B)' = A' \cap B'$

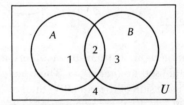

Figure 20

Proof

Step 1   What regions correspond to $(A \cup B)'$?

	Corresponds to regions
$A \cup B$	1, 2, and 3
So, $(A \cup B)'$	4

Step 2   What regions correspond to $A' \cap B'$?

	Corresponds to regions
A'	3 and 4
B'	1 and 4
So, $A' \cap B'$	4

Step 3   Since both $(A \cup B)'$ and $A' \cap B'$ correspond to region 4, we conclude that the theorem is true.

Exercise Set 2.5

1 Prove Theorem 1 again with a Venn diagram labeled as follows:

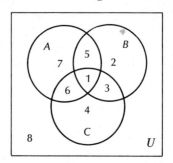

Figure 21

2 Prove Theorem 2. Use Figure 21.

3 Prove Theorem 3 with a Venn diagram labeled as follows:

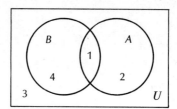

Figure 22

4 The method of proof for theorems about equal sets can be used to show that certain sets are *not* equal. The conclusion of the technique (comparing regions for each side of the equality) then results in two different sets of regions. Prove that $(A \cup B)'$ is not equal to $A' \cup B'$.

5 Prove that $(A \cap B)'$ is not equal to $A' \cap B'$.

6 Prove each of the following theorems:
   (a) Theorem 4: $(A \cap B)' = A' \cup B'$
   (b) Theorem 5: $A \cup (B \cup C) = (A \cup B) \cup (A \cup C)$
   (c) Theorem 6: $A \cap \emptyset = \emptyset$ Hint: Since the empty set contains no elements, the Venn diagram allowing for the set A and $\emptyset$ contains only two regions, "in A" or "not in A."
   (d) Theorem 7: $A \cup \emptyset = A$
   °(e) Theorem 8: $(A \cap B) \subseteq (A \cup B)$ Hint: This theorem says that one compound set is a subset of another. The method of proof can be altered to prove this theorem.
   (f) Theorem 9: $S \cup S = S$
   (g) Theorem 10: $S \cap S = S$
   (h) Theorem 11: $R = (R \cap T) \cup (R \cap T')$
   (i) Theorem 12: $A \cap (B \cup C)' = (A \cap B') \cap (A \cap C')$
   °(j) Theorem 13: $(A \cap B') \subseteq A$
   (k) Theorem 14: $G \cap G' = \emptyset$
   (l) Theorem 15: $A \cap (B \cap C)' = (A \cap B') \cup (A \cap C')$

2.6 MEMBERSHIP TABLES

Recalling that $A \cup B$ is the set of elements found in A or in B, or in both, we can tabulate all the possibilities for membership or nonmembership in A and B and for membership in the compound set $A \cup B$.

Table 2   Membership table for $A \cup B$

	A	B	$A \cup B$	A	B	$A \cup B$
Line 1:	in	in	in	$\in$	$\in$	$\in$
Line 2:	in	not in	in	$\in$	$\notin$	$\in$
Line 3:	not in	in	in	$\notin$	$\in$	$\in$
Line 4:	not in	not in	not in	$\notin$	$\notin$	$\notin$

Both parts of the table are the same. The one to the right uses "$\in$" for "is an element of" or "in," and uses "$\notin$" for "is not an element of" or "not in."

Remember that the Venn diagram for $A \cup B$ was shaded as in Figure 23. We have numbered the regions to correspond to the lines in our membership table, Region 1 is the region with elements in A and in B; region 2 is the region with elements in A but not in B, etc.

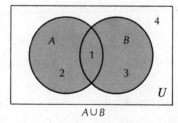

Figure 23 $A \cup B$

We can do the same thing for $A \cap B$, Table 3 and Figure 24.

Table 3   Membership table for $A \cap B$

	A	B	$A \cap B$	A	B	$A \cap B$
Line 1:	in	in	in	$\in$	$\in$	$\in$
Line 2:	in	not in	not in	$\in$	$\notin$	$\notin$
Line 3:	not in	in	not in	$\notin$	$\in$	$\notin$
Line 4:	not in	not in	not in	$\notin$	$\notin$	$\notin$

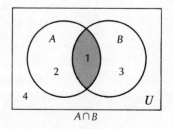

Figure 24

For A' we only need one subset, which contains those elements in U but not in A: Table 4 and Figure 25.

Table 4   Membership table for A'

	A	A'	A	A'
Line 1:	in	not in	$\in$	$\notin$
Line 2:	not in	in	$\notin$	$\in$

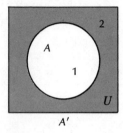

Figure 25

The three basic set operations can be combined and membership tables can be constructed for these new compound statements.

EXAMPLE 1

Determine the membership table for $(A \cup B)'$.
(First, the union is found and then the complement is derived from the union.)

A	B	$(A \cup B)$	$(A \cup B)'$
$\in$	$\in$	$\in$	$\notin$
$\in$	$\notin$	$\in$	$\notin$
$\notin$	$\in$	$\in$	$\notin$
$\notin$	$\notin$	$\notin$	$\in$

Before the double vertical line we have listed all the possibilities for membership and nonmembership in two subsets of a universal set. Under $A \cup B$ we have placed the appropriate membership symbol for union which we listed in Table 2. Note that the $\in$'s and $\notin$'s are placed under the operation symbol. Then we have determined the membership table for the complement of $A \cup B$, that is for $(A \cup B)'$. The table for $A \cup B$ runs $\in$ - $\in$ - $\in$ - $\notin$ line-by-line and so the table for $(A \cup B)'$ runs $\notin$ - $\notin$ - $\notin$ - $\in$ line-by-line. (It should be clear that if an element is in a certain set, then it is not in that set's complement; $\in$ changes to $\notin$ and $\notin$ to $\in$.)

Note the placement of the membership symbols under the complement sign in the column headed $(A \cup B)'$.

We can compress the table somewhat by not making a separate column for $A \cup B$.

A	B	$(A \cup B)'$	
∈	∈	∈	∉
∈	∉	∈	∉
∉	∈	∈	∉
∉	∉	∉	∈
		①	②

The membership table for $(A \cup B)'$ is in column 2. Column 1 was used to derive column 2.

We have in membership tables then, a kind of computational device. We move from column to column calculating new entries from previous ones. Study the following example carefully.

EXAMPLE 2

Find the membership table for $A' \cap B'$.

	A	B	A'	$\cap$	B'
	∈	∈	∉	∉	∉
	∈	∉	∉	∉	∈
Line 3:	∉	∈	∈	∉	∉
	∉	∉	∈	∈	∈
			①	③	②

Column 1 is found by reversing the entries under A. Column 2 is found by reversing the entries under B. Column 3 is found by combining columns 1 and 2 by way of intersection. (On line 3, for example, we have an element not in A but in B. So the element is in A' and not in B', which means it is not in the intersection of A' and B'.) Column 3 is the membership table for $A' \cap B'$.

EXAMPLE 3

Find the membership table for $A \cap (B \cup C)$.

	A	B	C	A	$\cap$	$(B \cup C)$
1.	∈	∈	∈	∈	∈	∈
2.	∈	∈	∉	∈	∈	∈
3.	∈	∉	∈	∈	∈	∈
4.	∈	∉	∉	∈	∉	∉
5.	∉	∈	∈	∉	∉	∈
6.	∉	∈	∉	∉	∉	∈
7.	∉	∉	∈	∉	∉	∈
8.	∉	∉	∉	∉	∉	∉
				①	③	②

The eight lines in this table correspond to the eight separate regions of a Venn diagram for three subsets, labeled correspondingly as follows:

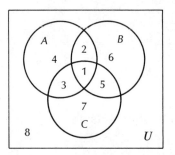

Figure 26

In column 1 we simply copied the entries under A. In column 2 we found the union of B and C by looking at the columns headed B and C and recalling the definition of union. In column 3 we found the intersection by looking at columns 1 and 2. Column 3 is the membership table for $A \cap (B \cup C)$.

Exercise Set 2.6

(Do this exercise before proceeding.)
1 Set up a Venn diagram labeled exactly the same as Figure 26. By the methods of Section 2.4, shade those regions corresponding to $A \cap (B \cup C)$. The result should be regions 1, 2, and 3 which correspond to lines 1, 2, and 3 of Example 3. Thus an element belongs to $A \cap (B \cup C)$ if one of three things holds:
   1. It is in A, in B, and in C (line 1 or region 1).
   2. It is in A and B but not in C (line 2 or region 2).
   3. It is in A and C but not in B (line 3 or region 3).

*2.7 PROOFS BY MEMBERSHIP TABLES

Theorems about sets can be proved by the method of Section 2.5 or by membership tables. In this section we investigate the latter technique. Suppose we wish to prove Theorem 3, $(A \cup B)' = A' \cap B'$. The proof consists of comparing the membership tables for the sets on each side of the equal sign. If the membership tables are the same, line-by-line, then the theorem is true. Table 5 shows the membership tables.

The membership table for $(A \cup B)'$ is in column 2 and for $A' \cap B'$ is in column 5. Since both are $\notin$- $\notin$- $\notin$- $\in$, Theorem 3 is proved. (These particular membership tables were constructed in Section 2.6 and explained thoroughly; you may wish to review that section if you have questions about the tables.)

Table 5 Membership table for Theorem 3

A	B	(A ∪ B)′		A′ ∩ B′		
∈	∈	∈	∉	∉	∉	∉
∈	∉	∈	∉	∉	∉	∈
∉	∈	∈	∉	∈	∉	∉
∉	∉	∉	∈	∈	∈	∈
		①	②	③	⑤	④

EXAMPLE

Prove Theorem 1, $A \cup (B \cap C) = (A \cup B) \cap (A \cup C)$.

Table 6 Membership table for Theorem 1

A	B	C	A ∪ (B ∩ C)			(A ∪ B) ∩ (A ∪ C)		
∈	∈	∈	∈	∈	∈	∈	∈	∈
∈	∈	∉	∈	∈	∉	∈	∈	∈
∈	∉	∈	∈	∈	∉	∈	∈	∈
∈	∉	∉	∈	∈	∉	∈	∈	∈
∉	∈	∈	∉	∈	∈	∈	∈	∈
∉	∈	∉	∉	∉	∉	∈	∉	∉
∉	∉	∈	∉	∉	∉	∉	∉	∈
∉	∉	∉	∉	∉	∉	∉	∉	∉
①	③		②			④	⑥	⑤

Since columns 3 and 6 are the same, the theorem is proved.

This second method of proof of theorems about set equalities may seem more difficult than the method of Section 2.5, but when you become proficient with the use of tables, it is quite rapid. Another advantage is this new method does not depend on Venn-diagram labeling.

Exercise Set 2.7

1 Using the sets $U = \{a, b, c, d, e, f, g\}$, $A = \{a, c, e, g\}$, and $B = \{d, e, f, g\}$, show that $(A \cup B)'$ is not the same as $A' \cup B'$.

2 (a) True or false: $A \cup B$ is never the same set as $A' \cup B'$. (See Example 1, Section 2.5)

   (b) True or false: $A \cup B = A' \cup B'$ is a theorem about sets.

3 Construct the membership table for
   (a) $(A \cap B)'$ (b) $A' \cap B'$ (c) $A' \cup B'$ (d) $(A \cup B)'$

4 Which pairs of sets in Exercise 3 are equal?

5 Construct the membership table for
   (a) $A \cap B'$ (b) $A \cap (B \cup B')$ (c) $(A \cup A')$
   (d) $A \cup (B \cap C)$ (e) $(A \cup B) \cap (A \cup C)$ (f) $A \cup (B \cup C)$
   (g) $(A \cup B) \cup C$ (h) $A \cap (B \cap C)$ (i) $(A \cap B) \cap C$

°6 Construct the membership table for $(R \cap T) \cup (R \cap T')$ and compare it to the membership table for R. Caution: Four lines are necessary for the first table, but only two are necessary for the second.

7 How many lines are necessary for a membership table involving
   (a) one set?   (b) two sets?   (c) three sets?   (d) four sets?

8 (a) How would you describe the pattern (to the left of the double line) that we have used for a table involving two sets? (See Example 1, Section 2.6.)

   (b) How would you describe the pattern (to the left of the double line) for the table given for three sets? (See Example 3, Section 2.6.)

9 Set up an organized table that could be used to find the membership table when four sets are involved. How many lines are necessary?

°10 Find the membership table for:
   (a) $(R \cap S) \cup (T' \cap V)$ (b) $(R \cup S) \cup (T \cup V)'$

11 (a) Find the membership table for $A \cap (B \cup C)'$.

   (b) Shade an appropriate Venn diagram to correspond to the membership table.

°12 Use membership tables to prove each of the following.
   (a) $A \cap (B \cup C) = (A \cap B) \cup (A \cap C)$
   (b) $(A \cap B)' = A' \cup B'$
   (c) $A \cup (B \cup C) = (A \cup B) \cup (A \cup C)$
   (d) $A \cap (B \cap C) = (A \cap B) \cap (A \cap C)$
   (e) $R \cup R = R$ (f) $S \cap S = S$ (g) $A = (A')'$

SUMMARY

This chapter introduces some basic mathematical ideas about sets. We were concerned about subsets of a universal set and the Venn diagrams used to represent relationships between the elements. The Fundamental Principle of Counting showed that we had created the most general Venn diagrams allowing for all possible intersections of subsets.

In Sections 2.3 and 2.4 we defined equal sets, the empty set, subset, union, intersection, and complement. These terms enable us either to compare sets or to create new compound sets. We then used a method of proof for theorems about equal sets. For example, $(A \cup B)'$ and $A' \cap B'$ are two different ways of expressing the same set. Thus $(A \cup B)' = A' \cap B'$ is a theorem about sets that holds for any choice of sets A and B.

The final two sections dealt with membership tables and proofs of set theorems using membership tables. Part of our emphasis was on the meaning of theorem and the idea of proof.

Before attempting the Review Test, you should be familiar with the following processes and ideas:

1. Well-defined sets as opposed to those that are not well-defined.

2. The use of $\in$ and $\notin$ as symbols relating to membership in sets.

3. Set-builder notation and the roster method for denoting sets.

4. Venn diagrams for one, two, or three subsets of a given universe.

5. The Fundamental Principle of Counting.

6. Distributing elements of given subsets of a given universe.

7. Definitions of equal, subset, union, intersection, and complement.

8. Given certain subsets of a universe, finding compound sets.

9. Constructing Venn diagrams for compound sets.

10. Proving statements about sets by using a labeled Venn diagram.

11. Constructing membership tables.

REVIEW TEST

1 (a) Is the set of powerful men well-defined?
  (b) Is the set of all snowmobiles well-defined?

2 (a) True or false: $t \in \{a, b, c, \ldots, z\}$
  (b) True or false: $\{a, b, c\} \notin \{a, b, c, \ldots, m\}$
  (c) True or false: $16 \in \{x \mid x \text{ is the square of a counting number}\}$

3 Let $A = \{1, 2, 3, 4\}$, $B = \{2, 4, 5, 6, 7\}$ and $U = \{1, 2, 3, \ldots, 10\}$. Set up a Venn diagram to show this and distribute the elements appropriately.

4 Complete the following table.

Number of subsets of U	Number of regions in most general Venn diagram
1	
2	4
3	8
4	
5	

5 In a class of 37 students the following information was discovered:
   (a) 15 had at one time or another owned a Chevy
   (b) 21 had owned a Ford
   (c) 4 had owned both a Chevy and a Ford
   How many had never owned either kind of car?

6 Let $U = \{a, b, c, d, e, f, g, h, i\}$, $A = \{a, e, i\}$, $B = \{a, b, c, d, e\}$, and $C = \{e, f, g, h\}$. Find each set below:
   (a) $A \cap B$ (b) $B \cup C$ (c) C'
   (d) $(A \cup B)'$ (e) $A \cap (B' \cup C)$

7 Shade those regions of a Venn diagram that correspond to each of the following:
   (a) $A \cap (B \cap C)$ (b) $(B \cup C) \cap A'$ (c) $R \cup S'$

8 Prove that $A \cup B' = (A' \cap B)'$.

9 Show the membership table for each of the following:
   (a) $R' \cup S$ (b) $A \cup (B \cap C)'$

10 Does Figure 27 allow for elements in
   (a) $A \cap C$?   (b) $(B \cup C)'$?
   (c) $(A \cap C) \cap B'$?   (d) $(B \cap C) \cap A'$?

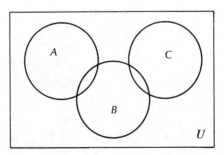

Figure 27

Benjamin Franklin (1706–1790) was one of the great universal minds during the American colonial period. From a humble beginning as a printer, he became a leading citizen of the world. He was instrumental in the Anglo-American debates that led to the drafting of the Declaration of Independence. We remember him for scientific discoveries as well (the lightning rod, the Franklin stove, and bifocal spectacles). Published in the *Pennsylvania Gazette* on October 30, 1735, the article below indicates another facet of Franklin's broad knowledge and intellect.

ON THE USEFULNESS OF THE MATHEMATICS
BENJAMIN FRANKLIN

Mathematics originally signifies any kind of discipline or learning, but now it is taken for that science which teaches or contemplates whatever is capable of being numbered or measured. That part of the mathematics which relates to numbers only is called *arithmetic*; and that which is concerned about measure in general, whether length, breadth, motion, force, &c., is called *geometry*.

As to the usefulness of arithmetic, it is well known that no business, commerce, trade, or employment whatsoever, even from the merchant to the shopkeeper, &c., can be managed and carried on without the assistance of numbers; for by these the trader computes the value of all sorts of goods that he dealeth in, does his business with ease and certainty, and informs himself how matters stand at any time with respect to men, money, or merchandise, to profit and loss, whether he goes forward or backward, grows richer or poorer. Neither is this science only useful to the merchant, but is reckoned the *primum mobile* (or first mover) of all mundane affairs in general, and is useful for all sorts and degrees of men, from the highest to the lowest.

As to the usefulness of geometry, it is as certain that no curious art or mechanic work can either be invented, improved, or performed without its assisting principles.

It is owing to this that astronomers are put into a way of making their observations, coming at the knowledge of the extent of the heavens, the

From Franklin, *The Works of Benjamin Franklin*; Chapter XVI (from the *Pennsylvania Gazette*, Oct. 30, 1735) pp. 417–422.

duration of time, the motions, magnitudes, and distances of the heavenly bodies, their situations, positions, risings, settings, aspects, and eclipses; also the measure of seasons, of years, and of ages.

It is by the assistance of this science that geographers present to our view at once the magnitude and form of the whole earth, the vast extent of the seas, the divisions of empires, kingdoms, and provinces.

It is by the help of geometry the ingenious mariner is instructed how to guide a ship through the vast ocean, from one part of the earth to another, the nearest and safest way and in the shortest time.

By help of this science the architects take their just measures for the structure of buildings, as private houses, churches, palaces, ships, fortifications, &c.

By its help engineers conduct all their works, take the situation and plan of towns, forts, and castles, measure their distances from one another, and carry their measures into places that are only accessible to the eye.

From hence also is deduced that admirable art of drawing sun-dials on any plane howsoever situate, and for any part of the world, to point out the exact time of the day, sun's declination, altitude, amplitude, azimuth, and other astronomical matters.

By geometry the surveyor is directed how to draw a map of any country, to divide his lands, and to lay down and plot any piece of ground, and thereby discover the area in acres, rods, and perches; the gauger is instructed how to find the capacities or solid contents of all kinds of vessels, in barrels, gallons, bushels, &c.; and the measurer is furnished with rules for finding the areas and contents of superficies and solids, and casting up all manner of workmanship. All these and many more useful arts too many to be enumerated here, wholly depend upon the aforesaid sciences—viz., arithmetic and geometry.

This science is descended from the infancy of the world, the inventors of which were the first propagators of human kind, as Adam, Noah, Abraham, Moses, and divers others.

There has not been any science so much esteemed and honored as this of the mathematics, nor with so much industry and vigilance become the care of great men, and labored in by the potentates of the world,—viz., emperors, kings, princes, &c.

Mathematical demonstrations are a logic of as much or more use than that commonly learned at schools, serving to a just formation of the mind, enlarging its capacity, and strengthening it so as to render the same capable of exact reasoning, and discerning truth from falsehood in all occurrences, even subjects not mathematical. For which reason, it is said, the Egyptians, Persians, and Lacedaemonians seldom elected any new kings but such as had some knowledge in the mathematics, imagining those who had not, men of imperfect judgments and unfit to rule and govern.

Though Plato's censure, that those who did not understand the 117th proposition of the 13th book of Euclid's Elements ought not to be ranked amongst rational creatures, was unreasonable and unjust; yet to give a man the character of universal learning, who is destitute of a competent knowledge in the mathematics, is no less so.

The usefulness of some particular parts of the mathematics in the common affairs of human life has rendered some knowledge of them very necessary to a great part of mankind, and very convenient to all the rest that are any way conversant beyond the limits of their own particular callings.

Those whom necessity has obliged to get their bread by manual industry, where some degree of art is required to go along with it, and who have had some insight into these studies, have very often found advantages from them sufficient to reward the pains they were at in acquiring them. And whatever may have been imputed to some other studies, under the notion of insignificancy and loss of time, yet these, I believe, never caused repentance in any, except it was for their remissness in the prosecution of them.

Philosophers do generally affirm that human knowledge to be most excellent which is conversant amongst the most excellent things. What science then can there be more noble, more excellent, more useful for men, more admirably high and demonstrative, than this of the mathematics?

I shall conclude with what Plato says in the seventh book of his *Republic* with regard to the excellence and usefulness of geometry, being to this purpose;

"Dear friend; you see then that mathematics are necessary, because by the exactness of the method we get a habit of using our minds to the best advantage. And it is remarkable that all men being capable by nature to reason and understand the sciences, the less acute, by studying this, though useless to them in every other respect, will gain this advantage—that their minds will be improved in reasoning aright; for no study employs it more, nor makes it susceptible of attention so much; and those who we find have a mind worth cultivating ought to apply themselves to this study."

3 SETS OF ORDERED PAIRS

L. Mac Chase/Charles Marut

So far, we have not been concerned with the order of the elements in a set. The set { a, b } is the same as the set { b, a }. What if we *want* to emphasize the order of elements?

In many sets order *is* important. An example might be a person's name; first and last names imply order. The person whose name is Michael Harvey is different from the person whose name is Harvey Michael. The spelling of words implies an order; changing the order changes the meaning. Each pair in the words "asp" and "sap," "dame" and "made," "steak" and "skate," contains the same letters, yet the words mean quite different things.

In this chapter we will concentrate on sets of two elements, or *ordered pairs*. (In advanced work mathematicians also consider ordered triples, ordered quadruples, ordered quintuples and even in general, ordered *n*-tuples.) By establishing sets of ordered pairs, we can organize certain properties and relationships.

The payoff in studying ordered pairs is that certain mathematics is clarified and simplified; some new mathematics may even be created. The beauty of mathematics lies in just such inventiveness and organization.

3.1 ORDERED PAIRS AND SETS

The fractions 3/4 and 4/3 are clearly different; the first one has a numerator of 3 and a denominator of 4, while the second has a numerator of 4 and a denominator of 3. We can create either of these fractions by drawing elements from the set { 3, 4 } and by specifying that the first element drawn will be the numerator and the second the denominator. Notice that words like "first" and "second" imply order. Thus two different fractions can be made by drawing from the set as specified, because of order.

In general, *ordered pairs* are symbols of the form (a, b), where parentheses are used and the two components are separated by a comma. The order in which the components a and b appear is important. While, in one sense, an ordered pair is a set of two elements, notice that we do not use braces to indicate ordered pairs, but parentheses. We also call the objects that are ordered *components* instead of *elements*.

We could create the fraction 3/4 by selecting first a 3 and then a 4 from the set { 3, 4 }, and could write the fraction as the ordered pair (3, 4). We could create the fraction 4/3 by selecting a 4 and then a 3, and could write this fraction as the ordered pair (4, 3).

We could write the set consisting of the two fractions 3/4 and 4/3 as the set of ordered pairs { (3, 4), (4, 3) }. Be sure to see this as a set of two ordered pairs and not as a set of four elements.

EXAMPLE 1

$\{ (2, 2), (1, 1), (1, 2) \}$ is a set consisting of three ordered pairs. Note the differences between this set and the set $\{ 1, 2 \}$.

EXAMPLE 2

If we toss two coins, each will fall either heads or tails. The set of words we use to describe this situation is $\{$ heads, tails $\}$. But if we wanted to indicate that the first coin landed tails and the second coin landed tails, we would use an ordered pair (tails, tails). Note that, for this example, order in tossing coins is important; we arbitrarily call one coin "first" and the other coin "second."

EXAMPLE 3

Denoting heads by H and tails by T, write down all the possible ways in which two coins may fall in a single toss of each:

$$\{ (H, H), (H, T), (T, H), (T, T) \}$$

This is a set consisting of four ordered pairs. There are four ways in which two coins can fall. (How does this follow from the Fundamental Principle of Counting? See page 40.)

EXAMPLE 4

Write down the set of all possible fractions that can be created from the elements of the set $\{ 1, 2, 3 \}$:

$$\{ 1/1, 1/2, 1/3, 2/1, 2/2, 2/3, 3/1, 3/2, 3/3 \}$$

or, in ordered pair notation,

$$\{ (1, 1), (1, 2), (1, 3), (2, 1), (2, 2), (2, 3), (3, 1), (3, 2), (3, 3) \}$$

We can express the set of Example 4, using set-generator notation as introduced in Chapter 2:

$$\{ (x, y) \mid x \in \{ 1, 2, 3 \} \text{ and } y \in \{ 1, 2, 3 \} \}$$

This would be read "the set of all ordered pairs (x, y) such that x is an element of the set $\{ 1, 2, 3 \}$ and y is an element of the set $\{ 1, 2, 3 \}$."

It should be clear by now that although the set $\{ 1, 2 \}$ equals the set $\{ 2, 1 \}$, the ordered pair $(1, 2)$ is not equal to the ordered pair $(2, 1)$. In general, two ordered pairs (a, b) and (c, d) are *equal* if the first components are the same and the second components are the same. That is, $(a, b) = (c, d)$ if $a = c$ and $b = d$.

EXAMPLE 5

$(1, 2) \neq (2, 1)$ since $1 \neq 2$ and $2 \neq 1$.

EXAMPLE 6

$(r, s) = (r, s)$ since both first components are the same and both second components are the same.

Exercise Set 3.1

1 (a) Express the set of fractions $\{\, 7/8, 5/3, 6/4 \,\}$ in ordered-pair notation.
  (b) Express the set of fractions $\{\, 8/7, 3/5, 4/6 \,\}$ in ordered-pair notation.

2 Tell whether each of the following is true or false.
  (a) $(7, 8)$ is an element of $\{\, 7, 8, 9 \,\}$.
  (b) Ordered pair $(7, 8)$ has components that are elements of the set $\{\, 7, 8, 9 \,\}$.
  (c) $(a, b) \in \{\, (a, b), (b, c), (c, d), (d, e), \ldots, (y, z) \,\}$.
  (d) $(a, c) \in \{\, (a, b), (b, c), (c, d), \ldots, (y, z) \,\}$.
  (e) $\{\, (x, y) \mid x \in N \text{ and } y \in N \,\}$ contains $(0, 0)$.
  (f) $\{\, (1, 5) \,\}$ contains one element, while $\{\, 1, 5 \,\}$ contains two elements.
  (g) $\{\, (1, 6),(2, 7),(3, 8),(4, 9),(5, 10),(6, 11) \,\}$ has a set of first components, which is $\{\, 1, 2, 3, 4, 5, 6 \,\}$, and a set of second components, which is $\{\, 6, 7, 8, 9, 10, 11 \,\}$.

3 (a) Does ordered pair $(7, 8)$ equal ordered pair $(8, 7)$?
  (b) Does $(3, 4) = (6, 8)$?
  (c) Does $(H, T) = (T, H)$? (See Example 3 in text.)
  (d) Is the set of ordered pairs $\{\, (a, b),(c, d) \,\}$ equal to the set of ordered pairs $\{\, (c, d),(a, b) \,\}$?

4 (a) What is the cardinality of the set of possible outcomes when we throw two coins?
  (b) What is the cardinality of the set $\{\, \alpha, \beta \,\}$?
  (c) What is the cardinality of the set $\{\, (\alpha, \beta) \,\}$?

5 Write the set of ordered pairs $\{\, (x, y) \mid x \in \{\, a, b \,\} \text{ and } y \in \{\, c, d, e \,\} \,\}$ without set-generator notation.

6 Write the set of ordered pairs $\{\, (1, a), (1, b), (1, c), (2, a), (2, b), (2, c) \,\}$, using set-generator notation.

3.2 CROSS PRODUCT SETS

Suppose $A = \{\, 1, 2, 3 \,\}$ and $B = \{\, 4, 5, 6, 7 \,\}$. A set of ordered pairs can be created by choosing the first element from A and the second from B.

Examples of such ordered pairs are $(1, 6)$ and $(3, 4)$. The set of *all* possible ordered pairs of the form (x, y), where x is an element of A and y is an element of B, is

$$\{ (1, 4), (1, 5), (1, 6), (1, 7), (2, 4), (2, 5),$$
$$(2, 6), (2, 7), (3, 4), (3, 5), (3, 6), (3, 7) \}$$

Such a set is called the *cross product set* or the *Cartesian product set*, *Cartesian* refers to the seventeenth-century philosopher and mathematician, Rene Descartes who first studied these types of sets. The name *cross product set* comes from the notation for the set and for the number of elements in the set, as we shall see.

If the cross product set is formed from sets A and B, it is denoted $A \times B$, provided that first elements come from A and second elements from B.

EXAMPLE 1

Let $A = \{ a, b \}$ and $B = \{ 1, 2, 3 \}$. Then $A \times B = \{ (a, 1), (a, 2), (a, 3), (b, 1),$ $(b, 2), (b, 3) \}$. But $B \times A = \{ (1, a), (1, b), (2, a), (2, b), (3, a), (3, b) \}$. Clearly $A \times B$ is not the same as $B \times A$.

EXAMPLE 2

Let $R = \{$ Jim, John, Rick $\}$ and $S = \{$ Sally, Sue $\}$. Then $R \times S = \{$ (Jim, Sally), (Jim, Sue), (John, Sally), (John, Sue), (Rick, Sally), (Rick, Sue) $\}$.

EXAMPLE 3

We can form the cross product set of a set with itself. Suppose $Q = \{ 1, 2,$ $3, 4, 5, 6 \}$. Then $Q \times Q = \{ (1, 1), (1, 2), (1, 3), (1, 4), (1, 5), (1, 6), (2, 1), (2, 2),$ $(2, 3), (2, 4), (2, 5), (2, 6), (3, 1), \ldots, (6, 6) \}$.

Notice in Example 1 that A contains two elements, B contains three elements, $A \times B$ contains six elements, and so does $B \times A$. In Example 3, Q contains six elements and $Q \times Q$ contains thirty-six elements. In general, if R contains p elements and S contains q elements, the cross product set, $R \times S$, always contains $p \times q$ elements. The reason for this is that in forming $R \times S$ we can select the first elements in p ways and the second elements in q ways. The Fundamental Principle of Counting assures us that we can form all possible ordered pairs in $p \times q$ ways.

More formally in terms of cardinality notation, if $n(A) = p$, and $n(B) = q$, then $n(A \times B) = pq$.

EXAMPLE 4

Let $Z = \{ a, b, c \}$ and $W = \{ 70, 80, 90, 100 \}$. Since $n(Z) = 3$ and $n(W) = 4$, it is true that $n(Z \times W) = 3(4) = 12$. Note also that $n(W \times Z) = 4(3) = 12$.

Exercise Set 3.2

1 Let $W = \{-1, 1\}$ and $V = \{-2, 2\}$.
   (a) Form $W \times V$.   (b) Form $V \times W$.   (c) Form $V \times V$.
   (d) Form $W \times W$.   (e) What is $n(V \times V)$?   (f) What is $n(W \times V)$?

2 Let $R = \{a, e, i, o, u\}$.
   (a) Form $R \times R$
   (b) How many ordered pairs are in the cross product of set R with itself?
        That is, count the elements in your answer to (a).
   (c) What is $n(R \times R)$?
   (d) Write $\{(x, y) \mid (x, y) \in R \times R$ and $x = y\}$, using the roster method.
   (e) Rewrite question (d) in words, using the fewest possible symbols.

3 (a) Two dice, one red and the other blue, are tossed. The *sample set* is the set
        of all possible outcomes. A 2 on the red die and a 6 on the blue die is a
        different outcome from a "blue 2 and a red 6." Calling the set of outcomes
        on any one die $S = \{1, 2, 3, 4, 5, 6\}$, write out all possible outcomes on
        rolling two dice, by creating $S \times S$.
   (b) How many possible ways are there to roll a sum of 7?
   (c) How many possible ways are there to roll at least one 3?
   (d) How many possible ways are there to roll a sum of at least 11?
   °(e) What is the probability that you roll doubles? That you roll at least a sum
        of 11? (While no mention has been made of probability, see if you can
        answer the question intuitively.)

4 Let $S = \{a, b, c, \ldots, j\}$. What is $n(S \times S)$?

°5 Let $S = \{a, b, c\}$. How many elements are in $S \times \emptyset$?

°6 What is the minimum cardinality of a cross product set?

7 Let $U = \{1, 2, 3, 4, 5, 6, 7, 8, 9, 10\}$ and $A = \{1, 3, 5, 7, 9\}$.
   (a) Find $n(U \times U)$.   (b) Find $A \times A'$.   (c) Find $n(A' \times A')$.

3.3 RELATIONS

Cross product sets are specific sets of ordered pairs. Any subset of a cross
product set is called a *relation*. If the cross product set is denoted as $A \times B$,
then we can denote a subset of $A \times B$ as the relation $R_{A \times B}$, or simply as R.

EXAMPLE 1

Let $A = \{1, 2\}$ and $B = \{1, 2, 3\}$. Then $A \times B = \{(1, 1),(1, 2),(1, 3),(2, 1),$
$(2, 2),(2, 3)\}$. *Any* subset of $A \times B$ is a relation. Here are some examples:
(a) $R_{A \times B} = \{(1, 1), (2, 2)\}$
(b) $R = \{(1, 2), (1, 3), (2, 3)\}$
(c) $R = \{(2, 3)\}$
(d) $R = \{\ \}$
(e) $R = A \times B$

Normally, we define relations that have some meaning, although meaning is *not* required. While the cross product set is the set of all possible ordered pairs, we are usually interested in a subset in which there is some relationship between the components. Thus in Example 1 (a) the components are equal and this subset contains *only* those ordered pairs whose components are equal. The first component is related by equality to the second component.

In Example 1 (b) the relationship is that the first component is less than the second component. Example 1 (c) shows that a relation may consist of only one ordered pair; perhaps there is no specific meaning implied in this example. Example 1 (d) shows that the empty subset of $A \times B$ is considered a relation, while Example 1 (e) shows that the set $A \times B$ is a relation itself. (A relation need not be a proper subset of a cross product set.)

The set of first components in the ordered pairs of a relation is called the *domain* of the relation, and the set of second components is called the *range* of the relation.

EXAMPLE 2

$A \times B = \{ (a, a), (a, b), (a, c), (b, a), (b, b), (b, c) \}$. Let $R = \{ (a, b), (a, c), (b, a), (b, c) \}$. Domain of $R = \{ a, b \}$; range of $R = \{ a, b, c \}$.

Suppose we have the following four line segments:

$$\underline{\hspace{3cm}}\ m$$
$$\underline{\hspace{3cm}}\ n$$
$$\underline{\hspace{3cm}}\ o$$
$$\underline{\hspace{3.5cm}}\ p$$

Let the segments m, n, o, and p be the elements of set M. Forming $M \times M$, we get

$$\{ (m, m), (m, n), (m, o), (m, p),$$
$$(n, m), (n, n), (n, o), (n, p),$$
$$(o, m), (o, n), (o, o), (o, p),$$
$$(p, m), (p, n), (p, o), (p, p) \}$$

Now let us form the relation R according to the following rule: $R = \{ (x, y) \mid (x, y) \in M \times M$ and x is shorter than $y \}$. This, in effect, *selects* a subset of the set of 16 ordered pairs by giving the verbal rule for the selection. The relation we want is the set of ordered pairs whose first component is shorter than the second. We can examine each ordered pair in the set $M \times M$ and ask "Is the first component shorter than the second?" If the answer is yes, that ordered pair belongs to the relation R. The result is

$$R = \{ (m, n), (m, o), (m, p), (n, o), (n, p), (o, p) \}$$

This may seem like a hard way to list lines that are shorter than others.

The beauty of the setup will become clear, though, if we stick with the procedure until we have clarified some other concepts. We will then be able to classify seemingly dissimilar relations according to properties that they have in common.

When an ordered pair (a, b) belongs to a relation, we will write $a \, R \, b$, and when (a, b) does not belong to a relation, we will write $a \, \cancel{R} \, b$. When $a \, R \, b$ holds, we can say "a is related to b."

EXAMPLE 3

$$S = \{ 1, 2, 3, 4, 9 \}$$

$$
\begin{aligned}
S \times S = \{ & (1, 1), (1, 2), (1, 3), (1, 4), (1, 9), \\
& (2, 1), (2, 2), (2, 3), (2, 4), (2, 9), \\
& (3, 1), (3, 2), (3, 3), (3, 4), (3, 9), \\
& (4, 1), (4, 2), (4, 3), (4, 4), (4, 9), \\
& (9, 1), (9, 2), (9, 3), (9, 4), (9, 9) \}
\end{aligned}
$$

We now define a relation as follows:

$$R = \{ (x, y) \mid (x, y) \in S \times S \text{ and } x = y^2 \}$$

In roster form it looks like this:

$$R = \{ (1, 1), (4, 2), (9, 3) \}$$

Here $(4, 2) \in R$ or $4 \, R \, 2$, which means "4 is the square of 2." Also $(3, 2) \notin R$ or $3 \, \cancel{R} \, 2$, which means "3 is not the square of 2."

Forming relations is a process of selection; we select those ordered pairs that stand in the relation and reject those ordered pairs that do not stand in the relation. In Example 3 we are organizing the search for those numbers in the set $\{ 1, 2, 3, 4, 9 \}$ that are the squares of other numbers in the set. Perhaps you knew the answer without using this highly structured procedure, but at least you must admit that this is a foolproof way to avoid missing any such numbers.

EXAMPLE 4

Using $S \times S$ as given in Example 3, we see that the relation

$$R = \{ (x, y) \mid (x, y) \in S \times S \text{ and } x + y \text{ is an odd number} \}$$

selects these ordered pairs:

$$
\begin{aligned}
\{ & (1, 2), (1, 4), (2, 1), (2, 3), (2, 9), (3, 2), \\
& (3, 4), (4, 1), (4, 3), (4, 9), (9, 2), (9, 4) \}
\end{aligned}
$$

Exercise Set 3.3

1 Let $A = \{ a, b, c \}$ and $B = \{ b, c, d, e \}$.
   (a) Form $A \times B$.
   (b) Select the subset of $A \times B$ (that is, write the ordered pairs of the relation) whose first component alphabetically precedes the second. (Here R stands for "alphabetically precedes.")
   (c) Is $b \, R \, b$? Does (b, b) belong to R?
   (d) Given the same $A \times B$, write the ordered pairs in the relation $R = \{ (x, y) \mid (x, y) \in A \times B$ and x is the same as $y \}$.

2 Let $N = \{ 1, 2, 3, 4, 5, \dots \}$ and consider $N \times N$.
   (a) What is a rule that describes the following relation? $\{ (1, 1), (2, 4), (3, 9),$ $(4, 16), (5, 25), \dots \}$.
   (b) Write the subset of $N \times N$ such that (x, y) is an element of $N \times N$ and $x = 2y$.
   (c) Write in roster form the relation $R = \{ (x, y) \mid (x, y) \in N \times N$ and x is exactly divisible by $y \}$. (Exactly divisible means that there is no remainder.)
   (d) What is the domain of $N \times N$? What is the range?

3 Let W be the set of all first names, and let Z be the set of all last names.
   (a) In words describe $W \times Z$.
   (b) In words describe $Z \times W$.

4 Let $A = \{ 1, 2, 3 \}$ and consider, or write down, $A \times A$.
   (a) What is $n(A \times A)$?
   (b) How many subsets of $A \times A$ are there?
   (c) How many relations can be formed from $A \times A$?
   (d) What ordered pairs in $A \times A$ belong to the relation "numerically precedes"?
   (e) What ordered pairs in $A \times A$ belong to the relation "is the same number as"?

3.4 REFLEXIVE, SYMMETRIC, AND TRANSITIVE RELATIONS

Only relations on the cross product set of a set with itself will be considered in this section. That is, we will consider subsets of $A \times A$, or $B \times B$, or $S \times S$, but will not deal with subsets of $A \times B$ or $S \times T$. Such relations may be reflexive, symmetric, or transitive—terms we now define.

DEFINITION 1   Reflexive Property

Let R be a relation on $A \times A$. If $a \, R \, a$ for every a in set A, then R is said to be *reflexive*. That is, R is reflexive if the ordered pair (a, a) is in R for every element $a \in A$.

EXAMPLE 1

If $A = \{1, 2, 3\}$, then consider $A \times A$. Let $R = \{(1, 1), (2, 2), (3, 3)\}$. Here R might be "is the same number as." Then R is reflexive, since $1 \in A$ and $(1, 1) \in R$, $2 \in A$ and $(2, 2) \in R$, and $3 \in A$ and $(3, 3) \in R$.

When we can name the relation, as we did in Example 1, reflexivity implies that every element stands in that relation to itself. Thus in Example 1, 1 is the same number as itself, 2 is the same number as itself, and 3 is the same number as itself.

EXAMPLE 2

Let R be the relation "weighs the same as" on the set of all people. Clearly R is reflexive, since every person weighs the same as himself.

DEFINITION 2   Symmetric Property

Let R be a relation on $A \times A$. If, whenever $a\ R\ b$, for some elements a and b in A, it is also true that $b\ R\ a$, then R is said to be *symmetric*. That is, R is symmetric if (b, a) is in R whenever (a, b) is in R.

EXAMPLE 3

Consider the relation $R = \{(1, 2), (2, 1), (3, 4), (4, 3)\}$. First, $(1, 2)$ is in R and so is $(2, 1)$. Second, $(3, 4)$ is in R and so is $(4, 3)$. Considering any element $(a, b) \in R$, we also find $(b, a) \in R$, so R is symmetric.

A relation is symmetric if, whenever one element is related to a second, the second element has the same relation to the first. "Weighs the same as" on the set of all people is a symmetric relation, since if person x weighs the same as person y, it must also be true that person y weighs the same as person x. However, "is taller than" on the set of people is *not* a symmetric relation; if person x is taller than person y, then y is not taller than x.

DEFINITION 3   Transitive Property

Let R be a relation on $A \times A$. If, whenever $a\ R\ b$ and $b\ R\ c$ for three elements $a, b, c \in A$, it is also true that $a\ R\ c$, then R is said to be *transitive*. That is, R is transitive if (a, c) is in R whenever (a, b) and (b, c) are in R.

EXAMPLE 4

If a relation happened to be $\{(l, m), (m, n), (l, n)\}$, it would be transitive; (l, m) and (m, n) are in the relation and so is (l, n).

Transitivity compares a first and a second ordered pair to a third one. The relation "is taller than" is transitive. Suppose person a is taller than person b, and further that person b is taller than person c. Then it is true that person a is taller than person c.

EXAMPLE 5

Let $N = \{1, 2, 3, 4, 5, \ldots\}$ and R be the relation "is less than." Some members of R are $(1, 2), (1, 3), (2, 9)$, and $(5, 6)$.

(a) Relation R is not reflexive since we do not find ordered pairs like $(1, 1)$ or $(2, 2)$ in the relation. That is, no number is less than itself.

(b) Relation R is not symmetric. The ordered pair $(2, 9) \in R$ but $(9, 2) \notin R$. That is, whenever a first number is less than a second number, the second number is not less than the first.

(c) Relation R is transitive. Note that $(1, 2)$ and $(2, 9)$ are in R and that $(1, 9)$ is also in R.

Exercise Set 3.4

1 Suppose $\{(a, a), (b, b), (a, b)\}$ is a relation R.
   (a) Which ordered pairs indicate that the relation is reflexive?
   (b) Note that $(a, b) \in R$. What other ordered pair would have to be added to R to make the relation symmetric?

2 Is the relation "is the same age as" on the set of all people reflexive? Symmetric? Transitive?

3 Consider each of the following relations on the set indicated and tell whether it is reflexive, symmetric, or transitive:
   (a) R_1: "is the same color as," on the set of all flowers in Mrs. Perkins's garden
   (b) R_2: "is the father of," on the set of all people
   (c) R_3: "is the same height as," on the set of students in this college
   (d) R_4: "is the same length as," on the set of all city streets
   (e) R_5: "costs the same as," on the set of all cars
   (f) R_6: "lives in the same water as," on the set of all fish
   (g) R_7: "swims faster than," on the set of all fish
   (h) R_8: "lives farther north than," on the set of all trees
   (i) R_9: "is divisible by," on the set of all natural numbers
   (j) R_{10}: $\{(x, y) \mid x \in N, y \in N, \text{ and } x \div y \text{ leaves a remainder of } 0\}$

4 Suppose $\{(1, 3), (3, 5)\}$ is a relation R.
   (a) The ordered pairs $(1, 1)$, $(3, 3)$, $(5, 5)$ would have to be added to the set to make R reflexive. What ordered pairs would have to be added to the set to make R reflexive and symmetric?
   (b) What ordered pairs would have to be added to the set to make R reflexive, symmetric, and transitive?

5 Consider the relation $\{(a, b), (b, c), (a, a), (b, b), (c, c), (b, a), (c, b), (a, c)\}$.
   (a) What ordered pair would have to be added to the set to make the relation symmetric?
   (b) After the addition of that ordered pair to the set, is the relation transitive?

6 Is $\{(1, 1)\}$ reflexive? Symmetric? Transitive?

°7 Let $S_1 = \{\ \ \}$ $S_2 = \{a\}$
    $S_3 = \{b\}$ $S_4 = \{c\}$
    $S_5 = \{a, b\}$ $S_6 = \{a, c\}$
    $S_7 = \{b, c\}$ $S_8 = \{a, b, c\}$
   The sets listed above are all the subsets of $\{a, b, c\}$. Let $S = \{S_1, S_2, S_3, S_4, S_5, S_6, S_7, S_8\}$. That is, S is the set of all subsets of $\{a, b, c\}$. The cross product set $S \times S = \{(S_1, S_1), (S_1, S_2), (S_1, S_3), \ldots, (S_2, S_1), (S_2, S_2), (S_2, S_3), \ldots, (S_3, S_1), (S_3, S_2), \ldots, (S_8, S_8)\}$.
   (a) How many ordered pairs are in $S \times S$?
   (b) Write in roster form the relation on $S \times S$, "is the same subset as."
   (c) Is this relation reflexive? Symmetric? Transitive?
   (d) Write in roster form the relation on $S \times S$, "is a subset of." For example, (S_2, S_5) belongs to the relation because $S_2 \subseteq S_5$. However, (S_6, S_3) does not belong to the relation because S_6 is not a subset of S_3.
   (e) Is the relation in (d) reflexive? How many ordered pairs in $S \times S$ so indicate?
   (f) Is the relation in (d) symmetric? Can you prove your answer to this question?
   (g) The relation in (d) is transitive. Cite one example showing this transitivity.

8 Let $L = \{5, 2, V, II\}$.
   (a) Write out $L \times L$ by completing this table:

	5	2	V	II
5	(5, 5)		(5, V)	
2		(2, 2)		
V	(V, 5)			
II				(II, II)

   (b) Write down the subset of $L \times L$ defined by the relation "stands for the same number as."
   (c) Which ordered pairs in your answer to (b) indicate that the relation is reflexive? Where are these ordered pairs located in the table?

(d) The relation is symmetric. Extend the table below to show all pairs that exhibit the symmetric property.

In the relation	Also in the relation
(5, 5)	(5, 5)
(5, V)	(V, 5)
(2, 2)	(2, 2)

(e) Describe positionally which ordered pairs in the table indicate symmetry.

(f) The presence of (II, 2), (2, II), and (II, II) indicates one instance of transitivity. Write down a different set of three ordered pairs that also show the transitive property.

°9 (a) Write down all 16 subsets of the set { 1, 2, 3, 4 }.

(b) Consider the relation "is equivalent to" for sets. For example, recall that { 1, 2, 3 } is equivalent to { 2, 3, 4 }. Now sketch five rectangles, each about 1 inch wide and 2 inches high. Place all the subsets that are equivalent to each other in the same rectangle.

(c) Is the relation reflexive? Is it symmetric? Is it transitive?

°10 Consider all the subsets of the set { 1, 2, 3, 4 } and the relation "has greater cardinality than." *Attempt* to draw rectangles as in exercise 9, separating the subsets into groups that stand in the relationship.

°11 (a) S = {a, b}. Consider S x S. Other than the subset {(a, a), (b, b)} there is only one relation that is reflexive, symmetric, and transitive. How many ordered pairs are there in that relation?

(b) S = {a, b, c}. Consider S x S. Is the cross product set itself a reflexive, symmetric, and transitive relation?

°12 (a) Complete the table below, in which the ordered pairs are elements of A x A, where A = { r, s, t, v }.

	r	s	t	v
r	(r, r)	(r, s)		
s	(s, r)	(s, s)		
t	(t, r)			
v	(v, r)			(v, v)

(b) Is the set of ordered pairs in the body of the table a reflexive, symmetric, and transitive relation?

3.5 EQUIVALENCE RELATIONS

Suppose 40 men and women sign up for a choir. Maestro Jones gives each person a voice test and classifies each as a soprano, alto, tenor, or bass. At the first rehearsal they all sit wherever they choose. But Maestro Jones likes to have his choir organized into sections, so he simply tells everyone to get up and move to the appropriate parts of the room, according to the following diagram:

Altos	Sopranos	Tenors	Basses

Figure 1

We can say that the choir is *partitioned* into four *classes*. The word *partition* means separated or grouped, and *class* is simply a term for one of the groups. In the alto class are all those singers who have an alto voice range; in the soprano class are all those who have a soprano voice range; and so on. The different classes of singers do not have any elements in common. Anyone in the alto class is not in the tenor class and vice versa.

Any two singers in the alto class have the same voice range. What we have is a relation on the set of 40 people, which might be called "has the same voice range as." Notice how nicely the 40 people are organized. Suppose Jack Williams is sitting in the tenor section and finds his friend Sam Smith sitting there also. The ordered pair (Williams, Smith) belongs to the relation "has the same voice range as." In fact all the tenors in pairs would belong to that relation. The same is true of the other classes; all pairs of altos belong to the relation.

A mathematician would say that Maestro Jones has partitioned the singers into classes according to whether or not they belong (in pairs) to the relation "has the same voice range as."

If Jones tried to arrange the singers into groups, using the relation "sings higher than," the room would soon turn into chaos. Suppose singer A finds that she sings higher than singer B, determining the ordered pair (A, B). Then we could *try* to put singers A and B into some class, represented schematically below.

A
B

That is, A belongs to the class because A sings higher than B. But then B does not belong in the same class as A, since B does not sing higher than A. So B must leave the class we tried to set up. Is A in a class by herself?

A

Since A cannot sing higher than A, we cannot even put A in the class. As a relation, "sings higher than" may stand between two people in one order, but the relation does not partition the set of singers.

Notice that Maestro Jones gets order when he says "All those singing alto sit here; all those singing soprano sit here; all those singing tenor sit here; all those singing bass sit here." But if he said "If you sing higher than your neighbor, both of you belong in section 1," complete disorder would result. The very first pair that stand in that relation both belong and do not belong in section 1!

On the set of singers the relation "has the same voice range" is reflexive, symmetric, and transitive. (Be sure you understand why.) On the same set, the relation "sings higher than" is not reflexive, is not symmetric, but is transitive.

A relation that is reflexive, symmetric, and transitive is called an *equivalence relation*. Equivalence relations partition the set from which they are formed into *equivalence classes* (or simply *classes*).

"Has the same voice as" is an equivalence relation that organizes the set of singers into four classes. "Sings higher than" is not an equivalence relation, and so we cannot think how to separate the singers into classes by using that relation.

The organization of any set into distinct (nonintersecting) subsets can be explained in terms of equivalence relations. Once these subsets are formed or determined, it is possible to give a distinctive name to a particular subset.

EXAMPLE 1

Imagine 250 AKC-registered dogs in the New York dog show to be set loose in Madison Square Garden. The relation "is the same breed as" is an equivalence relation (reflexive, symmetric, and transitive), and so it partitions the set of 250 dogs into classes. Even an amateur could pick out dogs that stand in the relation. An expert might know the *names* of the classes: Chihuahua, Great Dane, Shepherd, Pekingese, Greyhound, Collie, and so forth.

EXAMPLE 2

Assuming all men have one and only one nationality, the set of all men on earth is partitioned by the relation "belongs to the same nation as." If Green and Brown are both Canadians, they stand in that relation. Observe that the relation *is* an equivalence relation.

In Example 2 we could name the country after any one of its citizens. Thus Green belongs to "Green's country." Brown also belongs to Green's country. Membership in an equivalence class is so distinctive that any element of the class could be used to name the class.

Exercise Set 3.5

1 Suppose, as director of admissions in a college, you had to separate 1000 incoming freshmen into manageable groups for registration purposes.
   (a) Think of a relation that would partition the 1000 students into distinct groups. (There are many possibilities.)
   (b) What is an appropriate name for each of your groups?

2 The set of all U.S. telephone numbers is partitioned by "is in the same area code as."
   (a) Is this relation reflexive? Symmetric? Transitive?
   (b) What is the name of the class, determined by the relation, in which your telephone number belongs?

3 Tell which of the following are equivalence relations:
   (a) "Is the same size as," on the set of all shoes
   (b) "Wears more perfume than," on the set of all women
   (c) "Is related to," on the set of all people
   (d) "Smells better than," on the set of all flowers
   (e) "Is synonymous with," on the set of words in a dictionary
   (f) "Runs smoother than," on the set of all engines
   (g) "Is smarter than," on the set of all students

4 Describe the partitions or equivalence classes formed by the equivalence relations in Question 3.

5 The word "same" can be used to redefine any equivalence relation not written with that word. For example, "is related to" can be expressed "has the same ancestors as." Rewrite each of these equivalence relations, using the word "same":
   (a) "is synonymous with," on the set of words
   (b) "runs just like," on the set of engines
   (c) "is equivalent to," on a set of sets
   (d) "is equal to," on the set of fractions
   (e) "is as rich as," on the set of wealthy people
   (f) "is similar to," on the set of triangles
   (g) "has as many elements as," on a set of sets
   (h) "is exactly as old as," on the set of people
   (i) "is as happy as," on the set of all people
   (j) "is as smart as," on the set of all students

6 "Lives in the same county as" is an equivalence relation on the set of all people in your state. How many equivalence classes are thus formed? What is the name of the equivalence class in which you would be found?

7 There are four suits in a deck of cards: clubs, diamonds, hearts, and spades. Tell whether each of the following is true or false.
   (a) "Is the same suit as" is an equivalence relation on the set of cards.
   (b) The deck is partitioned by this relation.
   (c) A card drawn at random can be in different equivalence classes (given the same relation).

8 In some card games the cards are ranked A-K-Q-J-10-9-8-7-6-5-4-3-2 from highest to lowest. Is the relation "is the same rank as" an equivalence relation? If so, describe the classes and tell how many cards are in each class.

°9 Consider the relation $R = \{ (x, y) \mid x$ and y are natural numbers and $(x, y) \in R$ if x and y have the same remainder when divided by 5 $\}$. For example, (7, 12) belongs to R because the same remainder, 2, occurs when both 7 and 12 are divided by 5; the ordered pair (1, 6) belongs to R (same remainder, 1); the ordered pair (2, 7) belongs to R (same remainder, 2).

   (a) Write six ordered pairs that yield the same remainder, 0, when each element is divided by 5.
   (b) Do the same for remainders 1, 2, 3, and 4.
   (c) Set up a diagram like Figure 1, distributing the natural numbers into appropriate classes.
   (d) How many equivalence classes are there? What is an appropriate name for each?

°10 Consider $N^{+0} = \{ 0, 1, 2, 3, 4, 5, \ldots \}$. This is simply the set N with zero added. Imagine all possible ordered pairs in $N^{+0} \times N^{+0}$—examples: (2, 10), (0, 0), (1, 7). We define the relation R to mean "two ordered pairs (a, b) and (c, d) are equivalent if their differences, $b - a$ and $d - c$, are equal." Thus (2, 7) is equivalent to (6, 11), since $7 - 2 = 11 - 6$. Also, (2, 12) is equivalent to (3, 13). Also, (7, 2) is equivalent to (8, 3).

   (a) Write several ordered pairs (a, b) in which the difference $b - a$ is:
      1. 4   2. −4   3. 7   4. 10   5. 15
      6. 20   7. 100   8. 0   9. −8
   (b) Check that R is reflexive. That is, verify that (a, b) and (a, b) are equivalent, or what is the same, $(a, b) R (a, b)$.
   (c) Check that R is symmetric. That is, if $(a, b) R (c, d)$, then $(c, d) R (a, b)$.
   (d) Give an example of the transitive property that holds for this relation. That is, show one example of ordered pairs of numbers, $(a, b), (c, d),$ and (e, f) where if $(a, b) R (c, d)$ and if $(c, d) R (e, f)$, then $(a, b) R (e, f)$.
   (e) What are the equivalence classes formed by this relation? How many classes are there?
   (f) Place several other ordered pairs in the classes identified below:

(6, 12)	(12,6)	(1, 4)	(10,7)	(5, 1)
(14, 20)	(6, 0)	(9, 12)	(8, 5)	
(0, 6)				

Note   While not essential to this question, it should be observed that this relation is a subset of the cross product set $(N^{+0} \times N^{+0}) \times (N^{+0} \times N^{+0})$. That is, the relation is a subset of a cross product set of a cross product set with itself. We are not comparing single ordered pairs, but a pair of ordered pairs. Thus we say (5, 7) × (9, 11), which implies that the ordered pair ((5, 7), (9, 11)) belongs to the relation; each component of the ordered pair is itself an ordered pair. This tricky point is worth some study.

°11 Is the relation in question 12, Exercise Set 3.4, an equivalence relation?

3.6 FUNCTIONS

Besides cross product sets and relations, another kind of set of ordered pairs is important to mathematicians. A *function* is a set of ordered pairs (x, y) in which no two different ordered pairs have the same first element.

EXAMPLE 1

The set $\{(1, 2), (3, 4), (5, 6)\}$ is a function.

The set of first elements in the ordered pairs of a function is called the *domain* of the function, and the set of second elements is called the *range* of the function. In Example 1 the domain is $\{1, 3, 5\}$ and the range is $\{2, 4, 6\}$.

EXAMPLE 2

The set $\{(1, 2), (1, 3), (2, 7), (4, 9)\}$ is not a function, since the ordered pairs $(1, 2)$ and $(1, 3)$ have the same first element.

It is especially convenient to state the rule for forming the ordered pairs of a function in set-generator notation.

EXAMPLE 3

The set $\{(x, y) \mid x \text{ and } y \text{ are counting numbers and } y = 2x\}$ denotes a function. In roster form it becomes $\{(1, 2), (2, 4), (3, 6), (4, 8), \ldots\}$.

In set-generator notation we always show a typical member of the set before the vertical line. Thus in Example 3 a typical member is (x, y). The x and y are called variables. In discussing functions, y (or the second element in the typical ordered pair) is called the *dependent variable*, and x (or the first element) is called the *independent variable*.

Then the independent variable takes on the values given in the domain of the function. The dependent variable has for its values those elements in the range of the function.

The part of the rule for forming the function that cites the set from which the variables are to be chosen is often stated only for the independent variable. The rest of the rule tells how to get the second element, or it shows how the second element depends upon the first.

EXAMPLE 4

Consider the set $\{(x, y) \mid x \text{ is a counting number and } y = 2x\}$. This is the same function as in Example 3. The x varies over the set of counting numbers, and the y depends upon x. In particular, whatever x is, y is twice x.

The statement of how y is to depend on x is the most important part of the rule defining the function. Sometimes we speak of the function $y = 2x$ but imply ordered pairs (x, y) where x is to be chosen from some given set. It is also clear that y *depends* on x. Mathematicians sometimes say y depends upon x by saying "y is a function of x" and use what is called *functional notation*. In functional notation "y is a function of x" is written

$$y = f(x)$$

If it is known that $y = 2x$ defines the relationship between x and y, we have

$$f(x) = 2x$$

(This last equation is read "f of x equals $2x$." Note that $f(x)$ does *not* mean "f times x.")

EXAMPLE 5

Suppose $x \in N$ and $y = 2x$. We can get ordered pairs belonging to the function by choosing x from N and substituting in the formula $y = 2x$. Thus if x is 1, $y = 2(1)$ or 2, and the ordered pair $(1, 2)$ is determined. If x is 5, $y = 2(5)$, and the ordered pair $(5, 10)$ is determined.

EXAMPLE 6

Suppose $x \in N$ and $f(x) = 2x$. This defines the function of Example 5, using functional notation. The same ordered pairs must belong to the function. If x is 5, then $f(x) = 2(5) = 10$, so the ordered pair $(5, 10)$ is found. The expression "$f(x) = 2x$ evaluated at $x = 6$" is written more briefly

$$f(6) = 2(6)$$

The ordered pair $(6, 12)$ is determined.

EXAMPLE 7

Let $f(x) = x^2 - 2x$ and find $f(4)$.
$f(4) = 4^2 - 2(4) = 16 - 8 = 8$
The ordered pair $(4, 8)$ is determined.

EXAMPLE 8

Let $f(x) = 3x + 4$. Find $f(1)$, $f(0)$, $f(8)$, and $f(h)$.
(a) $f(1) = 3(1) + 4 = 7$
(b) $f(0) = 3(0) + 4 = 4$
(c) $f(8) = 3(8) + 4 = 28$
(d) $f(h) = 3h + 4$

Functions always denote dependency. In everyday language we might say "My mood is a function of the weather" or "The temperature is a function of the season." It is reasonable to assume that mood depends on weather and that the temperature depends on the season.

To a mathematician, functions usually result in a formula in which the dependency can be stated exactly.

Exercise Set 3.6

1 True or false:
  (a) Every relation is a function.
  (b) Every function is a relation.
  (c) Any set of ordered pairs can be considered a relation.
  (d) $\{(1, 5)\}$ is a function.
  (e) $\{\ \}$ is a function.
  (f) The domain is the set of second components in the ordered pairs of a function
  (g) Functions must contain at least two ordered pairs.
  (h) In $y = x - 7$, x is the dependent variable.

2 (a) Is $\{(1, 2), (2, 4), (2, 1)\}$ a function?
  (b) Is $\{(1, 2), (1, 4), (1, 1)\}$ a function?

3 Write each function below in roster form:
  (a) $\{(x, y) \mid x \in N \text{ and } y = x + 7\}$
  (b) $\{(x, y) \mid x \in \{1, 2, 3, 4\} \text{ and } y = x^2\}$

4 Your salary at a job in which you are paid $2.50 per hour is a function of the number of hours worked.
  (a) Write a formula expressing this dependency, using S for salary and n for the number of hours worked.
  (b) Find your salary if n is 40.
  (c) "S depends on n" can be written $S = f(n)$. Find $f(20)$.

5 Let $f(x) = 3x + 7$.
  (a) Find $f(1)$. What ordered pair is determined?
  (b) Find $f(0)$, $f(3)$, and $f(10)$.
  (c) Find $f(a)$, $f(b)$, and $f(c)$.
  (d) Calculate $f(2) + f(3)$.

6 Let $f(x) = x^2$.
  (a) Find $f(2)$ and $f(3)$.
  (b) Find $f(a)$ and $f(b)$.
  (c) Is $f(2) + f(3) = f(2 + 3)$ or $f(5)$?
  (d) Is $f(a) + f(b) = f(a + b)$?

7 The formula $C = 5/9 (F - 32)$ can be used to convert from degrees Fahrenheit (F) to degrees Celsius (C). It shows the Celsius temperature as a function of Fahrenheit temperature. We would write $C = f(F)$.
  (a) Find the Celsius equivalent to $212°$ Fahrenheit.
  (b) Calculate $f(212)$.
  (c) Calculate $f(32)$.
  (d) Calculate $f(50)$.
  (e) A comfortable room temperature is $20°$. Is this Celsius or Fahrenheit?

SUMMARY

The set $\{\,a, b\,\}$ is the same as the set $\{\,b, a\,\}$. Yet often we want to specify the order of appearance of two elements. So we have defined an ordered pair; the ordered pairs (a, b) and (b, a) are not the same.

The cross product set $A \times B$ is a set of ordered pairs formed from the sets A and B. Each element of A is associated with each and every element of B. Thus if A contains three elements and B contains ten elements, $A \times B$ contains thirty ordered pairs.

Any subset of a cross product set is called a relation. If an ordered pair (a, b) is in a relation R, we say "a is related to b" and write $a\ R\ b$. Some relations can be expressed verbally. Thus the relation "lives in the same state as" stands between any two Texans. Other relations can be expressed in set-generator form. In that form a formula is given to express the relation between the first and second components of the ordered pairs.

When a relation is formed from the cross product of a set with itself, we can ask if the relation is reflexive or symmetric or transitive. Any relation that has all three properties is called an equivalence relation.

Equivalence relations are powerful. They organize the set from which they are formed into distinct classes. The classes can be appropriately named.

Though we have approached the study of equivalence relations mathematically, the result applies to every field in which organization into categories or groups is undertaken. In biology the plant kingdom is organized into phylums, classes, and orders; the relation "belongs to the same order as" puts the lily, tulip, asparagus, onion, and daffodil all in the same organizational group. The history of music is organized into the Early Christian, Medieval, Renaissance, Baroque, Classical, Romantic, and Contemporary periods; the relation "belongs to the same period as" puts the music of Handel, Bach, Vivaldi, Corelli, and Scarlatti in the Baroque period.

In Section 3.6 we defined a function. Once again this is a particular set of ordered pairs. In functions we emphasize dependency of y on x in any (x, y) belonging to the function. Functional notation introduces a calculational aspect in finding ordered pairs. Functions are the basis of a great deal of mathematics.

In preparing for the Review Test, you should be familiar with the following:

1. The difference between an ordered pair and a set with two elements.

2. Forming $A \times B$, given sets A and B.

3. Forming $A \times A$, given set A.

4. Determining whether a relation is reflexive, symmetric, or transitive.

5. Determining whether a relation is an equivalence relation.

6. Naming equivalence classes.

7. Using the notation R, $R_{A \times B}$, and $a\ R\ b$.

8. Determining whether a set of ordered pairs is a function.

9. Expressing functions in set-generator form and converting to roster form.

10. Using functional notation.

REVIEW TEST

1 True or false:
   (a) $\{r, s\} = \{s, r\}$
   (b) $(r, s) = (s, r)$
   (c) (r, s) is an element of $\{r, s, t\}$
   (d) $(1, 2)$ is an element of $\{(1, 2), (2, 3), (3, 4)\}$
   (e) $(2, 4) = (1, 2)$

2 Let $A = \{r, s\}$ and $B = \{l, m, n\}$.
   (a) Form $A \times B$.
   (b) Form $B \times A$.
   (c) Form $A \times A$.
   (d) What is $n(A \times B)$?

3 Let $L = \{a, b, c\}$ and consider the relations discussed below to be subsets of $L \times L$.
   (a) Write the ordered pairs in the relation "alphabetically precedes."
   (b) Write the ordered pairs in the relation "is the same letter as."
   (c) Which one of these relations is an equivalence relation?
   (d) If R is the relation "alphabetically precedes," is $c\ R\ a$?
   (e) If R is the relation "is *not* the same letter as," is $a\ R\ b$?

4 Suppose $\{(1, 4), (4, 1)\}$ is a relation R.
   (a) Is R reflexive?
   (b) Is R symmetric?
   (c) Is R transitive?

5 Let R be the relation "is older than" on the set of people.
   (a) Is R reflexive?
   (b) Is R symmetric?
   (c) Is R transitive?

6 Consider the relation "has the same zip code as" on the set of people in the United States.
   (a) Is this an equivalence relation?
   (b) If so, describe the equivalence classes.

7 If freshman, sophomore, junior, and senior are names of classes at a college, what is an equivalence relation that creates this partitioning?

8 Tell which of the following are functions:
   (a) $\{ (a, b), (b, c), (c, d), \ldots, (y, z) \}$.
   (b) $\{ (1, 1), (1, 2), (1, 3), (1, 4) \}$.
   (c) $\{ (1, -1), (-1, 1) \}$.
   (d) $\{$ (New York, Albany), (Kentucky, Frankfort), (Colorado, Denver), (Oregon, Salem), (Texas, Austin) $\}$.

9 Suppose we have $y = f(x) = 2x^2 + 3$.
   (a) Find $f(1)$.
   (b) What ordered pair in the function is determined by (a)?

10 The charge for a long-distance telephone call depends upon the number of minutes.
   (a) Using C for charge and m for minutes, write a formula defining this function if every minute costs 35¢.
   (b) If the call lasts 2 minutes, what is the charge?
   (c) If the call lasts 1 hour, what is the charge?

Godfrey Harold Hardy (1877–1947) was the leading
English pure mathematician of his times. His greatest
contributions were in analysis, an area related to
calculus, but he was also interested in number
theory. His book *A Course of Pure Mathematics*
(1908), written for undergraduates, became the first
English exposition of number theory, functions, and
limits. As such, it revolutionized the teaching of
mathematics in English and American colleges. In the
following article Hardy defends the aesthetics of
mathematics. It is from *A Mathematician's Apology*,
written in his later years.

A MATHEMATICIAN'S APOLOGY
GODFREY H. HARDY

A mathematician, like a painter or a poet, is a maker of patterns. If his
patterns are more permanent than theirs, it is because they are made
with *ideas*. A painter makes patterns with shapes and colours, a poet with
words. A painting may embody an 'idea', but the idea is usually
commonplace and unimportant. In poetry, ideas count for a good deal
more: but, as Housman insisted, the importance of ideas in poetry is
habitually exaggerated: 'I cannot satisfy myself that there are any such
things as poetical ideas. . . . Poetry is not the thing said but a way of
saying it.'

Not all the water in the rough rude sea
Can wash the balm from an anointed King.

Could lines be better, and could ideas be at once more trite and more
false? The poverty of the ideas seems hardly to affect the beauty of the
verbal pattern. A mathematician, on the other hand, has no material to
work with but ideas, and so his patterns are likely to last longer, since
ideas wear less with time than words.

The mathematician's patterns, like the painter's or the poet's, must be
beautiful; the ideas, like the colours or the words, must fit together in a
harmonious way. Beauty is the first test; there is no permanent place in
the world for ugly mathematics. And here I must deal with a
misconception which is still widespread (though probably much less so

From *A Mathematician's Apology* by G. H. Hardy, New York: Cambridge University Press,
1967, pp. 84–91. Reprinted by permission.

now than it was twenty years ago), what Whitehead has called the 'literary superstition' that love of and aesthetic appreciation of mathematics is 'a monomania confined to a few eccentrics in each generation'.

It would be difficult now to find an educated man quite insensitive to the aesthetic appeal of mathematics. It may be very hard to *define* mathematical beauty, but that is just as true of beauty of any kind—we may not know quite what we mean by a beautiful poem, but that does not prevent us from recognizing one when we read it. Even Professor Hogben, who is out to minimize at all costs the importance of the aesthetic element in mathematics, does not venture to deny its reality. 'There are, to be sure, individuals for whom mathematics exercises a coldly impersonal attraction. . . . The aesthetic appeal of mathematics may be very real for a chosen few.' But they are 'few', he suggests, and they feel 'coldly' (and are really rather ridiculous people, who live in silly little university towns sheltered from the fresh breezes of the wide open spaces). In this he is merely echoing Whitehead's 'literary superstition'.

The fact is that there are few more 'popular' subjects than mathematics. Most people have some appreciation of mathematics, just as most people can enjoy a pleasant tune; and there are probably more people really interested in mathematics than in music. Appearances may suggest the contrary, but there are easy explanations. Music can be used to stimulate mass emotion, while mathematics cannot; and musical incapacity is recognized (no doubt rightly) as mildly discreditable, whereas most people are so frightened of the name of mathematics that they are ready, quite unaffectedly, to exaggerate their own mathematical stupidity.

A very little reflection is enough to expose the absurdity of the 'literary superstition'. There are masses of chess-players in every civilized country—in Russia, almost the whole educated population; and every chess-player can recognize and appreciate a 'beautiful' game or problem. Yet a chess problem is *simply* an exercise in pure mathematics (a game not entirely, since psychology also plays a part), and everyone who calls a problem 'beautiful' is applauding mathematical beauty, even if it is beauty of a comparatively lowly kind. Chess problems are the hymn-tunes of mathematics.

We may learn the same lesson, at a lower level but for a wider public, from bridge, or descending further, from the puzzle columns of the popular newspapers. Nearly all their immense popularity is a tribute to the drawing power of rudimentary mathematics, and the better makers of puzzles, such as Dudeney or 'Caliban', use very little else. They know their business; what the public wants is a little intellectual 'kick', and nothing else has quite the kick of mathematics.

I might add that there is nothing in the world which pleases even famous men (and men who have used disparaging language about mathematics) quite so much as to discover, or rediscover, a genuine mathematical theorem. Herbert Spencer republished in his auto-biography a theorem about circles which he proved when he was twenty (not knowing that it had been proved over two thousand years before by Plato). Professor Soddy is a more recent and a more striking example (but *his* theorem really is his own.)

❊   ❊   ❊

We may say, roughly, that a mathematical idea is 'significant' if it can be connected, in a natural and illuminating way, with a large complex of other mathematical ideas. Thus a serious mathematical theorem, a theorem which connects significant ideas, is likely to lead to important advances in mathematics itself and even in other sciences. No chess problem has ever affected the general development of scientific thought; Pythagoras, Newton, Einstein have in their times changed its whole direction.

The seriousness of a theorem, of course, does not *lie in* its consequences, which are merely the *evidence* for its seriousness. Shakespeare had an enormous influence on the development of the English language, Otway next to none, but that is not why Shakespeare was the better poet. He was the better poet because he wrote much better poetry. The inferiority of the chess problem, like that of Otway's poetry, lies not in its consequences but in its content.

There is one more point which I shall dismiss very shortly, not because it is uninteresting but because it is difficult, and because I have no qualifications for any serious discussion in aesthetics. The beauty of a mathematical theorem *depends* a great deal on its seriousness, as even in poetry the beauty of a line may depend to some extent on the signifi-cance of the ideas which it contains. I quoted two lines of Shakespeare as an example of the sheer beauty of a verbal pattern; but

After life's fitful fever he sleeps well

seems still more beautiful. The pattern is just as fine, and in this case the ideas have significance and the thesis is sound, so that our emotions are stirred much more deeply. The ideas do matter to the pattern, even in poetry, and much more, naturally, in mathematics; but I must not try to argue the question seriously.

4 CREATING NEW NUMBERS

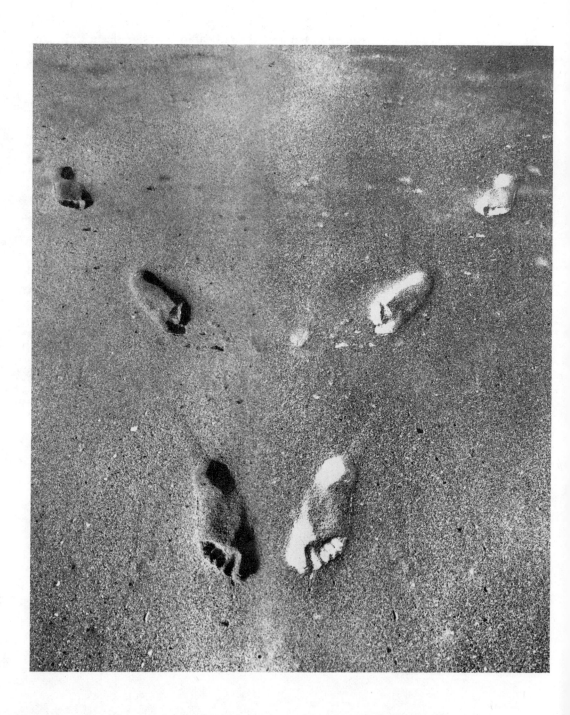

So far the only numbers we have discussed in any detail have been the counting or natural numbers. But what about fractions, decimals, negatives, and perhaps other kinds of numbers? Why are other numbers needed? Where did they come from?

We have remarked that it took thousands of years to develop our current decimal counting system. Yet negative numbers and fractions have been in common use for only three or four hundred years. The reason for their late development is that they depend upon the natural numbers for their meaning. Once our numeration system became widely adopted (about A.D. 1500), it was possible to extend the mathematics to include numbers like -3 or $7/8$.

In retrospect the mathematics seems easy. To get -3, just subtract 7 from 4 as if you were writing a check for \$7 on a balance of only \$4. To see why that couldn't be done 1000 years ago, we will use the mathematics of sets and ordered pairs to describe what probably was the natural evolution from a given set N (the counting numbers) to new creations: the integers, rational numbers, and irrational numbers.

4.1 BINARY OPERATIONS AND CLOSURE

After we learn to count, using the set $N = \{1, 2, 3, 4, \ldots\}$, we soon learn to add, subtract, multiply, and divide. These are *binary operations*, meaning that they are performed upon two numbers. While it is possible to add more than two numbers, the addition is actually done in stages and is dependent upon adding a pair of numbers at a time.

EXAMPLE 1

The sum $2 + 3 + 4$ is found in one of two ways. We think either $(2 + 3) + 4$ or $2 + (3 + 4)$, the parentheses indicating which pair of numbers to add first.

In ordinary arithmetic we learn to associate 5 with $2 + 3$, or 12 with 3×4, or 6 with $11 - 5$, or 3 with $15 \div 5$. The association of a third number with a given pair of numbers is the basis of a binary operation. The idea of a binary operation is given in the following definition:

DEFINITION 1

Let S be a set of elements. With every pair of elements in S, say a and b, associate a third element, c. The set of pairs, along with the associated third elements, is called a *binary operation* on the set S.

We sometimes indicate a binary operation that associates the element c with the elements a and b this way:

$$a \circ b = c$$

where $\circ$ can be read "operate."

The ordinary multiplication table (Table 1) indicates a binary operation on the set N. Note that the table is incomplete. For example, it does not show the product of 8 and 17. But we usually use this much of the table to find products that are not shown.

Table 1 Ordinary multiplication (partial table)

×	1	2	3	4	5	6	7	8	9
1	1	2	3	4	5	6	7	8	9
2	2	4	6	8	10	12	14	16	18
3	3	6	9	12	15	18	21	24	27
4	4	8	12	16	20	24	28	32	36
5	5	10	15	20	25	30	35	40	45
6	6	12	18	24	30	36	42	48	54
7	7	14	21	28	35	42	49	56	63
8	8	16	24	32	40	48	56	64	72
9	9	18	27	36	45	54	63	72	81

The entry 35 in the body of the table is associated by way of the operation × (times) with 7 and 5. We find 35 at the junction of row 7 and column 5. So $7 \times 5 = 35$. Observe that 5×7 is also associated with 35, but that this product is in a different location in the table.

Not all binary operations need to be on an infinite set, nor do they have to be familiar operations.

EXAMPLE 2

Let S be the set $\{a, b, c\}$ and $\circ$ be an operation on that set defined by Table 2.

Table 2
An abstract binary operation

∘	a	b	c
a	a	b	c
b	b	c	a
c	c	a	b

Table 3
A nonclosed binary operation

∘	a	b	c
a	b	c	d
b	c	b	a
c	d	a	b

Here we associate a with elements b and c by way of the operation ∘. Thus we have $b \circ c = a$. Similarly, $a \circ b = b$, and $c \circ c = b$, and so on. Note that there has been no *meaning* given to a, b, c, or the operation ∘. However, we have followed our definition of binary operation precisely.

While the binary operation in Example 2 has no meaning, there is some organization to Table 2. For comparison, we create another operation on the same set $\{a, b, c\}$. This one (Table 3) lacks a property that the previous binary operation had. Here some of the results of the operation are not in the set $\{a, b, c\}$. Observe that $c \circ a = d$ and $a \circ c = d$ and that d is not in $\{a, b, c\}$. We say that the operation ∘ is *not closed* with respect to the set.

All entries in the body of Table 2 are elements of the given set. When this occurs the binary operation is said to be *closed* with respect to the set. Table 2 has the property of *closure*. This is explained more precisely in the following definition:

DEFINITION 2

A binary operation $a \circ b$ on a set S is said to be closed with respect to that set (or closed *on* that set), if whenever c is associated with $a \circ b$, then c is also an element of S.

EXAMPLE 3

Let N be the set of natural numbers and consider the operation of ordinary multiplication. Since the product of any two natural numbers is also a natural number, we say that multiplication is closed on N. We can see intuitively that this means that if Table 1 were to be extended far enough, the result of any multiplication would be in the set N; that is, the only entries in the body of the table are natural numbers.

When a table is given that defines an operation, it is easy to detect the closure or nonclosure property. If the entries in the body of the table are only those in the given set, then the operation is closed. If some element in the body of the table is not in the given set, then the operation is not closed.

Exercise Set 4.1

1 Because binary operations involve only two elements at a time, an expression like $a + b + c$ must mean $(a + b) + c$ or $a + (b + c)$, where the parentheses indicate which addition is to be performed first. In each of the following, insert parentheses so that the first operation must clearly lead to the results given.
   (a) $5 + 3 + 1 = 5 + 4 = 9$
   (b) $5 + 3 + 1 = 8 + 1 = 9$
   (c) $6 \times 2 \times 2 = 12 \times 2 = 24$
   (d) $7 - 4 - 2 = 3 - 2 = 1$
   (e) $7 - 4 - 2 = 7 - 2 = 5$
   (f) $12 \div 3 \times 2 = 4 \times 2 = 8$
   (g) $12 \div 3 \times 2 = 12 \div 6 = 2$
   (h) Let $A = \{1, 2, 3, 4, 5\}$; $B = \{4, 5, 6, 7\}$; $C = \{2, 3, 4, 5, 6, 7, 8\}$. Then $A \cup B \cap C = \{1, 2, 3, 4, 5, 6, 7\}$.
   (i) Use the same sets as in (h). Now, $A \cup B \cap C = \{2, 3, 4, 5, 6, 7\}$.

2 (a) Let $S = \{a, b, c\}$. Define a binary operation $\circ$ on S (different from the example in the text) that is not closed.
   (b) Define a binary operation $\circ$ on S that is closed.
   °(c) How many closed binary operations on S can be defined? Hint: Use the Fundamental Counting Principle.

3 Is ordinary addition on N closed?

4 Let us set up notation that will shorten exercise 3, by rewriting it as, "Is $[+, N]$ closed?" Now answer each of these questions, explaining your answers briefly:
   (a) Is $[\times, N]$ closed?   (b) Is $[-, N]$ closed?   (c) Is $[\div, N]$ closed?
   (d) Let U be any universal set containing subsets S_1, S_2, S_3, S_4, and S_5. Let $S = \{S_1, S_2, S_3, S_4, S_5\}$. Is $[\cup, S]$ closed? Is $[\cap, S]$ closed?
   °(e) Let U be a universal set containing every subset that exists and let S be the set of all subsets. Is $[\cup, S]$ closed? Is $[\cap, S]$ closed?

5 Does the following table have closure?

$\circ$	a	b	c
a	a	a	a
b	a	a	a
c	a	a	a

6 (a) In Chapter 2 we introduced operations on sets. Which of the operations union, intersection, and complement are binary operations?
   (b) Does it matter whether we perform $A \cup B \cup C$ as $(A \cup B) \cup C$ or as $A \cup (B \cup C)$?

4.2 CREATING THE INTEGERS

In the set N subtraction is not closed. Some differences of natural numbers are natural numbers, but some are not. Certainly $8 - 5$ is a natural number, but $5 - 8$ is not.

When we say $8 - 5 = 3$ we mean that $8 = 5 + 3$ or that 8 is 3 more than 5. Every difference of natural numbers can be expressed as a corresponding addition. Thus $a - b = c$ means that $a = b + c$ or that a is c more than b. In the set of natural numbers this is understood only if a is greater than b.

Can we attach meaning to $a - b$ when a is smaller than b? For example, what is $5 - 8$? Let $5 - 8$ be some number, say x. Then $5 - 8 = x$ should mean $5 = 8 + x$. We invent the number (-3) to be the correct choice for x. Then if $5 = 8 + (-3)$ is to be consistent with the fact that $5 = 8 - 3$ in the set N, it is reasonable to define $8 + (-3)$ to mean $8 - 3$.

In general, $r + (-s)$ means $r - s$, where r is greater than s.

EXAMPLE 1

Invent a number to represent the difference $2 - 10$. Let $2 - 10 = x$. This will mean $2 = 10 + x$. Since it is a fact about natural numbers that $2 = 10 - 8$, we take x to be -8 so that $10 + (-8) = 10 - 8$.

We will not labor over every subtraction $a - b$, where a is smaller than b. The fact is, we can create a new number for every such subtraction by the method of Example 1.

These new numbers are called *negative integers* and can be denoted by I^-. The symbols used for the negative integers are the symbols used to represent the natural numbers, preceded by the sign $-$:

$$I^- = \{-1, -2, -3, -4, -5, \ldots\}$$

Some negative integer can be chosen from this set to represent the difference $a - b$ where a is smaller than b (assuming a and b are natural numbers).

If a is greater than b (and they are both natural numbers), then the difference $a - b$ is a natural number. To contrast this with the negative integers we can write such results by using the natural number preceded by the sign $+$. In this context, the natural numbers can be called the *positive integers* and can be denoted I^+:

$$I^+ = \{+1, +2, +3, +4, \ldots\}$$

The negative integers can be said to be the set of all differences $a - b$, where a is smaller than b. The positive integers are the set of all differences $a - b$, where a is greater than b.

What if a is equal to b $(a = b)$? That is, what about the difference $a - a$? We will call this the number 0.

DEFINITION 3
Zero is the number associated with the difference $a - a$, where a is any natural number.

Taken together, the sets I^- and I^+ and the number 0 constitute the set of integers, I. It is convenient to indicate the set I in roster form as follows:

$$I = \{\ldots, -4, -3, -2, -1, 0, +1, +2, +3, \ldots\}$$

The operation of subtraction on the set N can be partially tabulated. (See Table 4.) The elements in the body of the table are the "new" integers.

Observe that -2 occurs many times in the table: $-2 = 1 - 3$, $-2 = 2 - 4$, $-2 = 3 - 5$, $-2 = 4 - 6$, $-2 = 5 - 7$, and so on. Likewise, $+4$ occurs many times: $+4 = 5 - 1$, $+4 = 6 - 2$, and so on. We can say that -2 is the name of the equivalence class wherein all subtractions of natural numbers yield -2. The integers, then, can be thought of as a partitioning of the set of all ordered pairs of natural numbers; the partitioning is such that two ordered pairs belong to the same equivalence class if their differences are equal.

Table 4   Subtraction of natural numbers creates the integers

$-$	1	2	3	4	5	6	7	8	9
1	0	-1	-2	-3	-4	-5	-6	-7	-8
2	$+1$	0	-1	-2	-3	-4	-5	-6	-7
3	$+2$	$+1$	0	-1	-2	-3	-4	-5	-6
4	$+3$	$+2$	$+1$	0	-1	-2	-3	-4	-5
5	$+4$	$+3$	$+2$	$+1$	0	-1	-2	-3	-4
6	$+5$	$+4$	$+3$	$+2$	$+1$	0	-1	-2	-3
7	$+6$	$+5$	$+4$	$+3$	$+2$	$+1$	0	-1	-2
8	$+7$	$+6$	$+5$	$+4$	$+3$	$+2$	$+1$	0	-1

Note that the subtraction table organizes the ordered pairs of natural numbers. The difference $3 - 4$ corresponds to $(3, 4)$; $6 - 7$ corresponds to $(6, 7)$; $5 - 6$ corresponds to $(5, 6)$. Thus we can read the table backward to see which ordered pairs belong to the same equivalence class. First identify all differences that result in, say, -3. Then one equivalence class of ordered pairs is $\{ (1, 4), (2, 5), (3, 6), (4, 7), (5, 8), (6, 9), \ldots \}$. What better name to call this equivalence class than -3?

Another set of ordered pairs (a, b) that belong together because their difference $(a - b)$ is -7 is $\{ (1, 8), (2, 9), (3, 10), (4, 11), (5, 12), \ldots \}$. What better name to call this class than "-7"?

Likewise, "$+3$" is a good name for the equivalence class $\{ (4, 1), (5, 2), (6, 3), (7, 4), (8, 5), \ldots \}$.

We can visualize the equivalence classes better if we place together the ordered pairs belonging to the same class, as in Figure 1. The relation determining the partitioning is $(a, b) \, R \, (c, d)$ if $a - b = c - d$.

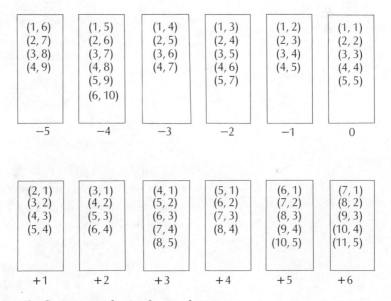

Figure 1   Some equivalence classes of integers

Now that we have created the integers, I, we can create binary operations on this new set in order to use integers in arithmetic. The operations are "ordinary" addition, subtraction, multiplication, and division. The term "ordinary" will mean "everyday" or "usual," unlike the abstract and meaningless operations we invented to explain closure in Section 4.1. Rather than explain why the ordinary operations give the results they do, we simply review them and give examples.

ADDITION

Addition of integers is closed and works like balancing a checkbook:

$$(+4) + (+5) = +9$$
$$(+4) + (-5) = -1$$
$$(+4) + (-3) = +1$$
$$(-4) + (+5) = +1$$
$$(-4) + (+3) = -1$$
$$(-4) + (-6) = -10$$

MULTIPLICATION

Multiplication of integers is also closed. The signs of the products of two integers are tabulated below:

a	b	$a \times b$
+	+	+
+	−	−
−	+	−
−	−	+

Thus $(-8)(+4) = -32$, $(+7)(+5) = 35$, and $(+3)(-4) = -12$.

SUBTRACTION

Subtraction of integers is also closed. Subtraction can be performed by converting each difference into a sum according to this rule: Change the subtraction sign to an addition sign and reverse the sign of the second integer; then perform the addition of the integers:

$$(+1) - (+4) = (+1) + (-4) = -3$$
$$(-7) - (+4) = (-7) + (-4) = -11$$
$$(-7) - (-2) = (-7) + (+2) = -5$$

DIVISION

Division is *not* closed on the set of integers. There is no integral answer to $+3 \div +4$ or $-7 \div +8$. However, when divisions are possible, the rules for signs are the same as they are for multiplication. Thus

$$+12 \div +4 = +3 \qquad +8 \div -2 = -4$$
$$-14 \div +7 = -2 \qquad -26 \div -13 = +2$$

Exercise Set 4.2

1 (a) Is addition closed with respect to I? That is, is $[+, I]$ closed?
   (b) Is $[\times, I]$ closed?
   (c) Is $[-, I]$ closed?
   (d) Give an example to show that $[\div, I]$ is not closed.

2 (a) Is $[+, I^+]$ closed?
   (b) Is $[-, I^+]$ closed?
   (c) Is $[\times, I^+]$ closed?

3 (a) Is $[+, I^-]$ closed?
   (b) Is $[-, I^-]$ closed?
   (c) Is $[\times, I^-]$ closed?

4 What subtraction is meant by each of the following?
   (a) $12 + (-6)$ (b) $20 + (-3)$ (c) $4 + (-1)$

5 What integer do we create to correspond to each of these differences?
   (a) $7 - 1$ (b) $30 - 17$ (c) $6 - 11$
   (d) $12 - 9$ (e) $9 - 9$ (f) $17 - 27$

6 All the integers are created by considering $[-, N]$, that is, by inventing new numbers to correspond to the operation subtraction on the set $\{1, 2, 3, 4, \ldots\}$. Can the same set of integers be constructed from the set $\{4, 5, 6, 7, \ldots\}$ or from the set $\{11, 12, 13, \ldots\}$?

7 The exercises below are included for practice with arithmetic operations on I. Find the members of I that are represented by the following.
   (a) $+2 + (-7)$ (b) $(-7) + (-4)$
   (c) $-7 + (-2) + (-8)$ (d) $-2 - (-2)$
   (e) $-6 + (-2) - (+7)$ (f) $14 - 7$
   (g) $+14 - (+7)$ (h) $(-7) \times (-7)$
   (i) $(-7)(+7)$ (j) $(-50) \div (-2)$
   (k) $(-7) \times (+4)$ (l) $(-2) \times (-6) - (+3)$
   (m) $[(-4) + (-3)] \times (-2)$ (n) $[-12 \div (-2)] \div (-3)$
   (o) $-1[+6 + (-2)(-3 + 4)]$ (p) $(-14 \div -7)(+4)$
   (q) $[-3 + (-4)][-5 - (-2)]$ (r) $[-3 - (-6)]^2$
   (s) $[-3 - (-4)][-3 + (-4)]$
   (t) $[(-2)(-3)(+4)(-1)(-6)] \div [2 + (-1)]$

8 One of the operations above is ambiguous; that is, two answers are possible. Which exercise is ambiguous?

°9 The ordered pair of natural numbers (a, b) can be used to denote the integer corresponding to $a - b$. Perform each of the following operations, where the ordered pairs are considered integers.
   (a) $(7, 4) \times (3, 1)$ (b) $(8, 10) + (10, 8)$ (c) $(15, 7) - (20, 30)$
   (d) $[(14, 2) \div (1, 5)] \times (11, 9)$ (e) $(22, 25) - [(27, 30) - (11, 5)]$

10 We have constructed the set I and have defined two related sets I^- and I^+.
   (a) Is $I^- \subseteq I$? Is $I^- \subset I$?   (b) Is $I^+ \subseteq I$?
   (c) What is $I^+ \cap I^-$?   (d) Find $I^+ \cup I^-$.

° 11 When 0 is added to the set I^+ or I^-, it is convenient to denote these new sets as $I^{+,0}$ and $I^{-,0}$.

$$I^{+,0} = \{ 0, +1, +2, +3, \ldots \}$$
$$I^{-,0} = \{ 0, -1, -2, -3, \ldots \}$$

The first set can be called the nonnegative integers, and the second set can be called the nonpositive integers.

(a) Is every integer either negative or positive?
(b) What is the complement of $I^{+,0}$, where I is taken as the universe?
(c) Similarly, what is the complement of $I^{-,0}$?
(d) Similarly, what is the complement of I^-? Of I^+?
(e) What is $I^{+,0} \cup I^{-,0}$?
(f) What is $I^{+,0} \cap I^{-,0}$?
(g) Is every integer either nonnegative or positive?

° 12 See Figure 1. Give three or four ordered pairs that would belong to the equivalence class whose name is

(a) +7   (b) +10   (c) −6   (d) −9

4.3 CREATING THE RATIONALS

Division of integers is not closed; $(+4) \div (-7)$ is not an integer. However, $(+12) \div (-3) = -4$. (For some integer divisions the results *are* integers.) We can say that $(+12) \div (-3) = -4$ because $+12 = (-3)(-4)$. In general $a \div b = c$ means $a = bc$. The idea is that every possible division of integers means a corresponding multiplication of integers.

The indicated division $(+4) \div (-7)$ does not result in an integer. But we can attach meaning to such divisions; in doing so we will create a new kind of number. Let $(+4) \div (-7) = x$. We would want the division to correspond to the multiplication in $(+4) = (-7)x$. We invent the number $+4/-7$ to be the x and call this new number a *rational*.

EXAMPLE 1

The indicated division $(+4) \div (-7)$ means the rational $+4/-7$. Likewise the division $(+12) \div (-3)$ will be written as the rational $+12/-3$.

In general, the quotient of any two integers $a \div b$ (except that b cannot be the integer 0) is defined as the rational number a/b. Later we will explain why b cannot be 0.

The distinction between $(+4) \div (-7)$ and $+4/-7$ may seem strange to you if you are accustomed to interchanging the $\div$ and / signs. We shall not belabor this point except to emphasize that we can create a rational to correspond to every division of two integers (except for division by the integer 0).

We can denote the set of all rationals a/b by the symbol R. A rational is commonly called a *fraction*. In the fraction a/b, we call a the *numerator* and b the *denominator*. In Table 5 you will see the pattern of rationals in tabular form. For now, the set of rationals is written as

$$R = \{ a/b \mid a \text{ and } b \text{ are integers, } b \neq 0, \text{ and } a/b \text{ corresponds to the indicated division } a \div b \}$$

Rationals with different numerators and denominators are certainly different rationals. Yet we often write $1/2 = 2/4$. (Both indicate division of positive integers.) Two rationals a/b and c/d are equivalent if $ad = bc$. Thus $1/2 = 2/4$ means "1/2 is equivalent to 2/4" since $1(4) = 2(2)$.

EXAMPLE 2

$$3/4 = 6/8 \qquad \text{since} \qquad 3(8) = 4(6)$$
$$2/3 \neq 4/5 \qquad \text{since} \qquad 2(5) \neq 3(4)$$

The sign $=$ standing between 1/2 and 2/4 *should* be read "is equivalent to," instead of "is equal to." However, it is common practice to use the latter.

The binary operations $+, -, \times, \div$ can all be defined on R. The rules for adding, subtracting, multiplying, and dividing rationals are stated as formulas.

The *sum* of a/b and c/d is $a/b + c/d = (ad + bc)/bd$.

Note that a, b, c, and d are all integers, and that $ad + bc$ and bd are therefore integers (since multiplication and addition of integers are closed with respect to I). Thus the sum of two rationals is a rational. That is, addition on R is closed.

EXAMPLE 3

$$2/3 + 4/6 = [2(6) + 3(4)]/3(6)$$
$$= (12 + 12)/18$$
$$= 24/18$$

The *difference* of two rationals is given by $a/b - c/d = (ad - bc)/bd$.

Once again, the formula for this difference yields a result in rational form. The numerator is $ad - bc$, and the denominator is bd. Subtraction on R is closed.

<u>EXAMPLE 4</u>

$$5/7 - 2/3 = [5(3) - 7(2)]/7(3)$$
$$= (15 - 14)/21$$
$$= 1/21$$

The *product* of two rationals is given by
$$a/b \cdot c/d = ac/bd.$$

Since a, b, c and d are integers, we know that ac and bd are integers, so multiplication on R is closed.

<u>EXAMPLE 5</u>

$$(+3)/(-5) \cdot (-4)/(-2) = (+3)(-4)/(-5)(-2)$$
$$= (-12)/(+10)$$

The *quotient* of two rationals is given by
$$a/b \div c/d = ad/bc.$$

In the formula above, a, b, c, and d are integers, and neither b nor c nor d can be 0. Since the quotient of two rationals is a rational, division on R is closed.

<u>EXAMPLE 6</u>

$$-3/-4 \div +2/-5 = (-3)(-5)/(-4)(+2)$$
$$= +15/-8$$

DIVISION BY ZERO

A rational number is created to correspond to every division $a \div b$ for integers a and b, except when $b = 0$. What happens if we try to divide by 0?

We will investigate $5 \div 0$. If we assume that $5/0$ is equivalent to some rational number a/b, we have

$$5/0 = a/b \tag{1}$$

Then these equivalent fractions would imply that

$$5(b) = 0(a) \tag{2}$$

It is a property of 0 that 0 times any number yields 0. So we then have

$$5(b) = 0 \tag{3}$$

Thus b must be 0:

$$b = 0 \tag{4}$$

Notice that there are no restrictions on a. Suppose we take $a = 0$. Then Equation (1) becomes

$$5/0 = 0/b \tag{5}$$

But Equation (4) requires $b = 0$. Then Equation (5) becomes

$$5/0 = 0/0 \tag{6}$$

However, $0/0 = 1/1$ since $0(1) = 0(1)$. Then we can write

$$5/0 = 1/1 \tag{7}$$

which means that $5(1) = 0(1)$. But this is clearly false ($5 \neq 0$).

Any attempt to divide by 0 ultimately results in false statements. To avoid this we do not allow division by 0.

EQUIVALENCE CLASSES OF RATIONALS

Treating the rationals as quotients of integers, it is possible to write the rationals in tabular form. The integers can be written

$$I = \{ 0, +1, -1, +2, -2, +3, -3, \ldots \}$$

It should be clear that all the integers (positive, negative, 0) are in this listing. We use this ordering of the integers as column and row headings in Table 5.

Table 5 The rationals as quotients of integers

÷	0	+1	−1	+2	−2	+3	−3	+4	−4
0	Division by 0 is impossible	0/+1	0/−1	0/+2	0/−2	0/+3	0/−3	0/+4	0/−4
+1		+1/+1	+1/−1	+1/+2	+1/−2	+1/+3	+1/−3	+1/+4	+1/−4
−1		−1/+1	−1/−1	−1/+2	−1/−2	−1/+3	−1/−3	−1/+4	−1/−4
+2		+2/+1	+2/−1	+2/+2	+2/−2	+2/+3	+2/−3	+2/+4	+2/−4
−2		−2/+1	−2/−1	−2/+2	−2/−2	−2/+3	−2/−3	−2/+4	−2/−4
+3		+3/+1	+3/−1	+3/+2	+3/−2	+3/+3	+3/−3	+3/+4	+3/−4
−3		−3/+1	−3/−1	−3/+2	−3/−2	−3/+3	−3/−3	−3/+4	−3/−4
+4		+4/+1	+4/−1	+4/+2	+4/−2	+4/+3	+4/−3	+4/+4	+4/−4

Of course, Table 5 is only a partial listing because there are infinitely many integers. But the pattern assures us that every rational would be listed if we were to extend the table far enough.

For example, in the second row we see $+1/+2$. This corresponds to the division $+1 \div +2$, which can be written as the ordered pair $(+1, +2)$. In this way we can think of every rational as an ordered pair.

We said that $a/b = c/d$ if $ad = bc$. In ordered-pair notation a/b becomes (a, b) and c/d becomes (c, d). Then the notion of equivalent rationals is written:

$$(a, b) = (c, d) \quad \text{if} \quad ad = bc$$

Now the rationals can be thought of as equivalence classes. Two rationals (a, b) and (c, d) belong to the same class if $ad = bc$.

EXAMPLE 7

Both $(+2, +3)$ and $(+4, +6)$ belong to the same class since $(+2)(+6) = (+3)(+4)$. In more familiar form we are merely saying that $2/3$ and $4/6$ belong to the same class.

The rationals are partitioned into classes by the *equivalent rationals* idea. In Figure 2 each box contains ordered pairs belonging to the same equivalence class, and each class is labeled by the "reduced and simplified form" of all the fractions in the box. Thus $+1$ is $+1/+1$ or $-3/-3$ or $+4/+4$ or $(+1, +1)$ or $(-3, -3)$ and so on.

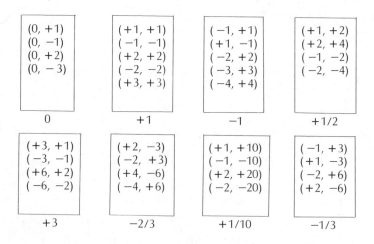

Figure 2   Some equivalence classes of rationals

Exercise Set 4.3

1 True or false:
   (a) The indicated division $+5 \div +13$ corresponds to the rational $+5/+13$.
   (b) The indicated division $+5 \div 10$ corresponds to the rational $+1/+2$.
   (c) $+3/+4$ is the same rational as $+6/+8$.
   (d) $+2/+3$ is equivalent to the rational $+4/+6$.
   (e) a/b, where a and b are any integers, is a rational.
   (f) In the rational a/b the numerator cannot be 0.

2 Consider the formula for the quotient of two rationals,

$$a/b \div c/d = ad/bc$$

   (a) What happens if $c = 0$?
   (b) Find $+2/-3 \div -4/+5$

3 (a) If $a, b, c,$ and d are integers, what kind of number is $ad + bc$?
   (b) Find the sum of $3/5$ and $2/9$.

4 (a) In the formula $a/b \cdot c/d = ac/bd$ can $c = 0$?
   (b) Find the product of $+2/-3$ and $-3/+2$.

5 (a) Why is it true that the rational a/a $(a \neq 0)$ is always equivalent to $+1/+1$?
   (b) Is $0/0$ in the same equivalence class as $+1/+1$?

6 The rational a/b can be written as the ordered pair (a, b).
   (a) Why is $(-2, 7)$ equivalent to $(-8, 28)$?
   (b) Why is $(3, 4)$ not equivalent to $(30, 31)$?

7 Write six other rationals belonging to the same equivalence class as each of the following:

   (a) $+4/+1$ (b) $(+3, +2)$
   (c) $(-2, -3)$ (d) $(-6, -5)$
   (e) $(7, -8)$ (f) $3/10$
   (g) $-3/10$ (h) $0/+4$
   (i) $(1, 10)$ (j) $(-50, 100)$

°8 When each of $a, b, c,$ and d is a positive integer, the rational a/b is less than the rational c/d if ad is less than bc. This can be written

$$a/b < c/d \quad \text{if} \quad ad < bc \quad (a, b, c, d \text{ all positive})$$

   (a) Is $4/6 < 5/7$? Why?
   (b) Which is smaller, $3/51$ or $2/17$?

°9 (a) Using the symbol $>$ for "greater than" and ideas similar to those in exercise 8, write the condition for $a/b > c/d$.
   (b) Which is larger, $9/15$ or $17/30$?

°10 Insert the correct symbol, $=$, $<$, or $>$, in each pair of rationals below.
   (a) $1/7$ $1/8$ (b) $-3/+5$ $+3/+5$
   (c) $3/51$ $1/17$ (d) $-2/+3$ $-3/+4$

4.4 THE IRRATIONALS

The natural numbers and integers are included in the set of rationals. Thus the natural number 5 can be considered to be a rational, perhaps as $+5/+1$. Also, the integer -7 can be considered to be the rational $-14/2$. In this way the set of rationals can be assumed to contain all the numbers we have studied so far.

In Section 4.3 we constructed the set of rationals from the set of integers, which in turn had been constructed from the set of natural numbers. We will not carry this device any further; that is, we will not attempt to create a new set of numbers from the rationals by considering ordered pairs of rationals. However, we will show the existence of another set of numbers, namely, the irrationals.

To show that numbers other than rationals exist, we will prove the theorem "there exists at least one number that is not rational." Before we do this, however, we need two preparatory ideas.

First, it must be clear that if the square of a natural number is even, then the natural number itself is even. This can been seen by inspecting Table 6. Observe that each square that is even is the square of a number that is also even. This is true no matter how far we extend the sets of natural numbers and their squares.

Table 6   Squares of natural numbers

Natural numbers	1	2	3	4	5	6	7	8	9	10	11
Squares	1	4	9	16	25	36	49	64	81	100	121
Even squares		↑		↑		↑		↑		↑	

Second, every even natural number can be expressed as twice some other natural number. The even natural numbers are $\{2, 4, 6, 8, \ldots\}$. Notice that $2 = 2(1), 4 = 2(2), 6 = 2(3), 8 = 2(4)$, and so on. Thus the set of evens can be written as

$$\{ 2K \mid K \text{ is a natural number} \}$$

We can now more easily follow the proof of Theorem 1.

Theorem 1   There exists at least one number that is not rational.

To prove the theorem we will prove that $\sqrt{2}$ (the square root of 2) is not rational. Let us assume that $\sqrt{2}$ *is* rational and investigate the results.

If $\sqrt{2}$ is rational, we can write it in the form a/b. We insist that this fraction be completely reduced. ("Complete reduction" of a fraction means that the numerator and denominator cannot be divided by the same number.) So we have

$$\sqrt{2} = a/b \text{ or } a/b = \sqrt{2} \tag{1}$$

then certainly,

$$a = b \cdot \sqrt{2} \tag{2}$$

Equation (2) says that a and $b \cdot \sqrt{2}$ are the same, so we expect their squares to be equal:

$$a^2 = (b \cdot \sqrt{2})^2 = b^2 \cdot 2 = 2b^2 \tag{3}$$

Then, since $a^2 = 2b^2$, we know that

$$a \text{ is even} \tag{4}$$

Thus a can be written as twice some natural number:

$$a = 2k \text{ for some natural number } k \tag{5}$$

Substituting $2k$ for a in (3), we have

$$(2k)^2 = 2b^2 \tag{6}$$

or equivalently,

$$4k^2 = 2b^2 \tag{7}$$

Then it must be true that

$$b^2 = 2k^2 \tag{8}$$

Then b^2 is even. Then

$$b \text{ is even} \tag{9}$$

Write b as twice some natural number, say L.

$$b = 2L \tag{10}$$

Now $\sqrt{2}$ was assumed to be the completely reduced fraction a/b. Since by Equation (5), a can be written as $2k$, and by Equation (10), b can be written as $2L$, we have shown that Equation (1) can be written

$$\sqrt{2} = 2k/2L \tag{11}$$

But this is not completely reduced since we can divide numerator $(2k)$ and denominator $(2L)$ by 2.

Our assumption leads to the contradiction that $\sqrt{2}$ is completely reduced and not completely reduced. So our assumption is wrong. It must be that $\sqrt{2}$ cannot be written in rational form.

We call a number that is not rational an *irrational number*. The following theorem assures us that there are many irrational numbers.

Theorem 2 There are at least as many irrationals as rationals.

Proof Let a/b be any rational. Assume that $a/b + \sqrt{2}$ is some rational, say c/d.

$$a/b + \sqrt{2} = c/d \quad \text{(assumption)} \tag{12}$$
$$a/b - c/d = -\sqrt{2} \quad \text{(simple algebra)} \tag{13}$$

On the left side of Equation (13) we see the difference of two rationals, which must be a rational. (Subtraction is closed with respect to R.) Then $a/b - c/d$ can be written in rational form, say r/s.

$$r/s = -\sqrt{2} \tag{14}$$
$$-(r/s) = \sqrt{2} \, \text{(multiplying both sides by } -1) \tag{15}$$

If r/s is rational, then so is $-(r/s)$. Equation (15) says $\sqrt{2}$ is rational, but we have already proved that $\sqrt{2}$ is not rational. Our assumption is wrong. Hence any rational $a/b + \sqrt{2}$ is irrational. Theorem 2 is proved.

There are infinitely many rationals. Theorem 2 asserts that there are just as many irrationals. Strange as it may seem, it has been shown that there are *more* irrationals than rationals. Somehow the infinitude of irrationals is greater than the infinitude of rationals. We will disscuss this in Chapter 11.

So far $\sqrt{2}$ has been shown to be an irrational number and we proved that $a/b + \sqrt{2}$ is an irrational. There are many ways to develop irrationals.

For example, if we multiply a known irrational by a nonzero integer, the result is irrational.

EXAMPLE 1

The products $+5 \cdot \sqrt{2}$, $-7 \cdot \sqrt{2}$, $14 \cdot \sqrt{2}$, and $-500 \cdot \sqrt{2}$ are all irrationals since they are the product of an integer and a number proved to be irrational.

Another way to get irrationals is to create the sum (or the difference) of an integer and a known irrational.

EXAMPLE 2

The sums $+7 + \sqrt{2}$ and $10 + \sqrt{2}$ and the difference $+7 - \sqrt{2}$ are all irrational numbers.

We can also start with a rational and add (or subtract) $\sqrt{2}$, and get an irrational.

EXAMPLE 3

The sums $2/3 + \sqrt{2}$ and $-2/7 + \sqrt{2}$ and the difference $-7/5 - \sqrt{2}$ are all irrational numbers.

In the exercises we ask you to show that $\sqrt{3}$ and $\sqrt{5}$ are irrational numbers. Thus, using the ideas in Examples 1 through 3 above, we see that all the numbers in Example 4 are irrational.

EXAMPLE 4

The numbers $-5\sqrt{3}$, $+10\sqrt{5}$, $-5 + \sqrt{3}$, $+10 + \sqrt{5}$, $-18 + \sqrt{3}$, $2/7 + \sqrt{5}$, and $(-4/+9) - \sqrt{5}$ are all irrationals.

So we can take either integers or rationals and, in combination with a known irrational, create new irrationals by adding, subtracting, or multiplying.

It would seem at first as if the sum or product of an irrational with an irrational would yield another irrational. That is true in some cases. The product $\sqrt{5} \cdot \sqrt{2}$ is irrational. However, the irrationals are not closed with respect to any of the ordinary arithmetic operations. Example 5 should convince you of this.

EXAMPLE 5

$$\sqrt{2} + (-\sqrt{2}) = 0$$
$$\sqrt{2} - \sqrt{2} = 0$$
$$\sqrt{2} \times \sqrt{2} = 2$$
$$\sqrt{2} \div \sqrt{2} = 1$$

We have demonstrated one example relative to each of the four operations that results in an answer that is not irrational. In Table 7 we show whether or not the basic arithmetic operations are closed on the sets of numbers we have dealt with. Irrationals and rationals will be contrasted in more detail in Section 4.5.

Table 7   Closure and sets of numbers

	Naturals	Integers	Rationals	Irrationals
Addition	closed	closed	closed	not closed
Subtraction	not closed	closed	closed	not closed
Multiplication	closed	closed	closed	not closed
Division	not closed	not closed	closed	not closed

Exercise Set 4.4

°1 (a) From Table 6 what relation can you find between the naturals 3, 6, 9, 12, and so on, and the squares 9, 36, 81, 144, and so on?

(b) Using this idea, prove that $\sqrt{3}$ is irrational in a fashion similar to the proof of Theorem 1.

°2 Prove that $\sqrt{5}$ is irrational.

3 The proof of Theorem 2 asserts that $a/b + \sqrt{2}$ is not a rational. What kind of number is $a/b + \sqrt{2}$?

4 Example 1 shows products of nonzero integers and $\sqrt{2}$ and asserts that they are irrational. Why do we stipulate "nonzero integer" in creating more irrationals this way?

5 Using the integers $+12$, -6, $+9$, and -15 and the irrational $\sqrt{5}$, write eight irrational numbers.

6 Using the rationals $+3/-4$ and $7/8$, write four irrational numbers.

7 True or false:

(a) The product of every integer and $\sqrt{2}$ is irrational.

(b) The sum of a rational and $\sqrt{3}$ is irrational.

(c) The sum of any two irrationals is irrational.

(d) Sometimes the quotient of two irrationals is rational.

(e) We can express $\sqrt{2}$ as a fraction whose numerator and denominator are integers.

8 The fact that the naturals are not closed under subtraction led us to develop the integers. What similar idea led us to develop the rationals?

°9 In the equation $x + 7 = 9$ we can choose a natural number for x that makes a true statement. (Namely, if x is 2, then $2 + 7 = 9$ is true.)

(a) Can we choose a natural number for x in the equation $x + 7 = 5$ that makes a true statement?

(b) Can we choose an integer for x in the equation $3x = 7$ to make a true statement?

(c) Can we choose a rational for x in the equation $x^2 = 3$ to make a true statement?

°10 Prove that $5 - \sqrt{2}$ is irrational. Hint: Assume that $5 - \sqrt{2}$ is rational and develop a contradiction of that assumption. You may use the fact that $a/b + \sqrt{2}$ is irrational from Theorem 2.

4.5 DECIMAL FORMS OF NUMBERS

In Chapter 1 we explored the meaning of units, tens, hundreds, and the like in the decimal representation of numbers. Such positional notation implies that we know where the decimal point is. The natural number 52 is usually written without a decimal point simply because all symbols to the right of the decimal point are zeros. However, 52 could be written as

$$52.0 \quad \text{or} \quad 52.00 \quad \text{or} \quad 52.000$$

When the decimal point is used it is common to refer to the number as a *decimal number* or simply as a *decimal*.

In this section we relate decimal numbers to the rationals and irrationals. The more interesting decimals are those in which the numerals to the right of the decimal point are not all zeros. In discussing a number like 54.9302 we can refer to the 54 as the *integer* (or integral) *part* and to the .9302 as the *noninteger part* or as the fractional part of the number. These terms seem obvious when 54.9302 is thought of as

$$54 + \frac{9302}{10000}$$

Every rational number can be written in decimal form merely by performing the division indicated by the fraction bar. (We will now think of 3/4 as "3 divided by 4.")

EXAMPLE 1

$$3/4 = 0.75$$
$$5/4 = 1.25$$
$$9/11 = 0.81818181\ldots$$
$$1/3 = 0.3333\ldots$$

The decimals 0.75 and 1.25 in Example 1 are called *terminating* decimals simply because they consist of only a finite number of symbols. A decimal like 0.17500000 . . . is also terminating. The zeros do not add value to the number, so it might as well have terminated after the 5.

The decimals 0.818181 . . . and 0.333 . . . are *nonterminating* decimals. The dots indicate that the representation goes on in the same manner—it does not terminate. Furthermore, these two decimals are *repeating* decimals. One or more of the digits repeats with a *cycle* of repetition.

EXAMPLE 2

0.818181 . . . has a cycle of repetition 2.
0.33333 . . . has a cycle of repetition 1.
5.123812381238 . . . has a cycle of repetition 4.

Every rational number a/b is either a terminating or a nonterminating repeating decimal. The truth of this is evident from inspection of the division indicated by the fraction bar.

EXAMPLE 3

Consider 7/8. The division of 7 by 8 is exact. That is, after a certain number of divisions there is no remainder, and so the division stops.

$$
\begin{array}{r}
0.875 \\
8\overline{)7.000} \\
6\,4 \\
\hline
60 \\
56 \\
\hline
40 \\
40 \\
\hline
0 \text{ (no remainder)}
\end{array}
$$

Thus $7/8 = 0.875$, which is a terminating decimal. It should be clear that in every case in which the division is exact, the decimal form of the rational is terminating. Other examples are $3/5 = 0.6$, $11/2 = 5.5$, $19/4 = 4.75$, and $1234/10000 = 0.1234$.

EXAMPLE 4

In the division indicated by 9/11 the pattern repeats after the second digit in the quotient:

$$
\begin{array}{r}
0.8181\ldots \\
11\overline{)9.0000} \\
8\,8 \\
\hline
20 \\
11 \\
\hline
90 \quad\longleftarrow \\
88 \\
\hline
20 \\
11 \\
\hline
9 \text{ and so on}
\end{array}
$$

The process of long division either terminates when there is no remainder, or continues without end. If it continues, a repeating pattern in the quotient must eventually occur. The key to seeing this is in looking at the pattern of subtraction. In Example 4 we know the quotient is the same as soon as we write "9" and bring down the 0. (The arrow points this out.) In dividing by 11 the only possible "subtractions" are 0, 1, 2, 3, 4, 5, 6, 7, 8, 9, or 10. Eventually one of these "subtractions" must recur in the division process.

It is also true that a terminating or repeating decimal is a rational. Given terminating decimals, it is easy to find the rational form.

EXAMPLE 5

The terminating decimal 0.37 means "thirty-seven hundredths," and so the rational form is 37/100.

It is interesting to find the rational corresponding to a nonterminating but repeating decimal.

EXAMPLE 6

Find the rational form of $0.1212 \ldots$. Let $x = 0.121212 \ldots$. Then $100x = 12.121212 \ldots$. Then $99x = 12$. (We get $99x$ by subtracting x from $100x$; we get 12 by subtracting $0.121212 \ldots$ from $12.121212 \ldots$.) If $99x = 12$, then $x = 12/99$. It happens that 12/99 reduces to 4/33. If you divide 4 by 33, you get $0.1212 \ldots$. Thus $0.1212 \ldots = 4/33$.

EXAMPLE 7

Find the rational form of $0.324324324 \ldots$. Let $x = 0.324324324 \ldots$. Then $1000x = 324.324324 \ldots$ So $999x = 324$ and $x = 324/999$. Thus $0.324324324 \ldots = 324/999$.

A strange result occurs when we apply this method to $0.49999 \ldots$.

EXAMPLE 8

Let $x = 0.4999 \ldots$. Then $10x = 4.999 \ldots$. So $9x = 4.5$. And $x = 4.5/9$ which is the same as 1/2. Thus $0.4999 \ldots = 1/2$. Of course, $1/2 = 0.5$ too. We can only conclude that $0.4999 \ldots = 0.5$.

The "strangeness" of the results of Example 8 lies in the fact that 0.4999 $\ldots$ and 0.5 differ in the tenths place ($4 \neq 5$), in the hundredths place ($9 \neq 0$), and in every other place. By our intuition it is probably true that $0.3 = 0.299 \ldots$, $0.25 = 0.24999 \ldots$, and $4.1 = 4.0999 \ldots$. You might like to verify each of these equalities.

Now we know the following two facts:

1. Every rational has a decimal form that either terminates or repeats.

2. If a decimal either terminates or repeats, then it can be converted to rational form.

If we call the set of terminating or nonterminating but repeating decimals T and the rationals R, fact 1 tells us that R is a subset of T. (See Figure 3 on the next page.)

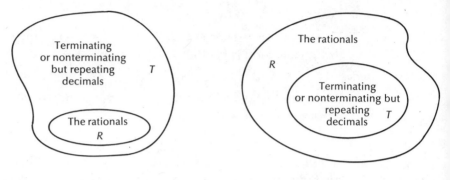

Figure 3
R is a subset of T

Figure 4
T is a subset of R

Fact 2 tells us that the terminating or repeating decimals are rationals, so T is a subset of R. (See Figure 4.) Together, the ideas illustrated in Figures 3 and 4 lead us to only one conclusion: $R = T$. Thus the set of rationals is the same as the set of terminating or nonterminating but repeating decimals.

It then follows that if any other kind of decimal exists, then that decimal is *irrational*. There are many ways to write decimals that are not terminating and not repeating. The following examples show how this can be done.

EXAMPLE 9

The decimal 0.121221222122221 . . . is a nonterminating and nonrepeating decimal. The extra 2 between each successive pair of ones destroys any possible cycle of repetition. If the cycle of repetition were of length 15, there would eventually be more than 15 twos to interrupt that cycle. Thus the decimal in this example is not repeating, and so it is not rational. It must then be irrational.

EXAMPLE 10

The number 0.41412412341234412345 . . . is irrational. Between successive pairs of fours there is an increasing sequence of counting numbers.

The rationals are the set of terminating or nonterminating but repeating decimals. We can define the irrationals as the set of nonterminating and nonrepeating decimals. These last two sentences can be taken as alternatives to our previous definitions of rational and irrational.

Exercise Set 4.5

1 What decimal is each of the following?
   (a) $7 + 9/100$ (b) $35 + 26/100$ (c) $23.9 + 3/10$
   (d) $1 + 237/1000$ (e) $1 - 2/100$ (f) $3/10 + 14/100$

2 Convert each of the following to their terminating decimal form:
   (a) $3/8$ (b) $13/4$ (c) $3/16$
   (d) $7/25$ (e) $17/20$ (f) $85/136$

3 Find the decimal form of each of the following rationals; extend divisions far
   enough to observe the cycling.
   (a) $7/11$ (b) $5/7$ (c) $4/9$ (d) $11/13$
   (e) $9/13$ (f) $15/17$ (g) $16/17$ (h) $13/24$

4 (a) Find the decimal equivalent to each of these rationals:

$$1/7, 2/7, 3/7, 4/7, 5/7, \text{ and } 6/7$$

   (b) What is the cycle of repetition of each?
   (c) Do you observe any pattern?

5 (a) Complete the following list through $16/17$, using a desk calculator if
       necessary.

$$1/17 = 0.0588235294117647\ldots$$
$$2/17 = 0.1176470588235294\ldots$$
$$3/17 = 0.176470588235294\ldots$$
$$4/17 = 0.2352941176470588\ldots$$
$$5/17 = 0.2941176470588235\ldots$$

   (b) What is the cycle of repetition of each decimal?
   (c) Do you observe any pattern in the list?

6 (a) Perform each of the divisions $1/11, 2/11, 3/11, \ldots, 10/11$.
   (b) What is the cycle of each?
   (c) What is the sum of the digits in each cycle?

°7 Assume that r/s has a decimal form that is repeating. What is the maximum
    possible cycle of repetition?

8 Find the rational form, r/s, of each of the following:
   (a) $0.6666666\ldots$ (b) $0.4444\ldots$
   (c) $0.7777777\ldots$ (d) $0.23232323\ldots$
   (e) $0.45454545\ldots$ (f) $0.3737373737\ldots$
   (g) $0.132132132\ldots$ (h) $0.647647647647647\ldots$
   (i) $1.424242\ldots$ (j) $0.825825825\ldots$
   (k) $0.34242424242\ldots$ (l) $0.001001001001001\ldots$

9 State a rule for shortening the work for examples like (a) through (h) in
   question 8.

10 Consider $0.49999\ldots = 0.5, 0.25 = 0.24999999\ldots$, and $0.71 = 0.709999\ldots$
    State a rule for finding the nonterminating equivalent for any terminating
    decimal.

11 Write three examples of decimals that are clearly irrational.

12 Show that $1.9999999\ldots = 2$.

*13 If x, y, and z are the digits of the repeating decimal $0.xyzxyzxyzxyz\ldots$, what is the fractional form of this number?

4.6 THE REAL NUMBER LINE

The rationals and irrationals taken together constitute the set of *real numbers*, or more briefly, the *reals*, denoted by the symbol $R^{\#}$. We will denote the irrationals by R; we have already used the symbol R to stand for the rationals.

Thus the set of reals is the union of the rationals and irrationals:

$$R^{\#} = R \cup R$$

Since the rationals are terminating or repeating decimals and the irrationals are the nonterminating and nonrepeating decimals, the reals must be the set of all decimal numbers.

EXAMPLE 1

The decimals 2.5, $1.333\ldots$, and $1.414114111411114\ldots$ are all reals; the first two are rational and the third is irrational.

The real numbers can be associated with the points of a line in a one-to-one way. That is, we can establish a one-to-one correspondence between the set of reals and the set of points on a line, as in Figure 5.

It should be clear that we can extend the line as far as we wish in either direction. We choose a point to correspond to 0, and select a unit length. Then we mark off the multiples of this length in both directions to determine points for $+1$, $+2$, $+3$, and so on, and for -1, -2, -3, and so on. These points can be constructed by setting a compass for the unit length and successively laying off unit lengths both to the right and to the left of the point chosen for 0.

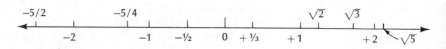

Figure 5 The real number line

Notice that we have also labeled points corresponding to some rationals and for the irrationals $\sqrt{2}$, $\sqrt{3}$, $\sqrt{5}$. Are we certain that these points exist on the line?

It is easy to locate rational points. (The expression *rational point* will mean "a point *corresponding* to a rational.") The segment from 0 to +1 can be divided into thirds, fourths, fifths, and the like by a construction from plane geometry. By setting a compass for, say 1/3, we can lay off four equal distances to the right of 0, ending with the rational point 4/3. We can set the compass for 1/2 and lay off three of these lengths to the left of 0, ending up with −3/2. In this way we can construct any rational, assuring ourselves that there is a point for every rational.

Finding points for the irrationals is more involved. First, we show a construction (Figure 6) that enables us to find the exact point corresponding to $\sqrt{2}$. Construct a perpendicular to the number line at +1. Lay off the unit length along this perpendicular. That is, make $AR = OR$. How long is OA? The Pythagorean theorem states that in a triangle with a right angle the square of the hypotenuse equals the sum of the squares of the legs. Then in our triangle $(OA)^2 = 1^2 + 1^2$. So $OA = \sqrt{1^2 + 1^2}$; $OA = \sqrt{2}$. We now set our compass for the distance OA and draw an arc of radius $\sqrt{2}$ until it intersects the number line at point P. The distance from O to P is $\sqrt{2}$, so point P corresponds to $\sqrt{2}$.

In the same way we can construct points on the number line for many other square roots. But many irrationals cannot be constructed. The irrational number π is shown here to 20 places:

$$\pi = 3.14159\ 26535\ 89793\ 23846\ldots$$

The dots are not intended to indicate what digit comes next; π is a nonrepeating nonterminating decimal and cannot be constructed with a straightedge and compass.

Yet certainly π is between 3.14 and 3.15. We could lay off three units, plus 0.1, plus 0.04, plus 0.001, plus 0.0005, and so on since these components of π can be thought of as rational lengths to be added. The

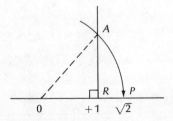

Figure 6   Constructing the length $\sqrt{2}$

fact that the decimal for π does not terminate would require that we have an infinity of time to complete the construction. However, we could come *as close as we wish* to the exact point for π. Even if we stopped at 3.14, we would be quite close to the correct point. At 3.14, the point corresponding to π is less than 2/1000 of a unit away! If we move to 3.15, the exact point corresponding to π has been passed.

For those irrationals that cannot be constructed we can use this method of successive moves to get as close as we wish. Our intuition tells us that a point for any irrational decimal can be approximated on the number line in this way.

The real number line contains points corresponding to every rational and irrational number. If a point is chosen at random on the line, it is conceivable that that point may correspond to some real number. We can approach the point from 0 by taking so many units, so many tenths, so many hundredths, and so on. The sum of these units and fractional parts is the decimal form of the real number.

Thus, given a real number, we can find a point on the line, and conversely, a point on the line corresponding to some real number. This is what it means to say that the real number line is in one-to-one correspondence with the set of real numbers.

Exercise Set 4.6

1 Given a point corresponding to $\sqrt{2}$ on the real number line, how would you find a point corresponding to $2\sqrt{2}$? $3\sqrt{2}$? $-2\sqrt{2}$? Are these points rational or irrational?
2 If the unit length is •————————• on a real number line, we can perform the following construction. Triangle ORA has a right angle at R and legs equal to 1. (Then $OA = \sqrt{2}$.) Triangle OAB has a right angle at A and legs equal to 1

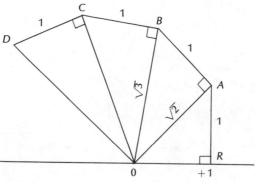

Figure 7

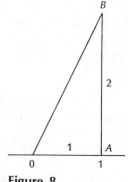

Figure 8

and $\sqrt{2}$. (Then $OB = \sqrt{3}$.) Construct two more triangles with right angles at B and C as indicated in Figure 7.

(a) Why is $OB = \sqrt{3}$?   (b) How long is OC?   (c) How long is OD?
(d) Can $\sqrt{n}$, where n is any natural number, be constructed?

3 Consider Figure 8, in which $OA = 1$ and $AB = 2$: How long is OB?

4 Given that $\sqrt{2} = 1.414\ldots$ and $\sqrt{3} = 1.732\ldots$,
(a) Find a rational number between $\sqrt{2}$ and $\sqrt{3}$.
°(b) Find an irrational number between $\sqrt{2}$ and $\sqrt{3}$.

5 Find an irrational number between 0.28 and 0.29

°6 We have $\pi/4 = 1 - 1/3 + 1/5 - 1/7 + 1/9 - 1/11 + 1/13 - \ldots$ (*All* terms must be taken to get $\pi/4$ exactly; this is, of course, impossible.) Find $\pi/4$ approximately by taking the first five terms and adding and subtracting as indicated. Approximate π from this. Compare your approximation with the value given in this section.

SUMMARY

This chapter introduced integers and rationals as sequential developments from the set of natural numbers. Integers were invented to include all differences of natural numbers. Then to make all quotients of integers possible (except the quotient indicating division by zero), the set of rationals was created. We can say that the rationals contain the integers, which in turn contain the natural numbers.

But we saw that a set of nonrationals, the irrationals, exists as well. Together the rationals and irrationals constitute the real numbers. There is, then, a sort of hierarchy of number sets as depicted in Figure 9.

Natural numbers exist as abstractions whether or not people use them or name them. Without an organized counting system using the natural numbers, civilization as we know it might not exist. Once that counting system *was* developed, growth in both commerce and technology followed. And when progress in mathematics made the need for other kinds of numbers apparent, they were developed.

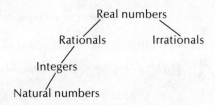

Figure 9   Hierarchy of number sets

Any decimal represents a real number. The rationals have a decimal representation that is terminating or nonterminating but repeating. The irrationals are nonterminating and nonrepeating in their decimal representation. Thus once we know that $\sqrt{2}$ is an irrational, we do not expect its decimal form (which starts $1.4142\ldots$) ever to terminate or to repeat.

It is possible to find a point on the number line corresponding to every real number. If the real number is an irrational in decimal form, we first measure the integer part away from the point for zero, then add the amounts indicated in each position to the right of the decimal point. Of course for such nonterminating decimals this is impossible, but we can *imagine* adding the correct number of 10ths, 100ths, 1000ths, and so on. We can get as close as we wish to the correct point for any irrational.

The real number line can be considered a model for the reals. Once we conceive of them in one-to-one correspondence with the points of a line, the line shows us much about the reals. For example, the real points are ordered in increasing fashion from left to right. It is easy to see why $+4$ is greater than -2; the point for $+4$ is more to the right than the point for -2. Here's another example: between any two reals there is another real; the midpoint of the segment joining any two real points always lies between them.

Before attempting the Review Test, you should be familiar with the following:

1. Binary operations defined in tabular form.

2. How to tell whether an operation is closed.

3. How subtraction is used to create integers.

4. How to tell whether two integers are in the same equivalence class.

5. The notation I, I^{+}, I^{-}.

6. The difference between $4 \div 3$ and $4/3$ in creating the rationals.

7. How to determine whether or not two rationals are equivalent.

8. The decimal form of rationals.

9. The decimal form of irrationals.

10. How to write a decimal that is clearly irrational.

11. How to change from rational form to decimal form.

12. How to change from decimal form to rational form.

13. The cycle of repetition of a repeating decimal.

14. Plotting points on a real number line.

*15. The integers and rationals as ordered pairs.

*16. Equivalence classes of integers and rationals.

REVIEW TEST

1 Insert parentheses in each of the following so that the binary operation leads to the given results.
   (a) $12 - 4 - 2 = 12 - 2 = 10$
   (b) $12 - 4 - 2 = 8 - 2 = 6$
   (c) $24 \div 6 + 2 = 24 \div 8 = 3$
   (d) $24 \div 6 + 2 = 4 + 2 = 6$

2 (a) What integer corresponds to the difference $5 - 9$?
   (b) What rational corresponds to the quotient $12 \div 7$?

3 True or false:
   *(a) As integers, $(7, 4)$ and $(15, 12)$ are in the same equivalence class.
   *(b) As rationals, $(+3, -5)$ and $(-6, +10)$ are in the same equivalence class.
   (c) $+2/-3$ is equivalent to $+4/-6$ because $(+2)(+4) = (-3)(-6)$.
   (d) The set of natural numbers is not closed with respect to subtraction.
   (e) The set of rationals is closed with respect to division.
   (f) The product of any two irrationals is always an irrational.
   (g) If an integer is not in I^+, then it is in I^-.

4 Find the decimal representation of each rational below.
   (a) 5/8   (b) 7/11   (c) 3/7

5 Write a decimal that is clearly irrational.

6 Find the rational form of each decimal below.
   (a) $0.525252\ldots$ (b) $0.7777\ldots$

7 From the word list below, complete these statements:
   (a) The decimal form of an irrational is _____.
   (b) If a decimal is _____ or _____ , then the number is rational.
   (c) If the division indicated by a/b is exact (no remainder), then the decimal is said to be _____.
   (d) A _____ decimal has no cycle of repetition.
       (i) terminating
       (ii) nonterminating and repeating
       (iii) nonterminating and nonrepeating

8 True or false:
   (a) A line segment can be divided into three parts by rule-and-compass construction.
   (b) The reals can all be expressed in decimal form.
   (c) There is only one way to express a given rational in decimal form.
   (d) We can construct a point on the real number line for every irrational.
   (e) On the number line a point exists for every irrational.
   (f) We can get as close as we wish to the corresponding point on the number line for the number $0.39999\ldots$, by the method of successive moves.
   (g) Since $0.39999\ldots = 0.4$, we can plot $0.39999\ldots$ exactly.

*9 Given a number line and a unit length, construct the point corresponding to $\sqrt{5}$.

In mathematics circles Alfred North Whitehead is
noted for his work with Bertrand Russell, *Principia
Mathematica*. This scholarly treatise establishes logic
as the foundation of mathematics. The thesis is that
all mathematical principles, from elementary
arithmetic to the most advanced theorems, emanate
from a few basic axioms. As an educator, Whitehead
communicated his philosophy clearly to his students.
This article from *An Introduction to Mathematics*
(1911) focuses on the symbolic language of
mathematics.

THE SYMBOLISM OF MATHEMATICS
ALFRED NORTH WHITEHEAD

We now return to pure mathematics, and consider more closely the
apparatus of ideas out of which the science is built. Our first concern is
with the symbolism of the science, and we start with the simplest and
universally known symbols, namely those of arithmetic.

Let us assume for the present that we have sufficiently clear ideas
about the integral numbers, represented in the Arabic notation by 0, 1,
2, . . . , 9, 10, 11, . . . , 100,101, . . . , and so on. . . . The interesting point
to notice is the admirable illustration which this numeral system affords
of the enormous importance of a good notation. By relieving the brain of
all unnecessary work, a good notation sets it free to concentrate on more
advanced problems, and in effect increases the mental power of the race.
Before the introduction of the Arabic notation, multiplication was
difficult, and the division even of integers called into play the highest
mathematical faculties. Probably nothing in the modern world would
have more astonished a Greek mathematician than to learn that, under
the influence of compulsory education, the whole population of
Western Europe, from the highest to the lowest, could perform the
operation of division for the largest numbers. This fact would have
seemed to him a sheer impossibility.

❋ ❋ ❋

Mathematics is often considered a difficult and mysterious science,
because of the numerous symbols which it employs. Of course, nothing

From *An Introduction to Mathematics* by Alfred North Whitehead, 1911, pp. 58–65. By
permission of The Clarendon Press, Oxford.

is more incomprehensible than a symbolism which we do not understand. Also a symbolism, which we only partially understand and are unaccustomed to use, is difficult to follow. In exactly the same way the technical terms of any profession or trade are incomprehensible to those who have never been trained to use them. But this is not because they are difficult in themselves. On the contrary they have invariably been introduced to make things easy. So in mathematics, granted that we are giving any serious attention to mathematical ideas, the symbolism is invariably an immense simplification. It is not only of practical use, but is of great interest. For it represents an analysis of the ideas of the subject and an almost pictorial representation of their relations to each other. If any one doubts the utility of symbols, let him write out in full, without any symbol whatever, the whole meaning of the following equations which represent some of the fundamental laws of algebra:

$$x + y = y + x \tag{1}$$
$$(x + y) + z = x + (y + z) \tag{2}$$
$$x \times y = y \times x \tag{3}$$
$$(x \times y) \times z = x \times (y \times z) \tag{4}$$
$$x \times (y + z) = (x \times y) + (x \times z) \tag{5}$$

. . . For example, without symbols, (1) becomes: If a second number be added to any given number the result is the same as if the first given number had been added to the second number.

This example shows that, by the aid of symbolism, we can make transitions in reasoning almost mechanically by the eye, which otherwise would call into play the higher faculties of the brain.

It is a profoundly erroneous truism, repeated by all the copy-books and by eminent people when they are making speeches, that we should cultivate the habit of thinking of what we are doing. The precise opposite is the case. Civilization advances by extending the number of important operations which we can perform without thinking about them. Operations of thought are like cavalry charges in a battle—they are strictly limited in number, they require fresh horses, and must only be made at decisive moments.

❋ ❋ ❋

It is interesting to note how important for the development of science a modest-looking symbol may be. It may stand for the emphatic presentation for an idea, often a very subtle idea, and by its existence make it easy to exhibit the relation of this idea to all the complex trains of ideas in which it occurs. For example, take the most modest of all symbols, namely, 0, which stands for the *number* zero. The Roman notation for numbers had no symbol for zero, and probably most mathematicians of the ancient world would have been horribly puzzled by the idea of the number zero. For, after all, it is a very subtle idea, not at all obvious. A great deal of discussion on the meaning of the zero of quantity will be found in philosophic works. Zero is not, in real truth, more difficult or subtle in idea than the other cardinal numbers. What do we mean by 1 or by 2, or by 3? But we are familiar with the use of these ideas, though we should most of us be puzzled to give a clear analysis of the simpler ideas which go to form them. The point about zero is that we do not need to use it in the operations of daily life. No one goes out to buy zero fish. It is in a way the most civilized of all the cardinals, and its use is only forced on us by the needs of cultivated modes of thought. Many important services are rendered by the symbol 0, which stands for the number zero.

The symbol developed in connection with the Arabic notation for numbers of which it is an essential part. For in that notation the value of a digit depends on the position in which it occurs. Consider, for example, the digit 5, as occurring in the numbers 25, 51, 3512, 5213. In the first number 5 stands for five, in the second number 5 stands for fifty, in the third number for five hundred, and in the fourth number for five thousand. Now, when we write the number fifty-one in the symbolic form 51, the digit 1 pushes the digit 5 along to the second place (reckoning from right to left) and thus gives it the value fifty. But when we want to symbolize fifty by itself, we can have no digit 1 to perform this service; we want a digit in the units place to add nothing to the total and yet to push the 5 along to the second place. This service is performed by 0, the symbol for zero. It is extremely probable that the men who introduced 0 for this purpose had no definite conception in their minds of the number zero. They simply wanted a mark to symbolize the fact that nothing was contributed by the digit's place in which it occurs. The idea of zero probably took shape gradually from a desire to assimilate the meaning of this mark to that of the marks, 1, 2, . . . , 9, which do represent cardinal numbers. This would not represent the only case in which a subtle idea has been introduced into mathematics by a symbolism which in its origin was dictated by practical convenience.

Thus the first use of 0 was to make the Arabic notation possible—no slight service. We can imagine that when it had been introduced for this purpose, practical men, of the sort who dislike fanciful ideas, deprecated the silly habit of identifying it with a number zero. But they were wrong, as such men always are when they desert their proper function of masticating food which others have prepared. For the next service performed by the symbol 0 essentially depends upon assigning to it the function of representing the number zero.

5 ORGANIZING MATHEMATICAL SYSTEMS

Mathematics is often called a study of numbers. While it is true that a large part of a mathematician's time is spent with numbers, numbers do not constitute mathematics. Mathematics cannot be defined in terms of facts about arithmetic, like 4(3) = 12, or in terms of solving an equation, like $4x = 12$.

All mathematicians use arithmetic facts; they may, also, for example, use procedures learned in algebra or calculus. Yet there is little (if any) mathematics in a statement like 4(3) = 12. It is mathematical to observe that 4(3) = 3(4), without caring what 4(3) or 3(4) may be. The mathematician is interested in *properties* and *relationships* that may be organized on a particular set.

In Chapter 4 we had little concern for calculation with the various numbers developed. We were quite concerned, however, with closure, a *property* that may or may not exist with respect to a certain operation on a set. In this chapter we will see how operations can be organized, and will classify their properties. There is some need to calculate or to "get an answer," but our goal is not to become experts at performing operations. Rather we hope to make progress toward understanding mathematical thinking.

Since mathematical properties do not apply exclusively to sets of numbers, we will sometimes find it useful to invent new sets and unusual operations on these sets. These may or may not have meaning outside of mathematics, but the sets and operations will exhibit the properties we describe.

5.1 MATHEMATICAL SYSTEMS

Before we deal with specific operations and properties, let's clarify what set and what operations are involved. Once a set and operation are chosen, we can ask, for example, if the operation is closed with respect to that set. Properties are studied relative to some mathematical system.

DEFINITION 1   Mathematical System

A *mathematical system* consists of a set of elements and one or more binary operations on that set.

If the set is S and the operation is ∗, the mathematical system is denoted [∗,S]. If there are two operations defined on S, say ∗ and ⊙, the system is denoted [∗, ⊙, S].

EXAMPLE 1

Let $T = \{ a, b, c, d \}$ and let the operation ! be defined by Table 1. This system would be denoted $[!, T]$.

Table 1   Operation ! on set T

!	a	b	c	d
a	a	b	c	d
b	b	c	d	a
c	c	d	a	b
d	d	a	b	c

In Example 1, there is no intended meaning for the operation !, although you may observe a pattern in the table. The system is abstract. We do not need to know the *meaning* of $c ! d$ in order to ask questions about the system. For example, is the system $[!, T]$ closed? Of course, the answer is "yes" since every entry in the table is in T. We don't care whether $c ! d = b$ or $c ! d = a$; closure is not dependent on specific answers.

We could have filled each of the 16 positions in Table 1 with any of a, b, c, or d and still have had a closed mathematical system. By the Fundamental Principle of Counting there must be 4(4)(4)(4)(4)(4)(4)(4)(4)(4)(4) (4)(4)(4)(4)(4) or 4,294,967,296 ways to define a closed system on T. We will see that only some of these four billion or so ways of defining a closed operation on T will exhibit other interesting properties.

EXAMPLE 2

A familiar mathematical system might be $[\times , N]$, where the operation is multiplication and the set is the counting numbers.

If a system is familiar, there may be no point in presenting a table to define the operation. In Example 1, however, we had to present a table to completely explain $[!, T]$. The table *defines* the operation ! on the set T.

5.2 THE COMMUTATIVE PROPERTY

Table 1 shows the property of commutativity. Notice that $b ! c = d$ and that $c ! b = d$; thus $b ! c = c ! b$. This is true for every pair of elements in T.

DEFINITION 2   Commutativity

Let [∘,S] be any mathematical system. If for every two elements a and b in S,

$$a \circ b = b \circ a$$

then the operation ∘ is said to be *commutative* on S.

The property of commutativity in a system tells us that we can reverse the order of the elements in performing the operation. A familiar example is [×, N]. Since $a \times b = b \times a$ for every a and b in N, we can say that [×, N] is a *commutative system*. The next example is more abstract.

EXAMPLE 1

Let S = { 0, 1, 2, 3 } and let × be defined by Table 2. The system [×,S] is commutative.

Table 2   Operation × on set S

×	0	1	2	3
0	0	0	0	0
1	0	1	2	3
2	0	2	10	12
3	0	3	12	21

The general commutative formula for any elements in a and b in a set and for an operation denoted by ∘ is

$$a \circ b = b \circ a$$

For Example 1, a can be chosen to be 0, 1, 2, or 3; there are four choices for b. By the Fundamental Principle of Counting then, there must be 4(4) or 16 cases of commutativity. To *prove* that [×,S] is commutative we could verify all 16 cases. One such case follows:

$$3 \times 2 \overset{?}{=} 2 \times 3 \tag{1}$$
$$12 \equiv 12 \tag{2}$$

Observe that we are checking commutativity for the elements 2 and 3. Step (1) questions whether the equality holds. The results in Step (2) are obtained from Table 2. Instead of =, Step (2) uses the sign ≡, which means "is the same as" and which in effect answers our question "yes." (We will use this device in checking cases from now on.)

Normally no one would want to go through 16 cases to check commutativity. But if we did, we would have a *proof* that the system is commutative. If S had 5 elements, there would be 25 cases to check; if there were 6 elements, there would be 36 cases to check. And, of course, no one could check all the cases for commutativity in [×,N]. How many cases would there be?

Commutativity or the lack of it is readily apparent when a table defines the operation. Consider the system [∘,S] where S = { a, b, c, d } and ∘ is defined in Table 3.

Table 3

A noncommutative system

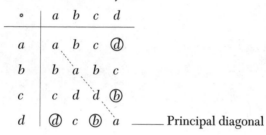

———— Principal diagonal

An imagined line from the upper left-hand corner of the table to the lower right-hand corner is called the *principal diagonal* of the table. Elements at right angles to the principal diagonal and at equal distances from it are said to be *symmetrical* with respect to the principal diagonal. The circled *d*'s are symmetrically placed and so are the circled *b*'s.

If the same elements are in every symmetric position of the table, then the operation is commutative. If one or more pairs of symmetric positions contain different elements, the operation is not commutative. In one pair of symmetrical positions in Table 3, we see a *d* and a *b*. Thus ∘ is not commutative. In particular, *c* ∘ *b* ≠ *b* ∘ *c*. This happens to be the only case that fails out of the 16 possible checks, but it is enough to destroy commutativity.

Observe that Table 2 is symmetric with respect to the principal diagonal. Therefore × is commutative on the set S.

Exercise Set 5.1

1 Consider the mathematical system [&, S], where S = { #, ☆, □ } and & is defined in Table 4.
(a) Is & closed on S? How can you tell?
(b) What elements are along the principal diagonal?
(c) Write a general commutative formula for &.

Table 4

Operation & on set S

&	#	☆	□
#	#	☆	□
☆	☆	□	#
□	□	#	☆

(d) How many cases of commutativity would you have to check to prove that & is commutative on S?

(e) Give two illustrations of commutativity checks.

(f) How many pairs of elements in the body of the table are in symmetrical positions relative to the principal diagonal?

2 In a set with ten elements, how many cases of commutativity could there be?

3 (a) In the system $[\times, N]$, is $\times$ commutative?

(b) In $[+, N]$, is $+$ commutative?

(c) In $[\div, N]$, is division closed? . . . Is division commutative?

(d) In $[-, I]$, is subtraction commutative?

(e) In $[\times, R]$, is multiplication commutative?

(f) In $[\times, R]$, is multiplication closed?

4 Addition on a 12-hour clock (Figure 1) can be thought of as sequential circular moves on a pointer about the face of the clock, starting with the pointer at 12. For example, the answer to $4 \oplus 3$ on a clock is obtained by moving the pointer from 12 to 4, and then moving the pointer 3 more hours to 7. Thus $4 \oplus 3 = 7$. All addition in this exercise is done the same way.

(a) What is $4 \oplus 8$?

(b) What is $4 \oplus 10$?

(c) What is $10 \oplus 4$?

(d) What is $9 \oplus 5$?

(e) What is $5 \oplus 9$?

(f) Is this kind of clock addition commutative?

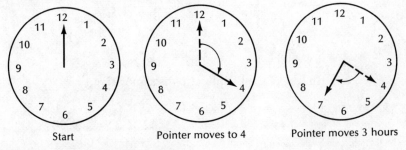

Start   Pointer moves to 4   Pointer moves 3 hours

Figure 1 12-hour clock arithmetic

*5 (a) In the clock addition of exercise 4 there is an implied mathematical system. What are the elements in the system?

(b) Make a table defining all possible "additions" on the elements of the system.

6 (a) Is forming the union of two sets commutative?
   (b) Is forming the intersection of two sets commutative?

7 Is the operation "x followed by y" commutative if x means putting on your shirt and y means putting on your tie?

°8 In the following proof the commutative property is used in two of the steps. Which steps?

Prove: $(a + b) + c = b + (c + a)$

$$
\begin{aligned}
(a + b) + c &= (b + a) + c && (1) \\
&= b + (a + c) && (2) \\
&= b + (c + a) && (3)
\end{aligned}
$$

5.3 THE ASSOCIATIVE PROPERTY

In ordinary arithmetic it is true that $(3 + 4) + 5 = 3 + (4 + 5)$. That is, to get the sum of 3, 4, and 5, we can group the 3 and 4 and then add 5, or we can group the 4 and 5 and add 3 to that result. The sum of these three numbers is obtained by adding 7 and 5, or 3 and 9. If a binary operation can be performed on three elements by grouping them in either of these two ways, that operation is associative.

DEFINITION 3   Associativity

Let $[\ *,S\ ]$ be a mathematical system. If for every three elements $a, b, c \in S$, we have

$$(a * b) * c = a * (b * c)$$

then $*$ is said to be *associative* on S. The system is called an associative system.

EXAMPLE 1

In $[\ \times,N\ ]$, multiplication is associative:

$$
\begin{aligned}
(3 \times 4) \times 5 &\overset{?}{=} 3 \times (4 \times 5) \\
12 \times 5 &\overset{?}{=} 3 \times 20 \\
60 &= 60
\end{aligned}
$$

EXAMPLE 2

Let $S = \{ a, b, c, d, e \}$ and define # as in Table 5. The system $[\,\#,S\,]$ is associative.

Table 5

An abstract associative system

#	a	b	c	d	e
a	a	b	c	d	e
b	b	c	d	e	a
c	c	d	e	a	b
d	d	e	a	b	c
e	e	a	b	c	d

To prove that [#,*S*] is associative in Example 2 would require 125 checks, since there are 5 elements and each position in the associative formula can be filled in 5 ways ($5 \times 5 \times 5 = 125$). One check would be

$$(b \ \# \ d) \ \# \ a \overset{?}{=} b \ \# \ (d \ \# \ a)$$
$$e \ \# \ a \overset{?}{=} b \ \# \ d$$
$$e \equiv e$$

Proof of associativity by cases is tedious, since there are so many cases. Of course, proof by cases for infinite systems is impossible. If the infinite system involves a familiar set like *I* or *N*, a moment's reflection will usually tell you whether or not associativity holds with respect to an ordinary operation.

EXAMPLE 3

In [$\div$,*N*], division is not associative:

$$(24 \div 6) \div 2 \overset{?}{=} 24 \div (6 \div 2)$$
$$4 \div 2 \overset{?}{=} 24 \div 3$$
$$2 \neq 8$$

In Example 3 we also know that [$\div$,*N*] is not a closed system. If a system is not closed, then it cannot be associative. Example 4 shows what the problem is.

EXAMPLE 4

Let $S = \{ 0, 1, 2 \}$ and $\circ$ be defined as in Table 6.

Table 6

Operation $\circ$ on set S

$\circ$	0	1	2
0	0	0	0
1	0	1	2
2	0	2	11

From the table we can find $2 \circ 2$, but we cannot find $1 \circ 11$.

$$(1 \circ 2) \circ 2 \overset{?}{=} 1 \circ (2 \circ 2)$$
$$2 \circ 2 \overset{?}{=} 1 \circ 11$$

This attempt to check for associativity fails because 11 is not in S.

A system must be closed to be associative. Systems that are closed, however, may *still* not be associative.

EXAMPLE 5

Table 7

A closed but nonassociative system

$\circ$	a	b	c
a	a	b	a
b	b	b	c
c	a	c	b

The operation is closed on the set $\{\, a, b, c \,\}$. However, $\circ$ is not associative in this case:

$$(a \circ c) \circ c \overset{?}{=} a \circ (c \circ c)$$
$$a \circ c \overset{?}{=} a \circ b$$
$$a \neq b$$

In every check of associativity there is an intermediate step due to the two different groupings. Thus in $(3 \times 5) \times 7 = 3 \times (5 \times 7)$ the intermediate step is $15 \times 7 = 3 \times 35$. These are quite different multiplications, but both yield 105. Commutativity requires no such intermediate step and is therefore usually more obvious than associativity. At any rate, commutativity is almost always easier to check for than associativity.

5.4 SYSTEM IDENTITIES

In the system $[\, \# , S \,]$ defined in Table 8, there is one element that operates on each of a, b, and c, and returns (or yields) a, b, and c, respectively. In Table 8 notice that

$$\begin{array}{lll} a \,\#\, a = a & \text{and} & a \,\#\, a = a \\ b \,\#\, a = b & \text{and} & a \,\#\, b = b \\ c \,\#\, a = c & \text{and} & a \,\#\, c = c \end{array}$$

Table 8
Operation # on set S

#	a	b	c
a	a	b	c
b	b	c	a
c	c	a	b

The element *a* is called the *system identity* (or the *set identity* or the *#identity*). It does not matter whether the elements of S operate on *a* or whether *a* operates on the elements of S. The identity always works in a commutative fashion.

DEFINITION 4   System Identity
Let S be a set on which an operation # is defined. If there is an element *i* in S such that, for any element *x* in S,

$$i \,\#\, x = x \qquad \text{and} \qquad x \,\#\, i = x$$

then *i* is called the *identity* element of S.

If we think of operations as combining two elements and yielding a third, the identity element combines with an element and yields that element. Schematically we can imagine this as follows:

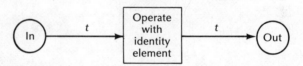

Figure 2   How the identity element works

If we think of the identity element as operating on any input element *t*, then the output is that same element *t*.

EXAMPLE 1
Consider [×,N]. Since $1 \times 1 = 1$, $1 \times 2 = 2$, $1 \times 3 = 3$, $1 \times 4 = 4$, and so on, and the process is commutative, we call 1 the *multiplicative identity* on N. Consider [+,N]. Since no natural number can be added to, say, 4 to yield 4, there is no *additive identity* in N. Consider [+,I]. Since for any integer, say *x*, we may add the integer 0 to yield *x*, 0 is the additive identity on *I*. $(-3) + 0 = (-3)$, $(+7) + 0 = +7$, $(-72) + 0 = -72$, $0 + 0 = 0$, and so on.

Table 9			
•	a	b	c
a	a	b	c
b	b		
c	c		

Table 10			
•	a	b	c
a		a	
b	a	b	c
c		c	

Table 11			
•	a	b	c
a			a
b			b
c	a	b	c

When a table is given that defines an operation on a set, it is easy to spot the identity, if one exists. We look for the entire set of row headings to be repeated in the body of the table, and for the entire set of column headings to be repeated in the body of the table. For a set with three elements there are three ways this can happen. In Table 9, a is the system identity; in Table 10, b is the system identity; in Table 11, c is the system identity. (The tables have been left partially completed so that the row-heading and column-heading "repeats" would stand out.)

Do you see that the element at the intersection of the repeated row and column headings is the identity? The identity, if there is one, always lies somewhere along the principal diagonal of the table. In Example 2 there is no identity since neither the row nor the column headings are repeated exactly in the table.

EXAMPLE 2

The system in Table 12 has no identity element.

Table 12

°	a	b	c
a	a	c	b
b	c	b	a
c	b	a	c

Remember that only one element of a system can be the identity element. Thus, in Example 2, it looks as if a operated on a to yield a, but since a does not operate on b to yield b, we know that a is not the identity element.

Exercise Set 5.2

1 Consider the system [#,T] (Table 13).
   (a) Is # closed on T?
   (b) Is # commutative on T?

Table 13

#	a	e	i	o	u
a	u	a	e	i	o
e	a	e	i	o	u
i	e	i	o	u	a
o	i	o	u	a	e
u	o	u	a	e	i

(c) Operation # is associative on T. Write an associative formula using variables other than $a, e, i, o,$ or u.

(d) Show two illustrations of associativity.

(e) What is the # identity?

(f) The element in the first row and first column is u. Change this element to o. Is the system still closed? Still commutative? The identity element still the same?

(g) Show that with the change in (f) above, the system is no longer associative.

2 Complete the following table so that the operation is closed and commutative, and so that s is the identity element.

∘	r	s	t
r			
s			
t			

Table 14

∘	a	b	c	d
a	a	b	c	d
b	b	a	c	d
c	c	d	a	b
d	d	c	b	a

°3 (a) Show that the system of Table 14 is not associative by finding a case that fails. Hint: Make use of the elements that indicate noncommutativity in the body of the table.

(b) Change the second row in Table 14 so that each element appears only once in each of the last two columns. Then recheck the case used to show nonassociativity in part (a).

4 In ordinary arithmetic,

(a) What is the additive identity?

(b) What is the multiplicative identity?

(c) Is there ever an identity under division? (Our definition of an identity requires that the identity be commutative.)

(d) Is there a subtractive identity?

5 Review "clock addition" as set up in question 4, Exercise Set 5.1.
   (a) Verify that $(7 \oplus 6) \oplus 10 = 7 \oplus (6 \oplus 10)$ on a 12-hour clock.
   (b) Give one other illustration of associativity.
   (c) Which number on a 12-hour clock acts as the additive identity?

6 Consider $[\, +,N \,]$.
   (a) Is there an additive identity?   (b) Is $+$ associative on N?

7 (a) Is there an additive identity element in the set of nonnegative integers?
   (b) Is there an additive identity element in the set of positive integers?

8 Tell which property—closure, commutativity, identity element, or associativity—accounts for each of the following in ordinary arithmetic.
   (a) $7 + 0 = 7$ (b) $87 \times 43 = 43 \times 87$
   (c) $(7 + 8) + 12 = 7 + (8 + 12)$ (d) $8 \times 1 = 8$
   (e) $(-2)\,[\,(-3) \times (-4)\,] = [\,(-2) \times (-3)\,] \times (-4)$

5.5 ELEMENT INVERSES

In a system $[\, \circ,S \,]$, where $S = \{\, r, s, t \,\}$, suppose r is the system identity. It may be true that some element operates on r to yield the identity r, that some element operates on s to yield the identity r, and that some element operates on t to yield the identity r. If all these things are true, then we say that an element inverse exists for each element in the system.

DEFINITION 5   Element Inverses
Let $[\, \circ,T \,]$ be a system with set identity e. If for each element a in T, there exists another element in T, say a^I, such that

$$a \circ a^I = e \text{ and } a^I \circ a = e$$

then a^I is called the *inverse* of a. We say that *element inverses* exist for each element in the system.

EXAMPLE 1

Table 15 shows a system with element inverses.

Table 15

$\circ$	r	s	t
r	r	s	t
s	s	t	r
t	t	r	s

Observe that *r* is the identity for the system. Element *r* is the ⁰ inverse of *r*, since *r* ⁰ *r* = *r*. Element *t* is the ⁰ inverse of *s*, since *s* ⁰ *t* = *r* and *t* ⁰ *s* = *r*. Element *s* is the ⁰ inverse of *t*, since *t* ⁰ *s* = *r* and *s* ⁰ *t* = *r*. (Here "⁰ inverse" is read "star inverse" or "operational inverse.")

If an element and its inverse operate on each other, the result is always the identity. To find the inverse of an element, we must first know the system identity. We then look in the table for that identity in each row. (Since we have required the inverse to operate commutatively, we could look for the identity element in each column as well.)

EXAMPLE 2

For the integers the additive identity is 0. Each integer has an additive inverse, since for any integer, say *i*, we can always find another integer that, when added to *i*, yields 0. Consider

$$+7 + (?) = 0$$

In place of the question mark we can write −7 to make the statement true. In searching for the additive inverse of −10 we are really solving the equation −10 + *x* = 0. Thus +10 is the additive inverse of −10.

If all the elements in a system have inverses, then the system identity appears in each row and column of the system table. Furthermore, if the inverses are commutative, as our definition of inverses requires, then the system identity appears either on the principal diagonal or in positions symmetrically placed about the principal diagonal. Observe what happens (Table 16) when this is not the case.

The identity in this system is *a*, but *a* is not symmetrically placed with respect to the principal diagonal. Now, *c* would appear to be the inverse of *b*, since *b* ⊙ *c* = *a*. But since *c* ⊙ *b* is not *a*, we see that *b* has no inverse. Also, *c* ⊙ *d* = *a*, but *d* ⊙ *c* ≠ *a*. Therefore *c* has no inverse. Likewise, *d* has no inverse. In this case *a* has an inverse since *a* ⊙ *a* = *a*; we can say that *a* is its own inverse. Any table exhibiting the identity property has at least one element with an inverse.

Table 16

⊙	*a*	*b*	*c*	*d*
a	*a*	*b*	*c*	*d*
b	*b*	*c*	*a*	*d*
c	*c*	*d*	*b*	*a*
d	*d*	*a*	*c*	*d*

EXAMPLE 3

Consider the system [$, T], where $T = \{i, j, k\}$ and $ is defined by Table 17. Note that k is the identity element. The inverses are tabulated in Table 18. The identity k is on the principal diagonal or it is placed symmetrically relative to it.

Table 17

$	i	j	k
i	j	k	i
j	k	i	j
k	i	j	k

Table 18

Element	Inverse	Reason
i	j	$i\,\$\,j = j\,\$\,i = k$
j	i	$j\,\$\,i = i\,\$\,j = k$
k	k	$k\,\$\,k = k\,\$\,k = k$

Exercise Set 5.3

1 Consider Table 19, which defines $\odot$ on the set $S = \{q, r, s, t, u\}$.
 (a) What is the system identity element?
 (b) What is the inverse of r? Why?
 (c) What is s^I? (See Definition 5 for the meaning of this notation.)
 (d) What is $s \odot s^I$?
 (e) True or False: $t \odot t^I = t^I \odot t = q$
 (f) True or False: $u \odot u^I = u \odot r = q$

Table 19

$\odot$	q	r	s	t	u
q	q	r	s	t	u
r	r	s	t	u	q
s	s	t	u	q	r
t	t	u	q	r	s
u	u	q	r	s	t

2 Complete the following table so that the operation is closed, so that it is commutative, so that it has x for an identity, and so that each element has an inverse.

#	w	x	y	z	a
w					
x					
y					
z					
a					

3 If we define $i^2 = -1$ so that $i = \sqrt{-1}$, then ordinary multiplication on the set $\{1, -1, +i, -i\}$ is closed, as shown in Table 20.

Table 20

×	+1	−1	+i	−i
+1	+1	−1	+i	−i
−1	−1	+1	−i	+i
+i	+i	−i	−1	+1
−i	−i	+i	+1	−1

(a) Why is $(-i) \times (+i)$ equal to $+1$? (Simply explain how this works.)
(b) What is the × identity?
(c) Complete the blanks in the following table:

Element	Inverse	Reason
+1	+1	$(+1) \times (+1) = +1$
−1		
+i	−i	$+i \times (-i) = (-i) \times i = +1$
−i		

4 (a) Complete the adjacent table so that 1 is the identity.
  (b) What is the inverse of −1?

#	1	−1
1		
−1		1

5 (a) Complete the adjacent table so that −1 is the identity.
  (b) What is the ☆ inverse of 1?

☆	1	−1
1	−1	
−1		

6 Complete the adjacent table so that the operation is closed and has a set identity and element inverses.

∘	a	b	c
a			
b			
c			

7 In $[\,+, N\,]$, does each element have an additive inverse? Explain.

8 On a 12-hour clock, we have previously defined addition so that $a \oplus b$ means b hours beyond a. Thus $11 \oplus 3 = 2$ and $9 \oplus 7 = 4$.
(a) What is the additive identity?
(b) What is the additive inverse of 3? Why?
(c) Make a table with two columns. In the first column, list the hours on a clock. In the second, list the corresponding inverses under addition.

5.6 THE DISTRIBUTIVE PROPERTY

In arithmetic the expression $3(5 + 7)$ means "three times the sum of five and seven." Notice that the expression involves both multiplication and addition. The so-called distributive property is that the product of 3 and $(5 + 7)$ can be written as the sum of $3(5)$ and $3(7)$. Here is a check that this is true:

$$3(5 + 7) \overset{?}{=} 3(5) + 3(7)$$
$$3(12) \overset{?}{=} 15 + 21$$
$$36 \equiv 36$$

Of course, this check shows only that the idea works in this case. However, it happens that in general, for any three natural numbers a, b, and c,

$$a(b + c) = ab + ac$$

This formula is a common way to state the distributive property. Notice that the a is distributed over the addition $b + c$. The single multiplication on the left becomes the sum of two multiplications on the right. The formula indicates that *multiplication is distributive with respect to addition* (for natural numbers).

EXAMPLE 1

Is $a(b + c + d) = ab + ac + ad$ for natural numbers a, b, c, and d? Once again we will give an example and leave the answer to our intuition. Let $a = 5, b = 2, c = 7, d = 10$. Then

$$5(2 + 7 + 10) \overset{?}{=} 5(2) + 5(7) + 5(10)$$
$$5(19) \overset{?}{=} 10 + 35 + 50$$
$$95 \equiv 95$$

Example 1 suggests that we can distribute multiplication over a sum of any number of terms. We can write a general distributive formula for multiplication with respect to addition for any real numbers, whether they be natural numbers, integers, rationals, or irrationals:

$$a(b_1 + b_2 + b_3 + \ldots + b_n) = ab_1 + ab_2 + ab_3 + \ldots + ab_n$$

However, you will seldom need such a formula.

EXAMPLE 2

Is multiplication distributive with respect to subtraction? This question must be answered by using a set of numbers for which subtraction is closed. We thus shift to the set of integers. Stated as a formula, our question is

$$a(b - c) \overset{?}{=} ab - ac$$

Try $a = +4$, $b = -2$, and $c = -3$:

$$(+4)\,[\,(-2) - (-3)\,]\ \overset{?}{=}\ (+4)(-2) - (+4)(-3)$$
$$(+4)\,[\,(-2 + 3)\,]\ \overset{?}{=}\ (-8) - (-12)$$
$$(+4)(+1)\ \overset{?}{=}\ -8 + 12$$
$$+4 \equiv +4$$

Once again the example does not *prove* that multiplication is distributive with respect to subtraction. It should be intuitive that any three integers could have been used in the formula. If you are unconvinced, try another set of three integers.

In the next example we show a pair of operations in which distributivity fails.

EXAMPLE 3

In $[\ +,\cdot,N\ ]$ is addition distributive with respect to multiplication? That is, does the following formula hold true in N?

$$a + b \cdot c = (a + b) \cdot (a + c)$$

Let's check one case:

$$7 + 3 \cdot 6 \overset{?}{=} (7 + 3) \cdot (7 + 6)$$
$$7 + 18 \overset{?}{=} 10 \cdot 13$$
$$25 \neq 130$$

This one case *proves* that addition is not distributive with respect to multiplication in the set N. If the answer to the question were yes, then *every* case would have to verify the idea.

We now formalize the definition of distributivity for any mathematical system having two operations.

DEFINITION 6   Distributivity

Let S be any mathematical system in which two operations $\circ$ and $\#$ are defined. If in the system $[\ \circ,\ \#,\ S\ ]$ we have

$$a \circ (b \ \# \ c) = (a \circ b) \ \# \ (a \circ c)$$

for every choice of three elements, a, b, c in S, then $\circ$ is *distributive* with respect to $\#$.

EXAMPLE 4

Let $S = \{0, 1, 2, 3\}$ and define the two operations as shown in Tables 21 and 22. In this abstract system, the two defined operations, $\oplus$ and $\otimes$, are not supposed to have any particular meaning. We give one illustrative check that the formula holds:

$$a \otimes (b \oplus c) \overset{?}{=} (a \otimes b) \oplus (a \otimes c)$$
$$3 \otimes (2 \oplus 3) \overset{?}{=} (3 \otimes 2) \oplus (3 \otimes 3)$$
$$3 \otimes 1 \overset{?}{=} 2 \oplus 1$$
$$3 = 3$$

Table 21

$\oplus$	0	1	2	3
0	0	1	2	3
1	1	2	3	0
2	2	3	0	1
3	3	0	1	2

Table 22

$\otimes$	0	1	2	3
0	0	0	0	0
1	0	1	2	3
2	0	2	0	2
3	0	3	2	1

In Example 4 how many checks would be necessary to constitute a proof that $\otimes$ is distributive with respect to $\oplus$? Since each of a, b, and c in the distributive formula can be chosen in four ways (to be either 0, 1, 2, or 3), there are 4(4)(4) or 64 cases to be checked to prove the stated distributivity. While some other method of proof might be fashioned, at least we can see how we might prove distributivity if we had enough time to go through all the cases.

EXAMPLE 5

Prove that in $[\oplus, \otimes, S]$ as set up in Example 4, $\oplus$ is *not* distributive with respect to $\otimes$. The distributive formula is written:

$$a \oplus (b \otimes c) = (a \oplus b) \otimes (a \oplus c)$$

One case is

$$3 \oplus (2 \otimes 3) \overset{?}{=} (3 \oplus 2) \otimes (3 \oplus 3)$$
$$3 \oplus 2 \overset{?}{=} 1 \otimes 2$$
$$1 \neq 2$$

Be careful not to assume, on the basis of examples given so far, that distributivity of a first operation with respect to a second operation works only one way. In one of the exercises that follow, two operations are defined on a set, and each is distributive with respect to the other.

Exercise Set 5.4

1 Perform each of the following indicated operations in two ways, using the notion of distributivity.

(a) $4(5 + 6)$ (b) $8(7 - 9)$
(c) $5(1/2 + 1/4)$ (d) $(1/2)(3 + 4 + 5)$

2 Give an example that shows that the statement "multiplication is distributive with respect to subtraction on the set N" fails to make sense because of closure difficulties.

3 (a) Give examples illustrating that multiplication is distributive with respect to addition on the set of rationals.

(b) Prove that division is not distributive with respect to addition on the set of rationals by giving one counter-example. (A counter-example is simply an example that shows that something is not true.)

4 Let # and ∘ be two operations defined on the set N with the following meanings: Operation a # b means select the larger of a and b, or select the number itself, if a and b are equal. Operation a ∘ b means select the first number (that is, select a). For example,

$$7 \# 8 = 8 \qquad 8 \# 7 = 8 \qquad 5 \# 5 = 5$$
$$4 \circ 10 = 4 \qquad 10 \circ 4 = 10 \qquad 4 \circ 4 = 4$$

Calculate each of the following:

(a) $10 \# 12$ (b) $15 \# 15$
(c) $25 \# 26$ (d) $4 \circ 8$
(e) $8 \circ 4$ (f) $5 \circ (6 \# 10)$
(g) $(5 \circ 6) \# (5 \circ 10)$ (h) $10 \# (12 \circ 4)$
(i) $(10 \# 12) \circ (10 \# 4)$ (j) $(5 \# 8) \# 10$
(k) $5 \# (8 \# 10)$

5 (a) For the operations defined in exercise 4 write a formula that expresses the fact that # is distributive with respect to ∘.

(b) For the same operations, write a formula that expresses the fact that ∘ is distributive with respect to #.

(c) Are the formulas you wrote in (a) and (b) true?

(d) Show three illustrations of each of the formulas in (a) and (b).

*6 (a) Prove that # is distributive with respect to ∘. (See exercise 4.) Hint: Work with the formulas and meanings of # and ∘. We cannot prove this fact by examples.

(b) Prove that ∘ is distributive with respect to #.

*7 See question 4 (j) and 4 (k).

(a) Write a formula stating that # is associative. Is # associative?

(b) Write a formula stating that ∘ is associative. Is ∘ associative?

(c) Prove that ∘ is associative.

(d) Try to prove that # is associative. Hint: Working in general we cannot determine whether a # b is a or b, but we can make two cases out of a # b. In case 1, a # b = a; in case 2, a # b = b.

5.7   THE GROUP—A SPECIAL SYSTEM

As we have seen, a given mathematical system may or may not have certain properties. Thus the natural numbers, with respect to multiplication, are closed, commutative, associative, and have an identity but no multiplicative inverses. Modern mathematics is greatly concerned with determining just which properties hold for a particular system. If a certain set of properties hold with respect to an operation (or operations) on a set, mathematicians may term the system a *group*, or a *ring*, or a *field*. These are technical words, each of which refers to a specific set of properties.

In this section we introduce groups. The theory of groups owes its initial development to Augustin-Louis Cauchy (1789–1857), Evariste Galois (1811—1832), and Niels Henrik Abel (1802—1829). Galois and Abel had rather tragic and all too short lives. Galois misspent his youth as a political activist, was arrested by authorities shortly after the French Revolution of 1830, and died in a duel at the age of twenty. Abel lived in extreme poverty and died of tuberculosis at 26. The unfinished work of these two mathematical geniuses has kept other mathematicians busy ever since. In the last 150 years group theory has been expanded immensely.

We now define what we mean by a group.

DEFINITION 7   Group

A *group* is a mathematical system [∘, S] in which the operation ∘ has the following properties:
1. The operation is *closed*.
2. The operation is *associative*.
3. A *set identity* exists.
4. *Element inverses* exist for each element.

You can see that a group does not have all the properties covered in this chapter. In particular, a group does not have to be a commutative system, nor is there a distributive property.

EXAMPLE 1

The system [+,N] is not a group. While addition of natural numbers is closed and associative, there is no set identity (or additive identity). Without an identity there can be no inverses for each element.

The lack of an additive identity and additive inverses in N suggests that we look at the integers.

EXAMPLE 2

The system $[+ ,I]$ does constitute a group. Addition of integers is closed and associative, and the integer 0 is the additive identity for the set. Each integer has an additive inverse. The additive inverse of $+7$ is -7; the additive inverse of -13 is $+13$; the additive inverse of 0 is 0. In general, if i is an integer, its additive inverse is $-i$, since $i + (-i) = 0$.

The 12-hour clock, with addition as defined in Exercise Set 5.1, is a group. Remember that we defined $a \oplus b$ to mean b hours beyond a. The operation is closed and associative; the additive identity is 12, since 12 hours beyond any hour is that hour. Also an additive inverse exists for each hour; it is always the difference between that hour and 12. Thus the additive inverse of 5 is 7, since $5 \oplus 7 = 12$.

"Clock addition" can be defined with any number of hours. Let us now create a 5-hour system in which we can get answers to addition by thinking of the model illustrated in Figure 3. The five divisions of the circle are labeled as in Figure 3. (Instead of a 5, we use a 0.) The operation $2 \oplus 4$ would mean 4 "hours" beyond 2, so $2 \oplus 4 = 1$. Table 23 is the complete addition table.

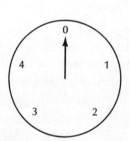

Figure 3   A model for 5-hour clock arithmetic

Table 23

$\oplus$	0	1	2	3	4
0	0	1	2	3	4
1	1	2	3	4	0
2	2	3	4	0	1
3	3	4	0	1	2
4	4	0	1	2	3

Clearly $\oplus$ is closed; all the elements in the body of the table are in the set $\{ 0, 1, 2, 3, 4 \}$. The operation $\oplus$ is associative. One case is

$$(3 \oplus 4) \oplus 2 \overset{?}{=} 3 \oplus (4 \oplus 2)$$
$$2 \oplus 2 \overset{?}{=} 3 \oplus 1$$
$$4 = 4$$

The element 0 is the $\oplus$ identity. Since 0 is either on the principal diagonal or symmetrically placed with respect to it, we are assured that each of the five "hours" has a $\oplus$ inverse. The inverses are tabulated in Table 24.

Table 24   Inverses for a 5-hour clock

Element	Inverse	Reason
0	0	$0 \oplus 0 = 0$
1	4	$1 \oplus 4 = 0$
2	3	$2 \oplus 3 = 0$
3	2	$3 \oplus 2 = 0$
4	1	$4 \oplus 1 = 0$

In the same way we could set up a 4-hour clock, a 6-hour clock, and so on. Each such clock arithmetic constitutes a group with respect to addition. Note that once a model like Figure 3 is given to develop a table like Table 23, we no longer need the model. All properties of the model can be verified by using results given in the table.

Some groups have the additional property of commutativity. In that case the group is called a *commutative group* or an *Abelian group* (after Niels Abel). The symmetry with respect to the principal diagonal of Table 23 is evidence that 5-hour clock addition is commutative, so that system would be a commutative group. That not all groups have to be commutative is clear if we consider the next example.

EXAMPLE 3

Consider the set $S = \{\, r, s, t, a, b, c \,\}$ and the operation $\#$ defined by Table 25. Closure is obvious; all entries in the table are in the given set S. Associativity, while not obvious, is a fact; there would be 216 cases to check in proving this. The identity element is r, and it is placed in symmetric positions relative to the principal diagonal, so inverses exist for each element. Thus the table defines the operation $\#$ so that $[\#, S]$ is a group. Yet $\#$ is not commutative. What is $a \# s$? What is $s \# a$? Can you find other examples showing that $\#$ is not commutative?

Table 25   A noncommutative group

$\#$	r	s	t	a	b	c
r	r	s	t	a	b	c
s	s	t	r	b	c	a
t	t	r	s	c	a	b
a	a	c	b	r	t	s
b	b	a	c	s	r	t
c	c	b	a	t	s	r

Don't hunt for a noncommutative group with fewer than six elements. It can be proved that all groups with fewer than six elements are commutative.

Example 3 illustrates a group in which the elements are abstract. That is, the table was given without reference to a model, and the elements do not seem to have any particular meaning or interpretation. Yet the example served our purpose, enabling us to show that noncommutative groups exist.

Cauchy, Galois, and Abel used group theory to help them solve problems. While a complete description of those problems cannot be undertaken here, we can say that group theory helped these mathematicians to determine the conditions under which an equation can be solved. This led to the solution of one age-old problem, namely, "Can any angle be trisected?" Today we have a proof that any angle *cannot* be trisected, and the proof rests upon the theory of groups.

In addition, group theory has applications in many areas of mathematics and physics, including geometry, crystallography, atomic theory, mechanics, and differential equations. While a pure mathematician may be interested only in the abstract, there are often interpretations of his work that have great impact on the real world.

Exercise Set 5.5

1 Consider the set of all rationals and the operation of addition.
    (a) Is addition closed on R?
    (b) Is addition commutative on R?
    (c) Is there an additive identity for all rationals?
    (d) Does each element have an additive inverse?
    (e) If they exist, list the additive inverses of $3/4$, $-7/5$, and $+15/13$.
    (f) Does addition on R constitute a group?
    (g) If so, is this an Abelian group?

2 Answer exercise 1 for the operation of multiplication.

3 Consider the set of nonzero rational numbers under multiplication. Do this set and operation constitute a group?

4 Make the table for ordinary multiplication on the set $\{1, -1\}$. Is each element its own multiplicative inverse? Verify that the group properties hold.

5 Show that addition on a 6-hour clock constitutes a group. Label the hours 0, 1, 2, 3, 4, 5.

6 Consider the set $\{1, a, a^2\}$ and the operation of multiplication as defined in Table 26.
    (a) Notice that there is closure, that there is an operational identity (what element is it?) and that each element has an inverse. What is the inverse of a^2?

Table 26

$\odot$	1	a	a^2
1	1	a	a^2
a	a	a^2	1
a^2	a^2	1	a

(b) Write an associative formula and present at least two illustrations to show that $\odot$ is associative on the set.

(c) How many checks of associativity would be necessary to constitute a proof that this operation is associative?

(d) Do this operation and the set constitute a commutative group?

7 Consider the model given in Figure 4, defining $\odot$. The operation $a \odot b$ means to rotate the figure counterclockwise through $a°$ and then to rotate it through $b°$, where a and b can be either $0°$, $60°$, $120°$, $180°$, $240°$, or $300°$. The $\odot$ symbol can be thought of as "followed by." Let the rotation through $0°$ be written as R_0, rotation through $60°$ be written as R_{60}, and so on. Example: $R_{60} \odot R_{60} = R_{120}$. Show that $\odot$ constitutes a group on this set of rotations.

Figure 4

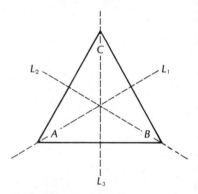

Figure 5 Position *I*

*8 On a piece of paper sketch and cut out an equilateral triangle and label the vertices *A*, *B*, and *C* as in Figure 5. If the triangle is rotated in a plane counterclockwise through $120°$, vertex *B* assumes the position first occupied by vertex *C*, vertex *C* goes to the position first occupied by *A*, and vertex *A* goes to the position first occupied by *B*. If we disregard the labels, we see that the figure is oriented exactly as it was originally. (Mathematicians say that the equilateral triangle remains *invariant* under a $120°$ counterclockwise rotation in a plane.) Let us call this rotation R_{120} (Figure 6). Starting from position *I*, a counterclockwise rotation of $240°$ results in Figure 7.

Starting again from position *I*, we can leave the orientation of the figure invariant by flipping it in space about the axis of line L_1, through $180°$. The result is shown in Figure 8. This can also be done about line L_2 with the result shown in Figure 9, and about line L_3 with the result shown in Figure 10.

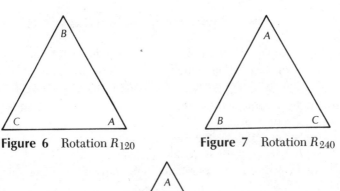

Figure 6   Rotation R_{120} **Figure 7**   Rotation R_{240}

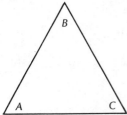
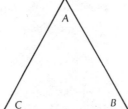
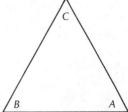

Figure 8   Position F_1 **Figure 9**   Position F_2 **Figure 10**   Position F_3

We can start at I, and perform first one rotation or flip, and then another rotation or flip; R_{120} followed by F_1 yields F_2. We write this as follows: $R_{120} \odot F_1 = F_2$. Notice that we choose I to be the identity move, which is a rotation of $0°$ or perhaps $360°$; at any rate I leaves the figure unchanged.

(a) Complete Table 27.
(b) Show that the six motions (called "the symmetries of an equilateral triangle") constitute a group. Assume associativity.
(c) Calculate $(F_1 \odot R_{120}) \odot F_3$.
(d) Calculate $F_1 \odot (R_{120} \odot F_3)$.
(e) Parts (c) and (d) of this exercise are evidence of what property?
(f) Is "the symmetries of an equilateral triangle" a commutative group?
(g) What is $F_1 \odot R_{120}$? What is $R_{120} \odot F_1$?

Table 27

$\odot$	I	R_{120}	R_{240}	F_1	F_2	F_3
I	I	R_{120}	R_{240}	F_1	F_2	F_3
R_{120}	R_{120}	R_{240}	I	F_2		
R_{240}	R_{240}					
F_1	F_1		F_2			R_{120}
F_2	F_2					
F_3	F_3					

*9 (a) In exercise 8, replace I by r, R_{120} by s, R_{240} by t, F_1 by a, F_2 by b, and F_3 by c.

    (b) Compare the result with Table 25 in the text. What do you notice?

    (c) Can the rotations and flips of an equilateral triangle be a physical model that generates the group of Example 3?

*10 (a) In a mathematics dictionary (or other appropriate source) find the word "isomorphism."

    (b) What is $R_{240} \odot F_3$? (See exercise 8.)

    (c) What is $t \# c$? (See Example 3 in text.)

    (d) How does isomorphism apply to exercise 8 and Example 3?

11 Write a brief report, after doing some library research, on Galois or groups or both. Suggested sources:

    (a) E. T. Bell, *The Development of Mathematics*. New York: McGraw-Hill, 1945.

    (b) E. T. Bell, *Men of Mathematics*. New York: Simon & Schuster, 1937.

    (c) Flora Dinkines, *Abstract Mathematical Systems*. New York: Appleton-Century-Crofts, 1964, pp. 24–28.

    (d) David Eugene Smith, *A Source Book in Mathematics*. New York: Dover (reprint), 1959, Vol. I, pp. 278–285.

    (e) Morris Kline, *Mathematical Thought from Ancient to Modern Times*. New York: Oxford Univ. Press, 1972, Chap. 31.

    (f) Jagjit Singh, *Great Ideas of Modern Mathematics: Their Nature and Use*. New York: Dover, 1959.

    (g) E. T. Bell, *Mathematics—Queen and Servant of Science*. New York: McGraw-Hill, 1951, Chap. 9.

    (h) Robert D. Carmichael, *Groups of Finite Order*. New York: Dover, 1956, Chap. 1.

    (i) Dirk J. Struik, *A Concise History of Mathematics*. New York: Dover, 1948, pp. 223–226.

SUMMARY

Mathematics organizes whatever it deals with. Some observers would go so far as to say that mathematics *is* organization.

In this chapter we have given some insight into what such statements might mean. The first few sections dealt with a few specific properties that a mathematical system may have. While numbers provide good examples of commutativity or associativity or identities, we can find examples outside of sets of numbers having such properties. Thus "putting on a shirt followed by putting on a tie" is not commutative, since the results are quite different if the tie is put on first.

Some of our examples were operations on sets of letters, where a table defined the "answers." Even the operation symbols were unfamiliar, appearing as a $\odot$ or a # or a ∘. We saw that "addition" on a 12-hour clock was a model exhibiting several of the properties.

The chapter culminated in a discussion of groups, which are mathematical systems having four specific properties. It is this part of the chapter that is evidence of the way the mathematician performs. The group properties apply to certain mathematical systems. Thus particular examples of groups will be known to have the property stated by the theorem. Working in the abstract, the mathematician discovers new facts and new relationships, then applies his discoveries to a model and ultimately to the real world.

Before attempting the Review Test, you should be familiar with the following:

1. Notation like [∘,S] for mathematical systems.

2. How to check for commutativity, given a table defining an operation.

3. The general commutative formula $a \circ b = b \circ a$ and how to check this in a given system.

4. How to calculate the number of checks necessary to prove commutativity, using the Fundamental Principle of Counting.

5. Clock addition for any clock, 12-hour, 4-hour, 6-hour, and so on.

6. The general associative formula $(a \circ b) \circ c = a \circ (b \circ c)$, making checks, and counting cases.

7. How to find system identitites.

8. How to find element inverses.

9. The distributive formula $a \circ (b \odot c) = (a \circ b) \odot (a \circ c)$, making checks, and counting cases.

10. The definition of a group and the process of determining whether or not a certain system constitutes a group.

REVIEW TEST

1 Consider the system [∘,S], where $S = \{ r, s, t, v \}$ and ∘ is defined as shown in Table 28.
(a) Is [∘,S] closed? Is it a commutative system?

Table 28

∘	r	s	t	v
r	r	s	t	v
s	s	t	v	r
t	t	v	r	s
v	v	r	s	t

(b) $A \circ (B \circ C) = (A \circ B) \circ C$ is an associative formula. Show a check of associativity by choosing $A = r$, $B = s$, and $C = v$.

(c) What is the system identity element?

(d) What is the ∘ inverse of t?

(e) Find $v \circ t$.

2 In 12-hour clock addition what is $10 \oplus 5$?

3 Complete the following table so that ∘ is closed and b is the system identity.

∘	a	b	c
a			
b			
c			

4 In ordinary arithmetic
   (a) What is the multiplicative identity?
   (b) What is the additive identity?
   (c) What is the additive inverse of $+4$?
   (d) What is the multiplicative inverse of 3?

5 Consider $[\times, N]$, where $\times$ is ordinary multiplication.
   (a) Is $\times$ associative in this system?
   (b) Is there a multiplicative identity in N? If so, what is it?

6 Consider $[\times, +, I]$, where I is the set of integers.
   (a) Is $+$ distributive with respect to $\times$?
   (b) Show a check of distributivity in the formula

$$a \times (b + c) = (a \times b) + (a \times c)$$

   by choosing $a = +3$, $b = -2$, and $c = -5$.
   (c) What is the multiplicative identity in I?
   (d) Do multiplicative inverses exist for each element?

7 From the set of words { commutativity, associativity, distributivity, identity, inverse } choose the appropriate one that accounts for each of the following:
(a) $3 \times (4 + 5) = (3 \times 4) + (3 \times 5)$
(b) $a + b = b + a$
(c) $3.5 + 0 = 0 + 3.5$
(d) $7 \times 1/7 = 1$
(e) $(3 + 7) + 10 = 3 + (7 + 10)$

8 Why is $[\times, N]$ not a group?

9 Tell why Table 29 constitutes a group. In particular, what is the $\oplus$ identity? Also list the inverse of each element.

Table 29

$\oplus$	1	2	3
1	1	2	3
2	2	3	1
3	3	1	2

10 (a) Make a table showing addition on a 3-hour clock. Use the numbers 0, 1, and 2.
(b) What is the identity?
(c) What is the inverse of 0? Of 1? Of 2?
(d) Is the system a group?
(e) Is it a commutative group?

Mathematicians and nonmathematicians have attempted to define mathematics. Though no brief definition can possibly connote all that the word means, the following passages show how some persons have tried to describe the spirit, the peculiarities, or the essence of mathematics.

MATHEMATICS IS...

Originally, mathematics was the collective name for geometry, arithmetic, and certain physical sciences (as astronomy and optics) involving geometrical reasoning. In modern use the word is applied, (a) in a strict sense, to the abstract science which investigates deductively the conclusions implicit in the elementary conceptions of spatial and numerical relations, and which includes as its main divisions geometry, arithmetic, and algebra; and (b) in a wider sense, so as to include those branches of physical or other research which consist in the application of this abstract science to concrete data. When the word is used in its wider sense, the abstract science is distinguished as *pure mathematics*, and its concrete applications (e.g. in astronomy, various branches of physics, the theory of probabilities) as *applied* or *mixed mathematics*.

The Oxford English Dictionary, 1933

Mathematics, like dialectics, is an organ of the inner higher sense; in practice it is an art like eloquence. Both alike care nothing for content, to both nothing is of value but the form. It is of no difference to mathematics whether it computes pennies or guineas, to rhetoric whether it is the true or the false that is being defended.

Goethe (1749–1832)

Mathematics is the queen of the sciences, and arithmetic the queen of mathematics. She often condescends to render service to astronomy and other natural sciences, but under all circumstances the first place is her due.

C. F. Gauss (1777–1855)

Mathematics is the science which draws necessary conclusions.

Benjamin Pierce (1809–1880)

Mathematics is the only true metaphysics.

William Thompson, Lord Kelvin (1824–1907)

Mathematics in general is fundamentally the science of self-evident things.

Felix Klein (1849–1925)

Mathematics is thought moving in the sphere of complete abstraction from any particular instance of what it is talking about. So far is this view of mathematics from being obvious, that we can easily assure ourselves that it is not, even now, generally understood. For example, it is habitually thought that the certainty of mathematics is a reason for the certainty of our geometrical knowledge of the space of the physical universe. This is a delusion which has vitiated much philosophy in the past, and some philosophy in the present.

Alfred North Whitehead (1861–1947)

Mathematics is a game played according to certain simple rules with meaningless marks on paper.

David Hilbert (1862–1943)

Mathematics may be defined as the subject in which we never know what we are talking about nor whether what we say is true.

Bertrand Russell (1872–1970)

Mathematics is no more the art of reckoning and computation than architecture is the art of making bricks or hewing wood, no more than painting is the art of mixing colors on a palette, no more than the science of geology is the art of breaking rocks, or the science of anatomy the art of butchering.

C. J. Keyser, 1907

Mathematics is not so much a body of knowledge as a special kind of language, one so perfect and abstract that—hopefully—it may be understood by intelligent creatures throughout the universe, however different their organs of sense and perception.

Mathematics, Life Science Library, 1963

Mathematics is the systematic treatment of magnitude, relationships between figures and forms, and relations between quantities expressed symbolically.

The Random House Dictionary of the English Language, 1967

6 LOGIC AND THE STRUCTURE OF THOUGHT

Logic has intrigued mathematicians and philosophers for centuries. Without agreement among mathematicians on certain logical principles, mathematical thinking as we know it would not be possible.

Throughout most of history logic in the western world has been Aristotelian logic—logic of the kind descirbed by Aristotle (384–322 B.C.). For example, one of the principles of logic stated by Aristotle is that a statement must be either true or false but cannot be both. Aristotle pointed out that logic provides the methods of proof in mathematics. He also claimed that logic stems from the thought processes of the human mind.

In the nineteenth century, mathematicians like Augustus De Morgan (1806–1871) and George Boole (1815–1864) began to apply the mathematician's use of carefully defined symbols and standard operations to logic, with the result that today logic is not so much a study of argumentation as it is a study of the representation of thought. In turn, since the early 1900s all of mathematics has been thought to be derivable from basic logic.

Logic certainly seems to reflect the workings of the mind, as Aristotle said it did. You don't need a course in formal logic to make the logical statement "If Fido falls in the creek, he will get wet." If you then hear that Fido has fallen in the creek, you instantly draw a conclusion: Fido is wet.

Mathematicians are, of course, concerned with the *form* of statements leading to conclusions. For example, in the Fido incident, we can let a stand for "Fido falls in the creek" and b stand for "he will get wet." Then the form looks like this:

1. If a, then b.　　(A given statement)
2. a.　　(a occurs or is true)
3. Therefore b.　　(b follows from 1 and 2)

Such a pattern, in which two *premises* are stated and a *conclusion* follows, is called an *argument*. You will learn other argument patterns later. For now it is enough to realize that a and b can be any simple sentences. Consider this argument:

If the moon is made of green cheese, I'll eat my hat.
The moon is made of green cheese.
Therefore I'll eat my hat.

As in the Fido argument, the conclusion is inescapable, yet we see that one of the components of the argument is false. However, the argument itself is *valid*.

In this chapter we introduce enough basic definitions and techniques to permit an appreciation of symbolic logic. Our goal is to show how mathematicians organize statements and how they determine whether or not arguments are valid.

6.1 BASIC NOTIONS

Logic deals with *simple statements* that can be judged true or false. The sentence "My dog has fleas" is an example, since we can determine whether the assertion is true or false. Neither of the following sentences is an acceptable simple statement:

> Are you mad at me?
> Get out of bed!

The first is interrogative (a question), and the second is imperative (a command). The question "Are they true or false?" is meaningless.

It is common practice to let lowercase letters like p, q, and r, stand for simple statements. Thus the notation

$$p: \text{It is raining.}$$

means that p stands for the sentence following the colon.

We will be combining two or more sentences with connectives like "and," "or," "if . . . then," and we will sometimes negate the meaning of a given sentence. These combined sentences are called *compound statements*.

EXAMPLE 1

Let p: It is Friday.
        q: I get paid.

Then "p and q" is the compound sentence "It is Friday and I get paid." Also "If p, then q" would be the compound sentence "If it is Friday, then I get paid." Another compound sentence might be "p or q"; that is, "It is Friday or I get paid."

To avoid the need for writing connectives and negations in words, we use the symbols defined in Table 1.

Table 1   Logic symbols and their meaning

Symbol	Meaning	Name of symbol
~	not	negation
$\wedge$	and	conjunction
$\vee$	or	disjunction
$\rightarrow$	if . . . then	implication
$\leftrightarrow$	if and only if	biconditional

EXAMPLE 2

Let p: John does his homework.
           q: John goes to the movies.

Then we can create the following sentences:
    $\sim p$: John does not do his homework.
    $p \wedge q$: John does his homework and goes to the movies.
    $p \vee q$: John does his homework or goes to the movies.
    $p \rightarrow q$: If John does his homework, then he will go to the movies.
    $p \leftrightarrow q$: John does his homework if and only if he goes to the movies.

Note in the sentence for $p \rightarrow q$ that we can change the tense without changing the symbols. A sentence like "If I get rich, then I can buy anything I want" could just as well be written "If I were rich, I could buy anything I wanted." Both sentences could be symbolized $p \rightarrow q$.

The connective "or" means "one or the other, or both." Thus $p \vee q$ in Example 2 means that John might do both. This is the inclusive sense of the word "or." On the other hand, in the sentence "You may have coffee or tea," the word "or" means "one or the other but not both." We will not make use of this exclusive meaning of "or."

Many compound sentences are a mixture of negations and connectives. The sentence "I take Latin and don't like it very much" can be thought of as a conjunction one of whose components is a negation.

Let p: I take Latin.
       q: I like it very much.

The second component of the conjunction must be

$$\sim q: \text{I don't like it very much.}$$

Thus the entire sentence can by symbolized

$$p \wedge \sim q$$

Study the following example.

EXAMPLE 3

Let p: I am tall.
       q: I can play basketball.
Then
    $\sim p \rightarrow \sim q$: If I am not tall, then I cannot play basketball.
    $\sim p \vee q$: I am not tall or I play basketball.
    $\sim(p \wedge q)$: It is false that I am tall and play basketball.

The symbols $\wedge$, $\vee$, $\rightarrow$, and $\leftrightarrow$ are called *binary connectives* since they connect (or operate on) two simple statements. We can speak a sentence in the form $p \wedge q$ or $p \rightarrow q$, but we would not say "$p \wedge$" or "$p \rightarrow$," since these would not be complete sentences. Negation, however, is a *unary* operation since it applies to a single symbol; it does not connect two symbols. Thus $p \sim q$ is meaningless.

EXAMPLE 4

(a) In $\sim p \wedge q$ only the p is negated.
(b) In $\sim (q \wedge p)$ the parenthesized statement is negated.
(c) In $p \rightarrow \sim q$ only the q is negated.
(d) In $\sim p \wedge \sim q$, p and q are each negated.
(e) In $\sim(p \vee q) \rightarrow r$ the disjunction $p \vee q$ is negated.

Negation applies to what comes immediately after the symbol. Thus $\sim p \wedge q$ is a conjunction and not at all the same as $\sim (p \wedge q)$, which is a negation. In the first case p is negated and in the second case the parenthesized expression is negated.

Be careful not to assume that $\sim (p \wedge q)$ and $\sim p \wedge \sim q$ are the same. "It is false that Joe plays tennis and golf" is not the same idea as "Joe does not play tennis and Joes does not play golf." We will explain this further as we go on.

Exercise Set 6.1

1 Using the underlined letters as symbols, translate each of the following sentences into symbolic form. (Do not use the symbol $\sim$ in this question.)
  (a) Tomatoes grow on <u>b</u>ushes and cherries grow on <u>t</u>rees.
  (b) Roses are <u>r</u>ed and violets are <u>b</u>lue.
  (c) Roses are <u>r</u>ed or violets are <u>b</u>lue.
  (d) The Democrats always <u>c</u>ampaign hard and never <u>w</u>in an election.
  (e) If the Democrats <u>c</u>ampaign, then they won't <u>w</u>in.
  (f) You will <u>p</u>ass chemistry if you <u>u</u>nderstand it.
  (g) If you <u>u</u>nderstand chemistry, then you will <u>p</u>ass the course.
  (h) If Sam <u>m</u>akes the team, then he was <u>p</u>hysically fit.
  (i) If Sam is <u>p</u>hysically fit, then he will <u>m</u>ake the team.
  (j) Sam will <u>m</u>ake the team if he is <u>p</u>hysically fit.
  (k) Sam <u>m</u>akes the team if and only if he is <u>p</u>hysically fit.

2 Let p: Alice is tall.
      q: Betty is overweight.
  Translate each of the following into English:
  (a) $p \wedge q$ (b) $\sim p$ (c) $\sim q$
  (d) $\sim (p \wedge q)$ (e) $\sim p \wedge \sim q$

3 Let *r*: Fido has fleas.
     s: Fido scratches.
   Translate each of the following into English:
    (a) $r \lor s$ (b) $r \rightarrow s$ (c) $\sim(r \land s)$
    (d) $\sim r \land \sim s$ (e) $\sim(r \lor s)$ (f) $\sim r \lor \sim s$
    (g) $\sim(r \rightarrow s)$ (h) $\sim s \rightarrow \sim r$
    (i) $(r \land s) \lor (\sim r \land \sim s)$ (j) $(r \rightarrow s) \lor (\sim r \land \sim s)$

4 Supply a conclusion that follows from each of the following premises:
    (a) If I smoke, then I will not stay healthy.
       I smoke.
       Conclusion:
    (b) If John works hard, then he will get rich.
       John works hard.
       Conclusion:
    (c) If John works hard, then he will get rich.
       John doesn't get rich.
       Conclusion:
    (d) If you keep your marks up, you will graduate.
       If you graduate, then you will get a good job.
       Conclusion:

5 Why is each of the following not an acceptable statement in logic?
    (a) What time is it?
    (b) Don't overeat or you will gain weight.
    (c) Everything I say is false.
    (d) Water, water everywhere, and not a drop to drink!
    (e) You *would* have to track in mud!
    (f) Woe is me.

6 Symbolize each of the following sentences:
    (a) I told you not to be late.
    (b) If today is Tuesday, then yesterday was Monday.
    (c) If today had been Tuesday, then yesterday would have been Monday.
    (d) If today were Monday, then tomorrow would be Tuesday.
    (e) If that is a cat, then it catches mice.
    (f) It catches mice if it is a cat.
    (g) Every cat catches mice.
    (h) If I am 10 years old, I can vote.

6.2 TRUTH TABLES—NEGATION AND CONJUNCTION

A sentence like "New York City is the capital of Georgia" is clearly false; the sentence "New York City is north of the Mason-Dixon line" is true. With our basic connectives "and" ($\land$), "or" ($\lor$), and "if . . . then" ($\rightarrow$) we can combine these ideas; that is, we can create compound sentences from them, and ask whether or not the combination is true.

1. New York City is the capital of Georgia, and New York City is north of the Mason-Dixon line.

2. New York City is the capital of Georgia, or New York City is north of the Mason-Dixon line.

3. If New York City is the capital of Georgia, then New York City is north of the Mason-Dixon line.

Sentence 1 turns out to be false; sentence 2 turns out to be true; sentence 3 (surprisingly) turns out to be true. In this section and in Section 6.3 we will see how the truth value of a compound statement is determined from the truth values of the simple statements involved.

A *truth table* lists all the possible truth values for given simple statements (or components). It can then be used to determine the truth value of a compound statement. In the first few truth tables that follow, the assignments of all possible truth values to the components lie to the left of the double line.

NEGATION

A simple statement p can be either true or false. If p is true, its negation $\sim p$ is false. If p is false, its negation $\sim p$ is true. We can summarize these ideas in a truth table for negation, Table 2. Since there are only two truth values for a statement p, it is clear that we need only two lines to assign the possible truth values to the component p.

We can apply negation to compound statements as well. If, for example, $p \wedge q$ (the conjunction of p and q) is known to be true, then $\sim (p \wedge q)$ (the negation of the conjunction of p and q) would be false.

Table 2
Truth table for negation

	p	$\sim p$
Line 1	T	F
Line 2	F	T

(The double vertical line separates the assignment portion from the analysis portion.)

CONJUNCTION

If two simple statements p and q are conjoined, forming $p \wedge q$, the truth values of the result depend upon the individual truth values of p and q. Since p can be true or false (two possibilities) and q can be true or false (two possibilities), there are four possibilities for assigning first p true or

Table 3   Truth table for conjunction

	p	q	$p \wedge q$
Line 1	T	T	T
Line 2	T	F	F
Line 3	F	T	F
Line 4	F	F	F

false, and then q true or false. Thus we need four lines to assign all the truth value possibilities to the components p and q, as in Table 3.

We can see from the table that a conjunction is false if either component is false, or if both are false. Conjunction is true only if both components are true. Our rationale for this is given below. (We do not *prove* the results listed in the table. The table merely defines the truth values for conjunctions.)

To see how the table defining the truth values for $p \wedge q$ makes sense, suppose John makes the following remark:

I have a nickel and I have a dime.

We can take p to be "I have a nickel" and q to be "I have a dime." You search John's pockets, and find that indeed he has a nickel and he has a dime. Certainly you would judge John's remark to be true, a conclusion that agrees with line 1 of Table 4, which corresponds to line 1 of Table 3 (line 2 corresponds to line 2 of Table 3, and so on). In the case of line 2, you find that John has a nickel but doesn't have a dime, so his remark is false. In the case of line 3, you find that John does not have a nickel but does have a dime, so his remark is false. In the case of line 4, you find that John has neither a nickel nor a dime, so you certainly conclude that his remark is false.

We now have the truth tables for negation and conjunction (Tables 2 and 3). These tables can be used to derive other truth tables for compound statements that may involve both negation and conjunction. That is, we can derive truth tables for statements like $p \wedge \sim q$ or $\sim p \wedge q$ or $\sim (p \wedge q)$ simply by knowing the tables for negation and conjunction.

Table 4   Justifying the truth table for $p \wedge q$

	Nickel?	Dime?	Nickel and Dime?
Line 1	yes (T)	yes (T)	yes (T)
Line 2	yes (T)	no (F)	no (F)
Line 3	no (F)	yes (T)	no (F)
Line 4	no (F)	no (F)	no (F)

The process of using the basic table results to derive truth tables for other compound statements is called *truth-table analysis.*

We now show a step-by-step procedure for getting the truth table for $p \wedge \sim q$.

Step 1   The compound statement $p \wedge \sim q$ contains two components p and q, so there must be four lines listing all the combinations of T and F for the components. We set up a truth table listing the possibilities:

p	q	
T	T	
T	F	
F	T	
F	F	

Step 2   Since we want the truth table for $p \wedge \sim q$, we label the columns to the right of the double vertical line as follows:

p	q		$p \wedge \sim q$
T	T		
T	F		
F	T		
F	F		

The statement $p \wedge \sim q$ should be thought of as the conjunction of p and not q. The symbol $\wedge$ is the connective, while the symbol $\sim$ applies only to q. For clarity we might have written $p \wedge (\sim q)$. Thus we wish to know the truth values of the conjunction, given the truth-value assignments to p and q.

Step 3   Line-by-line p is assigned T–T–F–F, so we can write these values under p in the compound statement. At the same time q is assigned T–F–T–F, so $\sim q$ becomes F–T–F–T.

p	q		p	$\wedge \sim q$
T	T		T	F
T	F		T	T
F	T		F	F
F	F		F	T

Notice that we place the derived $\sim q$ values under the $\sim$ sign. On line 1, q is assigned T, so in the compound statement $\sim q$ will be F. Each of these $\sim q$ truth values comes from Table 2, defining negation.

Step 4   Looking at line 1 in the table of step 3, we see that p is T and $\sim q$ is F. We want to derive the truth value of $p \wedge \sim q$ under these conditions. That is, we want the truth value of a conjunction whose first component is T and second component is F. Table 3, the basic table for conjunction, tells us that such conjunctions are false. We write F under the $\wedge$ sign:

	p	q	p	$\wedge$	$\sim q$
Line 1	T	T	T	F	F
	T	F	T		T
	F	T	F		F
	F	F	F		T

Step 5   Now we complete the table line by line in the same fashion:

Table 5
Truth table for $p \wedge \sim q$

p	q	p	$\wedge$	$\sim q$
T	T	T	F	F
T	F	T	T	T
F	T	F	F	F
F	F	F	F	T
		1	3	2

We will say we have *constructed* the truth table for the conjunction $p \wedge \sim q$. Let's review the procedure: Carry over the truth values assigned to p (column labeled 1 in Table 5). Determine the truth values for $\sim q$ (column 2). Determine the truth values for the conjunction by looking at columns 1 and 2. The truth table for $p \wedge \sim q$ is F–T–F–F (column 3).

<u>EXAMPLE 1</u>

By truth-table analysis find the truth table for $\sim (p \wedge q)$.
    Since this is the negation of a conjunction, we first find the truth values for $p \wedge q$ (column 1). Then we negate each of these values in column 2. The negation $\sim (p \wedge q)$ has a truth table F–T–T–T.

Table 6
Truth table for $\sim (p \wedge q)$

p	q	$\sim$	$(p \wedge q)$
T	T	F	T
T	F	T	F
F	T	T	F
F	F	T	F
		2	1

By now it should be clear that we always use the same order of assignments of truth values to components. We list these standard assignments in Table 7.

Observe the pattern for truth tables involving two and three components. Once you are aware of the truth-value assignment portion of a truth table, it is unnecessary to use a double vertical line to separate assignments from analysis.

Table 7   Standard order of truth-value assignments

One component	Two components		Three components		
p	*p*	*q*	*p*	*q*	*r*
T	T	T	T	T	T
F	T	F	T	T	F
	F	T	T	F	T
	F	F	T	F	F
			F	T	T
			F	T	F
			F	F	T
			F	F	F

Exercise Set 6.2

1 True or false:
   (a) If a statement is true, its negation is false.
   (b) If q is false, then $\sim q$ is true.
   (c) If $p \wedge q$ is false, then $\sim (p \wedge q)$ is true.
   (d) Conjunctions are true only if both components are true.
   (e) A statement in the form $\sim p$ is false.
   (f) A statement with three components has six lines in its truth table.
   (g) Conjunctions are false if either component is false.

2 Complete the following truth tables:
   (a)

p	*q*	$\sim p \wedge q$		
T	T	F	T	
T	F	F	F	
F	T	T	T	
F	F	T	F	
		1	3	2

   (b)

p	*q*	$\sim p \wedge \sim q$		
T	T	F		
T	T	F		
F	T	T		
F	T	T		
		1	3	2

(c)

p	q	r	$(p \wedge q) \wedge \sim r$	
T	T	T	T	F
T	T	F	T	T
T	F	T	F	F
T	F	F	F	T
F	T	T		
F	T	F		
F	F	T		
F	F	F		
			1	3 2

(d)

p	q	r	$\sim p \wedge (q \wedge \sim r)$
			1 5 2 4 3

3 Find the truth tables for each of these compound statements.
   (a) $\sim (\sim p)$
   (b) $q \wedge p$
   (c) $r \wedge s$
   (d) $\sim p \wedge q$
   (e) $p \wedge p$
   (f) $\sim (\sim p \wedge \sim q)$
   (g) $p \wedge (q \wedge r)$
   (h) $(p \wedge q) \wedge r$
   (i) $\sim p \wedge (\sim q \wedge r)$
   (j) $\sim [(p \wedge q) \wedge r]$

4 Identify each of the following as a negation or as a conjunction of components.
   (a) $\sim p \wedge q$
   (b) $\sim (p \wedge q)$
   (c) $\sim (p \wedge q) \wedge r$
   (d) $\sim [(p \wedge q) \wedge r]$
   (e) $q \wedge (p \wedge \sim r)$

5 Under what conditions is the statement $p \wedge q \wedge r \wedge s$ true? (That is, for what truth values of p, q, r, s is the statement true?)

6 (a) If p is true and q is true, what is the truth value of $p \wedge (\sim q)$?
   (b) If p is false and q is true, what is the truth value of $\sim p \wedge (p \wedge \sim q)$?
   (c) If p is false, what is the truth value of $p \wedge (\sim p \wedge q \wedge r)$?

6.3 DISJUNCTION, IMPLICATION, AND THE BICONDITIONAL

DISJUNCTION

Table 8 defines the truth values for $p \vee q$ (p or q) for all possible assignments to p and q.

To make the table seem reasonable, we can argue from the meaning of the word "or." Since we are using the inclusive sense of the word, disjunctions will be true whenever one or the other (or both) of p and q is true. We can construct a sentence like "I have a nickel or I have a dime" and argue (Table 9) as we did previously for conjunction, to see the reasonableness of the table definition.

Table 8
Defining truth values
for disjunction

p	q	$p \vee q$
T	T	T
T	F	T
F	T	T
F	F	F

Table 9
Justifying the table for disjunction

Nickel?	Dime?	Nickel or Dime?
yes (T)	yes (T)	yes (T)
yes (T)	no (F)	yes (T)
no (F)	yes (T)	yes (T) .
no (F)	no (F)	no (F)

EXAMPLE 1

Determine whether the sentence "John likes chemistry or John likes math" is true or false, given the facts that "John likes chemistry" is false and "John likes math" is true.

Let c: John likes chemistry.

       m: John likes math.

Then the given disjunction is $c \vee m$.

The table below lists all the truth-value possibilities for c and m. Given that c is false and m is true, we are directed to line 3 of the table, where we see that the disjunction is true.

c	m	$c \vee m$
T	T	T
T	F	T
F	T	T
F	F	F

As we did with conjunction and negation, we can now construct truth tables involving disjunction and other connectives.

EXAMPLE 2

Show the truth-table analysis of $\sim p \vee \sim t$.
Column 1 is obtained from the negation table; so is column 2. Column 3 is obtained from columns 1 and 2 by way of the disjunction table (Table 8).

p	t	$\sim p \vee \sim t$		
T	T	F	F	F
T	F	F	T	T
F	T	T	T	F
F	F	T	T	T
		1	3	2

EXAMPLE 3

Show the truth-table analysis of $\sim (p \vee t)$.

Once you see $\sim (p \vee t)$ as a negation of a disjunction, it is clear that the truth table of $\sim (p \vee t)$ is listed under the $\sim$ sign. Columns 1 and 2 are simply copied from the assignments at the left. The disjunction table (Table 8) is used to get column 3. Column 4 is the negation of each entry in column 3. Thus the truth table for $\sim (p \vee t)$ is F–F–F–T.

p	t	$\sim$	$(p$	$\vee$	$t)$
T	T	F	T	T	T
T	F	F	T	T	F
F	T	F	F	T	T
F	F	T	F	F	F
		4	1	3	2

IMPLICATION

Consider the sentence "If Smith wins the lottery, then he will buy a new car," where p is "Smith wins the lottery" and q is "he will buy a new car." If Smith *does* win the lottery and does buy a new car (p true and q true), it is reasonable to call the given sentence true (line 1, Table 10). If Smith wins the lottery and does not buy a new car (p true and q false), then the given sentence is false (line 2).

However, if Smith does not win the lottery, then either does or does not buy a new car, the given statement is not false; that is, we consider it to be true. So when p is false and q is either true or false, we take $p \rightarrow q$ to be true (lines 3 and 4).

Table 10

Truth values for implication

	p	q	$p \rightarrow q$
Line 1	T	T	T
Line 2	T	F	F
Line 3	F	T	T
Line 4	F	F	T

Lines 3 and 4 are somehow hard to take, but some examples will make the defined truth values more reasonable. The tabular definition of implication truth values is consistent with all our other work in logic.

EXAMPLE 4

Let p: We go to the moon again.
        q: Technology prospers.

If I say to you $p \rightarrow q$, that is, "If we go to the moon again, then technology will prosper," and if we do *not* go to the moon again, then whether or not technology prospers, you cannot say I lied to you. If we do *not* go to the moon again, the implication itself, $p \rightarrow q$, is still true.

Implications begin with a hypothetical situation, the "if" part (called the *antecedent*). If the antecedent is true, the truth or falsity of the implication depends totally on the conclusion, the "then" part (called the *consequent*). In fact, most important implications have true antecedents; looking at lines 1 and 2 of the table defining truth values for implication, we are not bothered by the assignments T and F.

Examples 5 and 6 show truth tables of compound statements involving implication, negation, and other connectives.

EXAMPLE 5

Show the truth-table analysis for $\sim p \rightarrow q$.

p	q	$\sim p \rightarrow$	q
T	T	F T	T
T	F	F T	F
F	T	T T	T
F	F	T F	F
		1 3	2

Answer: T–T–T–F

EXAMPLE 6

p	q	$(p \rightarrow q) \wedge \sim q$
T	T	T F F
T	F	F F T
F	T	T F F
F	F	T T T
		1 3 2

Answer: F–F–F–T

In a study of logic, implication is probably the most important connective. Section 6.4 will deal with implication in more detail.

THE BICONDITIONAL

The sentence "Sue goes to college if and only if she wins a scholarship" is a biconditional. The connective "if and only if" is abbreviated by the symbol $\leftrightarrow$. Table 11 defines the truth values for $p \leftrightarrow q$.

Table 11
Truth table for the biconditional

p	q	$p \leftrightarrow q$
T	T	T
T	F	F
F	T	F
F	F	T

The biconditional $p \leftrightarrow q$ is true whenever both components have the same truth values, either both true or both false.

The biconditional is sometimes called "double implication," which accounts for the choice of the double headed arrow. Our sentence above is the same as this much longer sentence:

If Sue goes to college, then she won a scholarship,
and if Sue wins a scholarship, then she will go to college.

Letting p stand for "Sue goes to college" and q stand for "Sue wins a scholarship," we write the longer form of $p \leftrightarrow q$ as

$$(p \rightarrow q) \wedge (q \rightarrow p)$$

Let's compare the truth tables for these compound statements. Columns 1 and 4 in Table 12 have the same truth table values, so we say they are *equivalent* statements. While the meaning of $p \leftrightarrow q$ suggests that $(p \rightarrow q) \wedge (q \rightarrow p)$ *ought* to be the same thing, the truth table analysis in Table 12 proves that the statements *are* equivalent. There are many pairs of statements that have the same truth table values.

Table 12 Truth values for two forms of the biconditional

p	q	$p \leftrightarrow q$	$(p \rightarrow q)$	$\wedge$	$(q \rightarrow p)$
T	T	T	T	T	T
T	F	F	F	F	T
F	T	F	T	F	F
F	F	T	T	T	T
		1	2	4	3

EXAMPLE 7

In Example 5 we found the truth table for $\sim p \rightarrow q$ to be T—T—T—F. Since this is the same as the truth table for $p \vee q$, we conclude that $\sim p \rightarrow q$ is equivalent to $p \vee q$.

EXAMPLE 8

Let p: We fail to attend.
       q: We have a good time.

Then $\sim p \rightarrow q$: If we do not fail to attend, then we will have a good time. Also, $p \vee q$: We fail to attend or we have a good time. If you think about the last two sentences, you can see that they mean the same thing.

We are saying that ordinary disjunction $(p \vee q)$ is replaceable by implication as $\sim p \rightarrow q$ because the two forms are equivalent. Yet there are times when clear English demands the connective "or," and so we will not abandon our use of it. It is also true that the connective "and" can be expressed in terms of "if . . . , then." From this point of view, the logic we are studying can be called the study of "if . . . , then" sentences.

EXAMPLE 9

Express "I will buy peanuts or I will buy popcorn" in the "if . . . , then" form. Letting p be the first component and q be the second, we write our sentence in the form $p \vee q$, which we have seen is equivalent to $\sim p \rightarrow q$. The "if . . . , then" form is then "If I do not buy peanuts, then I will buy popcorn."

Exercise Set 6.3

1 Determine whether each of the following compound sentences is true or false. (Use ordinary assumptions about the truth or falsity of components.)
   (a) Some apples are red and some are green.
   (b) "Dana" is a man's name or "Dana" is a woman's name.
   (c) $2 + 2 = 5$ and $2 + 2 = 4$
   (d) $2 + 2 = 5$ or $2 + 2 = 4$
   (e) $2 + 2 = 4$ or $3 + 2 = 5$
   (f) Some roses are red or China landed the first man on the moon. (Note that there need not be any relationship between the statements in a compound sentence.)
   (g) If coffee contains caffeine, then Senator Bigglesbottom will be the next President.
   (h) If Bigglesbottom becomes the next President, then coffee contains caffeine.
   (i) If circles have sharp corners, then $2 + 2 = 5$.
   (j) If circles have sharp corners, then $2 + 2 = 4$.

2 (a) Find the truth table for $\sim p \vee q$.
  (b) To which of the following basic connectives is $\sim p \vee q$ equivalent: $p \wedge q$, $p \vee q$, or $p \rightarrow q$?
  (c) Write a sentence in words of the form $\sim p \vee q$ and then put it in the appropriate basic connective form.

3 (a) Why does a compound statement containing two components have four lines in its truth table?
  (b) How many lines would be in the truth table for $p \vee (q \wedge s)$?
  (c) If a compound statement contained four simple statements, how many lines would be needed in its truth table?

4 (a) Find the truth table for $p \vee (q \wedge s)$.
  (b) Is this a conjunction or a disjunction?

5 (a) Find the truth table for $(p \vee q) \wedge s$.
  (b) Is this a conjunction or a disjunction?
  (c) Are $p \vee (q \wedge s)$ and $(p \vee q) \wedge s$ equivalent?

6 (a) Show that $\sim(p \wedge q)$ and $\sim p \vee \sim q$ are equivalent statements.
  (b) Show that $\sim(p \vee q)$ and $\sim p \wedge \sim q$ are equivalent statements.

7 Find the truth table of each of the following:
  (a) $p \rightarrow \sim q$
  (b) $\sim p \rightarrow \sim q$
  (c) $p \rightarrow (p \vee r)$
  (d) $(p \rightarrow q) \vee (p \rightarrow \sim q)$
  (e) $(p \rightarrow q) \wedge (p \rightarrow \sim q)$

8 Show that $p \leftrightarrow q$ and $q \leftrightarrow p$ are equivalent.

9 Show by truth-table analysis that these two sentences are *not* equivalent:
  (a) If you study hard, then you will pass Chemistry.
  (b) If you pass Chemistry, then you studied hard.

10 Show whether or not the following sentences are equivalent:
  (a) If we send in troops, then we will restore peace.
  (b) If peace is not restored, then we didn't send in troops.

11 Show that $\sim(\sim p)$ and p are equivalent.

12 Find the truth table for $(\sim p \vee r) \rightarrow (p \vee \sim s)$

13 Show that each of the following pairs of statements are equivalent:
  (a) $p; p \vee p$
  (b) $p; p \wedge p$
  (c) $\sim q; \sim q \vee \sim q$
  (d) $p \vee q; q \vee p$
  (e) $p \wedge q; q \wedge p$
  (f) $p \wedge (q \wedge r); (p \wedge q) \wedge r$
  (g) $p \vee (q \vee r); (p \vee q) \vee r$
  (h) $p \wedge (q \vee r); (p \wedge q) \vee (p \wedge r)$
  (i) $p \vee (q \wedge r); (p \vee q) \wedge (p \vee r)$
  (j) $p \rightarrow q; \sim q \rightarrow \sim p$

6.4 FORMS OF IMPLICATIONS

In the implication $p \rightarrow q$, the antecedent is p and the consequent is q. By changing the form of the original implication, other related implications can be created. These are

1. $p \rightarrow q$ Given statement
2. $q \rightarrow p$ The *converse* of the statement
3. $\sim p \rightarrow \sim q$ The *inverse* of the statement
4. $\sim q \rightarrow \sim p$ The *contrapositive* of the statement

The converse is formed by interchanging the antecedent and consequent of the given statement. The inverse is formed by negating both the antecedent and the consequent. The contrapositive is formed by interchanging *and* negating the antecedent and the consequent. Some examples will help clarify these forms.

EXAMPLE 1

Statement: If I leave my car out, then it will rain.
Converse: If it rains, then I left my car out.
Inverse: If I do not leave my car out, then it will not rain.
Contrapositive: If it does not rain, then I did not leave my car out.

EXAMPLE 2

Statement: If $x + 5 = 7$, then $x = 2$.
Converse: If $x = 2$, then $x + 5 = 7$.
Inverse: If $x + 5 \neq 7$, then $x \neq 2$.
Contrapositive: If $x \neq 2$, then $x + 5 \neq 7$.

EXAMPLE 3

Statement: $\sim p \rightarrow q$
Converse: $q \rightarrow \sim p$
Inverse: $p \rightarrow \sim q$
Contrapositive: $\sim q \rightarrow p$

Table 13 shows the truth tables of any implication $p \rightarrow q$, its converse, its inverse, and its contrapositive. (You should verify that these are indeed the correct truth tables.)

Observe that $p \rightarrow q$ and $\sim q \rightarrow \sim p$ have the same truth tables, and that $q \rightarrow p$ and $\sim p \rightarrow \sim q$ have the same truth tables. Thus a statement and its contrapositive are equivalent. The converse of a statement and the inverse of that statement are equivalent.

The most important practical result of this is that a statement and its converse are *not* equivalent.

Table 13   Truth tables of implication and related forms

p	q	Statement $p \rightarrow q$	Converse $q \rightarrow p$	Inverse $\sim p \rightarrow \sim q$	Contrapositive $\sim q \rightarrow \sim p$
T	T	T	T	T	T
T	F	F	T	T	F
F	T	T	F	F	T
F	F	T	T	T	T

EXAMPLE 4

Consider the following statement and converse.

Let $p \rightarrow q$: If your average is 94, then you pass Math.
      $q \rightarrow p$: If you pass Math, then your average is 94.

Certainly these sentences do not mean the same thing.

Too often a person uses a statement and means its converse, or means *both* the statement and its converse.

EXAMPLE 5

Sue's mother says to her "If you wash the dishes, you can go out." Sue, being savvy about logic, goes out but does not wash the dishes. Later, an argument takes place. Sue's mother says "I told you that you had to wash the dishes if you wanted to go out! You have deliberately disobeyed me." Sue replies, "No, that's not right. You said that if I washed the dishes, then I could go out. You didn't say what I could or could not do if I did *not* wash the dishes." Do you see that Sue's mother thought she had said the converse of the original statement?

What might Sue's mother have said that would be logically correct and express what she meant? One correct remark would be "If you wash the dishes, then you can go out *and* if you want to go out, then you must wash the dishes." The form of this is $(p \rightarrow q) \wedge (q \rightarrow p)$.

Another possibility is "If you wash the dishes, then you can go out, and conversely." This is the same as the sentence in the preceding paragraph, but briefer.

A more likely remark that emphasizes the thought is: "If you wash the dishes, then you can go out and if you do *not* wash the dishes, then you cannot go out." (Imagine Sue's mother arching her eyebrows and shaking her finger!) The form of this is $(p \rightarrow q) \wedge (\sim p \rightarrow \sim q)$. Table 14 shows the truth tables of these acceptable forms and contrasts them with $p \rightarrow q$.

Table 14 Forms of Sue's mother's remark

p	q	$p \to q$	$(p \to q) \wedge (q \to p)$			$(p \to q) \wedge (\sim p \to \sim q)$				
T	T	T	T	T	T	T	T	F	T	F
T	F	F	F	F	T	F	F	F	T	T
F	T	T	T	F	F	T	F	T	F	F
F	F	T	T	T	T	T	T	T	T	T
		1	2	4	3	5	9	6	8	7

Columns 4 and 9 indicate that the last two compound statements are equivalent. Note that they are not equivalent to $p \to q$, column 1. You should remember that T–F–F–T is the truth table for $p \leftrightarrow q$, which is the same as $q \leftrightarrow p$. Sue's mother might have said "You can go out if and only if you wash the dishes."

It is probably hopeless to try to correct the misuse of statements and converses in everyday speech. The converse is implied in many sentences without actually having been stated. In fact advertisers frequently extol the virtues of their product by having you jump to the converse (or to the inverse) of the remark they actually make.

EXAMPLE 6

"Tirpsy users are beautiful people" is a hypothetical advertising slogan. A person hearing this is supposed to think "If I use Tirpsy, then I will be beautiful." Of course, you might not use Tirpsy and still be beautiful. The advertiser is hoping you are hearing an implied converse: "If I want to be beautiful, then I must use Tirpsy." He might also be suggesting the inverse: "If I don't use Tirpsy, then I won't be beautiful." Neither the converse nor the inverse is equivalent to the given slogan.

Exercise Set 6.4

1 Write the converse, inverse, and contrapositive of each of the following implications:

(a) $\sim a \to \sim b$ (b) $(a \vee b) \to c$ (c) $p \to \sim q$

(d) If it is warm, then the flowers will bloom.

(e) If I don't waste time, I'll be there by noon.

(f) If it rolls, it does not gather moss.

°2 (a) Let $p \to q$ be a given implication. Write its inverse.

(b) Write the inverse of the inverse.

(c) How is this related to the given statement?

°3 (a) Consider $p \rightarrow q$. Write its converse.
    (b) Write the contrapositive of the converse.
    (c) True or false?: The contrapositive of the converse of a given implication is the inverse of the implication.

°4 (a) What is the inverse of the contrapositive of a given implication?
    (b) What is the converse of the inverse of a given implication?

5 Show that the contrapositive of the statement $\sim p \rightarrow q$ is equivalent to the statement.

6 (a) True or false?: $\sim(p \rightarrow q)$ is equivalent to $\sim p \rightarrow \sim q$
   (b) True or false?: $\sim(q \rightarrow p)$ is equivalent to $\sim q \rightarrow \sim p$

7 A statement $p \rightarrow q$ and its contrapositive $\sim q \rightarrow \sim p$ are equivalent.
   (a) Find the truth table for $(p \rightarrow q) \rightarrow (\sim q \rightarrow \sim p)$
   (b) Find the truth table for $(\sim q \rightarrow \sim p) \rightarrow (p \rightarrow q)$
   (c) Find the truth table for $(p \rightarrow q) \leftrightarrow (\sim q \rightarrow \sim p)$

8 The President says "If we send in troops, then we will maintain peace."
   (a) Does this mean "We will maintain peace only if we send in troops?"
   (b) Write the converse of the President's remark.
   °(c) Write the sentence in (a) in the "if . . . , then" form.

9 A definition can always be written in the "if . . . , then" form, and always implies that the converse holds too. Example: Recall the definition of a subset: If A is a subset of B and B is a subset of A, then set A equals set B. In symbolic form, $(A \subseteq B \wedge B \subseteq A) \rightarrow (A = B)$. The converse also holds, although it has not been stated. This is a convention that mathematicians have adopted and applies only to definitions.
   (a) Write the converse of the definition above. Many definitions use the word "means." Example: "Isomorphic" means "same form". In the "if . . . , then" form this might be "If two things are isomorphic, then they have the same form."
   (b) Look up dictionary definitions for the following words and write them in the "if . . . , then" form.
        1. lanky   4. bird
        2. obsequious   5. obscure
        3. mantilla   6. triangle
   (c) Why is the following not a good definition of the word "dog"? A dog is a four-legged animal with a tail.

°10 (a) Complete the following table, where S stands for a statement in the $p \rightarrow q$

°	S	C	I	C_p
S	S	C		
C			C_p	
I	I			
C_p				

form (and also means not to change it), C stands for a statement in the form $q \to p$ (and also means to interchange antecedent and consequent), I stands for the statement $\sim p \to \sim q$ (and also means to negate antecedent and consequent), C_p stands for the statement $\sim q \to \sim p$ (and means to interchange and negate antecedent and consequent), and $C \circ I$ means to start with C and then to form the inverse; so we go from $q \to p$ to $\sim q \to \sim p$. Thus $C \circ I = C_p$.

(b) What statement acts as the identity element?

(c) Is $\circ$ commutative?

(d) Does each statement have an inverse? (Do not confuse the two meanings of the word "inverse.")

(e) Assuming associativity, is this system a group?

6.5 TAUTOLOGIES AND VALID ARGUMENTS

The most interesting compound statements can sometimes be true and sometimes be false. Those that are either always true or always false are usually not worth much. Let's look at some examples.

EXAMPLE 1

"It is raining or it is not raining." The form of this statement is $p \vee \sim p$. The truth table is as follows:

p	p	$\vee$	$\sim p$
T	T	T	F
F	F	T	T
	1	3	2

The result, column 3, indicates that the remark is always true. Observe that if I say "It is raining or it is not raining" I have told you nothing. I made no assertion; there is no way to dispute my remark. It is a conversation stopper. We will call a sentence whose truth table is always true a *tautology*.

EXAMPLE 2

"I win the game and I lose the game." The form of this sentence is $p \wedge \sim p$ and the truth table is as follows:

p	p	$\wedge$	$\sim p$
T	T	F	F
F	F	F	T
	1	3	2

The result, column 3, indicates that the sentence is always false. Such a statement is called a *contradiction*. Certainly the speaker of such a sentence has contradicted himself, and there is something wrong with his logic.

EXAMPLE 3

"If I pass Math, then I will graduate." Here the form is $p \rightarrow q$ and the truth table is T—F—T—T, so the remark is neither a tautology nor a contradiction.

In conversation we cannot answer tautologies; nor can we allow contradictions. Rational people avoid tautologies and contradictions, which means that their declarative compound sentences make some assertion that becomes true or false when the component truth values are known. Because of this, almost every sentence we utter that makes any sense leads to other remarks or thoughts and then possibly to a conclusion.

EXAMPLE 4

First remark: "If I pass Math, then I will graduate."
Second remark: "If I graduate, then I can go to State University."
A "correct" third remark that follows from the first two might be "If I pass Math, then I can go to State University."

Let p: I pass Math, q: I will graduate, and r: I can go to State University. The sequences of sentences in Example 4 can be symbolized as follows:

$$\begin{array}{ll} \text{Premise 1:} & p \rightarrow q \\ \text{Premise 2:} & q \rightarrow r \\ \hline \text{Conclusion:} & \therefore p \rightarrow r \end{array}$$

Above the line are two sentences (called *premises*) that lead to the third sentence (called the *conclusion*). The symbol $\therefore$ is read "therefore." The sequence of sentences is called an *argument*. We will refer to *valid* arguments and *invalid* arguments. A valid argument is one that is rational, meaning that the conclusion does follow from the premises. We will see that the argument in Example 4 is valid.

EXAMPLE 5

Premise 1: If I pass Math, then I will graduate.
Premise 2: If I graduate, then I can go to State University.
Conclusion: Therefore if I do not pass Math, then I cannot go to State University.

If p, q, and r are defined as in Example 4, we can symbolize the argument as follows:

$$p \rightarrow q$$
$$q \rightarrow r$$
$$\overline{\therefore \sim p \rightarrow \sim r}$$

We will present a method below for determining that this argument is invalid, that is, that the conclusion cannot be reached from the premises given.

To determine the validity or nonvalidity of arguments, we assume that the premises are all true. In a valid argument, the conclusion is true whenever the premises are true. In an invalid argument, the conclusion is false in at least one case in which the premises are true.

We are led, then, to investigate the truth tables of the premises and the conclusions. We now do this for Examples 4 and 5. Those rows in which the premises are both true are boldface in Table 15; we are interested in the conclusion on those lines only.

Lines 1, 5, 7, 8 are the only lines in which both premises ($p \rightarrow q$ and $q \rightarrow r$) are true. The conclusion ($p \rightarrow r$) for Example 4 is true on those lines, so the argument is valid. The conclusion ($\sim p \rightarrow \sim r$) for Example 5 is false on lines 5 and 7, so the argument is invalid.

Table 15   Comparing proofs of validity and invalidity

	p	q	r	Premise 1 $p \rightarrow q$	Premise 2 $q \rightarrow r$	Conclusion, Example 4 $p \rightarrow r$	Conclusion, Example 5 $\sim p \rightarrow \sim r$
Line 1:	T	T	T	**T**	**T**	**T**	**T**
Line 2:	T	T	F	T	F		
Line 3:	T	F	T	F	T		
Line 4:	T	F	F	F	T		
Line 5:	F	T	T	**T**	**T**	**T**	**F**
Line 6:	F	T	F	T	F		
Line 7:	F	F	T	**T**	**T**	**T**	**F**
Line 8:	F	F	F	**T**	**T**	**T**	**T**

EXAMPLE 6

Consider this argument:
   Premise 1: If Fido falls in the creek, then he will get wet.
   Premise 2: Fido falls in the creek.
   Conclusion: Fido gets wet.

The form is as follows:

$$c \to w$$
$$c$$
$$\text{———}$$
$$\therefore \quad w$$

		Premise 1	Premise 2	Conclusion
c	*w*	$c \to w$	*c*	*w*
T	T	T	T	T
T	F	F	T	
F	T	T	F	
F	F	T	F	

We are interested only in the first line of the table because this is where premise 1 *and* premise 2 are true. Since the conclusion is true on that line, the argument is valid.

A second method for determining validity is based upon the discovery that a certain compound statement derived from the argument is a tautology. We could show that this method derives from the first one, but we will state it simply as a rule.

TAUTOLOGY METHOD FOR DETERMINING VALIDITY
Given an argument with premises and conclusion,
1. First, form the conjunction of all premises.
2. Second, form an implication, using this conjunction as antecedent and the conclusion as consequent.
3. Third, find the truth table of the implication.
4. If the implication is a tautology, then the argument is valid. If the implication is not a tautology, then the argument is invalid.

In Example 6, the premises are $c \to w$ and c, while the conclusion is w.
Step 1   Form $(c \to w) \land c$, the conjunction of the premises.
Step 2   Form $[(c \to w) \land c] \to w$, the required implication.
Step 3

c	*w*	[(*c*	$\to$	*w*)	$\land$	*c*]	$\to$	*w*
T	T		T		T	T	T	T
T	F		F		F	T	T	F
F	T		T		F	F	T	T
F	F		T		F	F	T	F
			1		3	2	5	4

Step 4   Column 5 indicates that the implication is a tautology, so the argument is valid.

EXAMPLE 7

Show that $p \to q$
             $q \to r$
_____
             $\therefore p \to r$

is valid, by discovering the appropriate tautology.

p	q	r	$[(p$	$\to$	$q)$	$\wedge$	$(q$	$\to$	$r)]$	$\to$	$(p$	$\to$	$r)$
T	T	T		T		T		T		T		T	
T	T	F		T		F		F		T		F	
T	F	T		F		F		T		T		T	
T	F	F		F		F		T		T		F	
F	T	T		T		T		T		T		T	
F	T	F		T		F		F		T		T	
F	F	T		T		T		T		T		T	
F	F	F		T		T		T		T		T	
				1		3		2		5		4	

We have shown (column 5) that the conjunction of the premises in implication with the conclusion is a tautology. Therefore the argument is valid.

It is true that the implication formed is rather complex. Compared with the work we did for Examples 4, 5, and 6, the method seems tedious. Yet many people like this method of determining validity because it is mechanical.

Notice that validity is not dependent upon the actual truth values of the premises. In order to judge an argument valid, we must *assume* the premises to be true and then be able to derive the truth of the conclusion. The form of the argument may or may not be valid.

EXAMPLE 8

You can prove that the following argument is valid without knowing what "glub" and "snortle" mean. Is the first premise true?

Premise: If it is a glub, then it snortles.
Premise: It doesn't snortle.
Conclusion: Therefore it isn't a glub.

Exercise Set 6.5

1 Show that each of the following is a tautology:

(a) $p \to p$ (b) $\sim p \vee (\sim p \to q)$ (c) $(\sim p \to q) \to (p \vee q)$
(d) $(a \to b) \to (\sim b \to \sim a)$ (e) $(p \vee q) \to (\sim p \to q)$

2 (a) True or false: The negation of a tautology is a contradiction.
  (b) True or false: The negation of a contradiction is a tautology.

3 Show that each of the following is a contradiction:
  (a) $\sim(p \rightarrow p)$ (b) $[(r \rightarrow s) \wedge r] \wedge (\sim s)$ (c) $(r \wedge s) \wedge (\sim s)$

4 In question 13 of Exercise Set 6.3, ten pairs of equivalent statements are given. Write obvious tautologies that follow as a result of these equivalent statements.

5 Show that each of the following argument forms is valid.

(a) $p \rightarrow q$
      $\underline{\sim q}$
     $\therefore \sim p$

(b) $p \vee q$
      $\underline{\sim p}$
     $\therefore q$

(c) $\underline{p \wedge q}$
     $\therefore p$

(d) p
      $\underline{q}$
     $\therefore p \wedge q$

(e) $p \rightarrow q$
      $\underline{q \rightarrow r}$
     $\therefore p \rightarrow r$

(f) $\underline{p}$
     $\therefore p \vee q$

(g) $p \vee q$
      $\sim p$
      $\underline{q \rightarrow r}$
     $\therefore r$

(h) $(p \wedge q) \rightarrow \sim r$
      $\underline{q \wedge r}$
     $\therefore \sim p$

(i) $s \rightarrow b$
      $\sim b$
      $\underline{\sim s \rightarrow a}$
     $\therefore a$

(j) $p \rightarrow \sim q$
      $\sim r \rightarrow \sim s$
      $p \vee \sim r$
      $\underline{q}$
     $\therefore \sim s$

*6 Prove that if the conjunction of two statements is assumed true, then their disjunction follows.

7 Prove that the following are invalid arguments:

(a) $p \rightarrow q$
      $\underline{q}$
     $\therefore p$

(b) $p \rightarrow q$
      $\underline{\sim p}$
     $\therefore q$

(c) $p \rightarrow q$
      $\underline{\sim p}$
     $\therefore \sim q$

(d) $p \vee q$
      $\underline{\sim p}$
     $\therefore \sim q$

8 Prove each of the following arguments either valid or invalid.

(a) Either John buys a boat or Ted cannot go fishing. John buys a boat. Therefore Ted can go fishing.

(b) If we are quick enough, we will win the bid. If we get the bid, we can resell at a profit. Therefore if we are quick enough, we can resell at a profit.

(c) If Smith sells his car, then he can buy some stock. If Smith does not sell his car, then he will remain poor. Smith does not buy stock. Therefore he will not remain poor.

(d) Money is the root of all evil or the salvation of the poor. If money is the salvation of the poor, then it is not the root of all evil. Money is the root of all evil. Therefore money is not the salvation of the poor.

(e) Same premises as in (d). Conclusion: Therefore money is the salvation of the poor.

SUMMARY

The goal in our study of logic was to develop a method for determining validity of verbal arguments. To do that we had to symbolize basic statements and define truth tables for compound statements. We dealt with negation, conjunction, disjunction, implication, and the biconditional.

There was particular emphasis on implication, the "if ..., then" compound statement, and we saw how to form the converse, inverse, and contrapositive of given implications. The notion of equivalent statements was important in recognizing two different sentence forms that mean the same thing.

Behind all our work in logic was truth-table analysis, which enabled us to discover tautologies and contradictions and provided us with a tabular method of determining validity of arguments. There were two methods of testing for validity. The first involved checking the conclusions in a truth table for which all the premises were true. If the conclusion was true in all cases, the argument was valid. The second method involved checking to see if the conjunction of the premises, in implication with the conclusion, was a tautology.

While we are all rational and logical people, it is hard to imagine a study of logic without symbolization and without truth-table analysis. Some of the arguments in this chapter are quite tricky. To try to prove that they are valid by conversational arguing would be difficult. So the symbolic method of modern-day logic is mentally liberating in the sense that it lets us rely upon mechanical means to come to a conclusion. Sophisticated as Aristotle was in his logic, he did not have the simple techniques of truth-table analysis available to us today.

Because of the advent of symbolic Logic in the nineteenth century, tremendous progress has been made in the advancement of mathematics. Some believe that the greatest advance of all is the computer. Present-day computers are primarily "logic machines" (they even have "and" and "or" circuits!). They are the culmination of more than 100 years of mathematical research into logic.

Before attempting the Review Test, be sure you are familiar with the following:

1. The basic symbols $\sim$, $\wedge$, $\vee$, $\rightarrow$.

2. Truth tables for $\sim p, p \wedge q, p \vee q, p \rightarrow q, p \leftrightarrow q$.

3. Translating sentences into symbolic form.

4. Translating symbolic statements into English.

5. Finding the truth table for expressions like $p \wedge \sim q$ or $\sim p \vee \sim q$ or $\sim p \rightarrow (q \wedge \sim p)$ or $(p \wedge \sim q) \rightarrow (\sim r \vee p)$.

6. Showing that two statements are equivalent by truth-table analysis.

7. Tautologies and contradictions.

8. Forming the inverse, converse, and contrapositive of a given implication.

9. Testing the validity of a given argument by truth-table analysis.

REVIEW TEST

1 Given p: I study hard.
   q: I pass the course.
   r: I will be promoted.

Write each of the following in verbal form:

(a) $p \rightarrow q$

(b) $(p \wedge q) \rightarrow r$

(c) $p \vee \sim q$

(d) $\sim r \rightarrow (\sim p \wedge \sim q)$

2 Using any symbols you wish, translate each of the following into symbolic form:

(a) If John is old enough, he can make the team.

(b) Alice will get her license if she passes the road test.

(c) There will be enough fuel if the weather stays warm or new supplies are discovered.

3 Construct the truth table for $(\sim p \vee q) \rightarrow \sim r$

4 Write the converse of $\sim p \rightarrow (q \wedge r)$

5 Write the inverse of *If the employees get a raise, then they will not strike.*

6 Show that $\sim(p \wedge q)$ is not equivalent to $\sim p \wedge \sim q$.

7 Determine whether the following is a tautology or a contradiction or neither.

$$[(\sim p \rightarrow q) \wedge \sim q] \rightarrow p$$

8 Show that the following argument is invalid.

$$p \vee q$$
$$\underline{\sim q}$$
$$\therefore \sim p$$

9 Determine whether or not the following argument is valid: If it snows, then we can go skiing. If it doesn't snow, then we can go to the movies. We don't go skiing. Therefore we can go to the movies.

Lewis Carroll (pseudonym for Charles Lutwidge Dodgson) is most often remembered for his contributions to children's literature. However, Carroll was also a mathematician, noted primarily for his work in logic. The stories he wrote were intended for the amusement and entertainment of young children who were his friends. The excerpts below, from *Alice in Wonderland* and *Through the Looking Glass*, reflect not only Carroll's charm, but also his concern with logic and meaning. The apparent nonsense of the story line conveys the need for precision in the use of language.

Say What You Mean and Mean What You Say!
LEWIS CARROLL

"Come, we shall have some fun now!" thought Alice. "I'm glad they've begun asking riddles—I believe I can guess that," she added aloud.

"Do you mean that you think you can find out the answer to it?" said the March Hare.

"Exactly so," said Alice.

"Then you should say what you mean," the March Hare went on.

"I do," Alice hastily replied; "at least—at least I mean what I say— that's the same thing, you know."

"Not the same thing a bit!" said the Hatter. "Why, you might just as well say that 'I see what I eat' is the same thing as 'I eat what I see'!"

"You might just as well say," added the March Hare, "that 'I like what I get' is the same thing as 'I get what I like'!"

"You might just as well say," added the Dormouse, which seemed to be talking in its sleep, "that 'I breathe when I sleep' is the same thing as 'I sleep when I breathe'!"

"It *is* the same thing with you," said the Hatter, and here the conversation dropped. . . .

From "A Mad Tea-Party" in *Alice in Wonderland*

From *The Complete Works of Lewis Carroll*, New York: Random House, 1936. [My title —F. N. M.]

"In that case we start afresh," said Humpty Dumpty, "and it's my turn to choose a subject—" ("He talks about it just as if it was a game!" thought Alice.) "So here's a question for you. How old did you say you were?"

Alice made a short calculation, and said "Seven years and six months."

"Wrong!" Humpty Dumpty exclaimed triumphantly. "You never said a word like it!"

"I thought you meant 'How old *are* you?'" Alice explained.

"If I'd meant that, I'd have said it," said Humpty Dumpty.

From "Humpty Dumpty" in *Through the Looking Glass*

✿   ✿   ✿

"When *I* use a word," Humpty Dumpty said, in rather a scornful tone, "it means just what I choose it to mean—neither more nor less."

"The question is," said Alice, "whether you *can* make words mean so many different things."

"The question is," said Humpty Dumpty, "which is to be master—that's all."

Alice was too much puzzled to say anything; so after a minute Humpty Dumpty began again. "They've a temper, some of them—particularly verbs; they're the proudest—adjectives you can do anything with, but not verbs—however *I* can manage the whole lot of them! Impenetrability! That's what *I* say!"

"Would you tell me, please," said Alice, "what that means?"

"Now you talk like a reasonable child," said Humpty Dumpty, looking very much pleased. "I meant by 'impenetrability' that we've had enough of that subject, and it would be just as well if you'd mention what you mean to do next, as I suppose you don't mean to stop here all the rest of your life."

"That's a great deal to make one word mean," Alice said in a thoughtful tone.

"When I make a word do a lot of work like that," said Humpty Dumpty, "I always pay it extra."

"Oh!" said Alice. She was too much puzzled to make any other remark.

"Ah, you should see 'em come round me of a Saturday night," Humpty Dumpty went on, wagging his head gravely from side to side, "for to get their wages, you know."

(Alice didn't venture to ask what he paid them with; and so you see I can't tell *you*.)

From "Humpty Dumpty" in *Through the Looking Glass*

"You are sad," the Knight said in an anxious tone: "let me sing you a song to comfort you."

"Is it very long?" Alice asked, for she had heard a good deal of poetry that day.

"It's long," said the Knight, "but it's very, *very* beautiful. Everybody that hears me sing it—either it brings the *tears* into the eyes, or else—"

"Or else what?" said Alice, for the Knight had made a sudden pause.

"Or else it doesn't, you know. The name of the song is called '*Haddocks' Eyes.*'"

"Oh, that's the name of the song, is it?" Alice said, trying to feel interested.

"No, you don't understand," the Knight said, looking a little vexed. "That's what the name is *called*. The name really *is* '*The Aged Aged Man.*'"

"Then I ought to have said, 'That's what the *song* is called'?" Alice corrected herself.

"No, you oughtn't: that's quite another thing! The *song* is called '*Ways and Means*': but that's only what it's *called*. you know!"

"Well, what *is* the song, then?" said Alice, who was by this time completely bewildered.

"I was coming to that," the Knight said. "The song really *is* '*A-sitting on a Gate*': and the tune's my own invention."

So saying, he stopped his horse and let the reins fall on its neck: then, slowly beating time with one hand, and with a faint smile lighting up his gentle foolish face, as if he enjoyed the music of his song, he began.

From "It's My Own Invention" in *Through the Looking Glass*

7 NUMBER THEORY

An introduction to the theory of numbers deals primarily with the counting or natural numbers, that is, the set

$$N = \{\, 1, 2, 3, 4, 5, 6, \ldots \}$$

This set has many interesting properties, some of which are merely curiosities and some of which have opened up whole new fields of mathematical research.

For example, did you know that the numbers 123,123; 482,482; and 563,563 are exactly divisible by 13? In general, any six-digit number of the form *abc,abc* is divisible by 13.

Here's an interesting fact about the odd numbers, grouped as follows:

1,   3, 5,   7, 9, 11,   13, 15, 17, 19,   21, 23, 25, 27, 29, . . .

The first odd number, 1, is 1^3. The sum of the next two is 2^3; that is, $3 + 5 = 2^3$. The sum of the next three is 3^3; that is $7 + 9 + 11 = 3^3$. Also, 13 + 15 + 17 + 19 = 4^3, and so on. So we have a progression in which the *first* number equals 1 cubed, the next *two* numbers add up to 2 cubed, the next *three* numbers add up to 3 cubed, and next *four* numbers add up to 4 cubed, and so on. This unusual property of numbers was discovered by Nichomachus in the first century A. D.

Strange properties of numbers have always fascinated mathematicians and nonmathematicians alike. Pierre de Fermat (1601–1665), today regarded as having been a mathematician of the first order, spent his entire career as a French government official. He was probably the world's greatest number theorist and is credited, with Blaise Pascal (1623—1662), with providing the foundation for probability theory. Fermat must have filled his leisure time with doodlings and calculations with numbers.

The history of mathematics is full of stories about number theorists. Euclid (about 300 B.C), whom we revere primarily as a geometer, proved theorems about numbers. In fact his monumental work *The Elements* contained three chapters on the theory of numbers. Even earlier (about 540 B.C.), Pythagoras discovered numerical relationships between the musical tones of the scale. He seems to have worshipped the properties of numbers; to him every part of nature stood in some harmonious numeric relationship to the whole universe. One of the most famous of all mathematical quotes is attributed to Pythagoras:

"Number rules the universe."

Since 1700 much more number theory has been written and developed than in all previous time. From one point of view every mathematician is a number theorist, since numbers provide such clear examples of mathematical properties. But the fascination of numbers *per se* guarantees that the field will always have its specialists.

7.1 PATTERNS AND PROGRESSIONS

Probably the greatest force motivating work in number theory is the discovery of patterns in sequences, or successions, of numbers. Even on a simple intuitive level, many people discover some pattern in a set of numbers.

EXAMPLE 1

Suppose you had to add up the first 25 counting numbers:

$$1 + 2 + 3 + 4 + 5 + 6 + 7 + 8 + 9 + 10 + 11 + 12 + 13 + 14 + 15$$
$$+ 16 + 17 + 18 + 19 + 20 + 21 + 22 + 23 + 24 + 25$$

If this sum is simply regrouped as follows, it can be found mentally without much work.

$$25 + (24 + 1) + (23 + 2) + (22 + 3) + (21 + 4) + (20 + 5) + (19 + 6)$$
$$+ (18 + 7) + (17 + 8) + (16 + 9) + (15 + 10) + (14 + 11) + (13 + 12)$$

Each of the terms in parenthesis has a sum of 25, so we can visualize the sum as $25 + 12(25)$. Since twelve 25's is 300, it is easy to see that the answer is 325.

Many people, in adding columns of figures, have learned to reorder and regroup numbers, say in sums of 10, to simplify the work. Thus in the above example, the reordering and regrouping to sums of 25 is not very unusual. Yet the process is completely general; we can add up the first n numbers by separating out the last one and regrouping the others so that the sum of each group is n. It can be seen that there are $(n - 1)/2$ of these groups. Thus the sum of the first n counting numbers is $n + \dfrac{(n - 1)}{2} \cdot n$.

It is possible to show that this formula simplifies to $\dfrac{n(n + 1)}{2}$. An intuitive observation has led us to a general formula for the sum of the first n counting numbers.

Now that the formula is known, we can let $n = 25$ to get the sum in Example 1.

$$\frac{n(n + 1)}{2} = \frac{25(26)}{2} = 25(13) = 325$$

EXAMPLE 2

Find the sum $1 + 2 + 3 + \ldots + 200$. The formula $\dfrac{n(n + 1)}{2}$, where $n = 200$, will give the answer:

$$\frac{200(201)}{2} = 100(201) = 20{,}100$$

Let's look now at progressions of numbers in which each number is derived from the one that precedes it, according to some pattern. If we can detect the pattern, we can supply as many numbers in the progression as we wish. It will be interesting to find a formula that governs the pattern and enables us to find any term, say the 50th, without supplying all of the first 49 terms.

EXAMPLE 3

What is the next term (number) in the sequence 1, 4, 7, 10, 13, ... ? Upon inspection we see that each term is three more than the preceding one. Thus the next number is $13 + 3 = 16$.

The terms in the sequence of Example 3 differ from each other by a constant. Such sequences of numbers are called *arithmetic progressions*. The difference between any two successive terms must always be the same. This difference is generally denoted by d and is called the *common difference*. In Example 3, $d = 3$. Observe the following pattern in that sequence.

$$
\begin{aligned}
2\text{nd term:} \quad & 4 = 1 + 3 \\
3\text{rd term:} \quad & 7 = 4 + 3 = (1 + 3) + 3 = 1 + 2(3) \\
4\text{th term:} \quad & 10 = 1 + 3(3) \\
5\text{th term:} \quad & 13 = 1 + 4(3) \\
6\text{th term:} \quad & 16 = 1 + 5(3)
\end{aligned}
$$

What is the nth term? The pattern suggests it should be $1 + (n - 1)3$.

DEFINITION 1   Arithmetic Progression

An arithmetic progression, with first term a and common difference d, has for its nth term, $a + (n - 1)d$. Given a particular a and d, an *arithmetic progression* may be defined as the set of numbers $a + (n - 1)d$, taken in order.

It should be observed that n takes on values 1, 2, 3, 4, 5, ... In particular if $n = 1$, the nth term formula gives $a + 0 \cdot d = a$, which *is* the first term.

EXAMPLE 4

Consider the progression 5, 12, 19, 26, ... What is the next term, and what is a formula for the nth term? Clearly $d = 7$, so the next term is 33. By Definition 1, the nth term is $5 + (n - 1)7$ or $7n - 2$.

EXAMPLE 5

Write the first few terms of the arithmetic progression with $a = 10$ and $d = 12$. What is the 50th term? The progression is 10, 10 + 12, 10 + 12 + 12, 10 + 12 + 12 + 12, and so on, or 10, 22, 34, 46, ... The 50th term is given by

$$10 + (50 - 1) \cdot 12 = 10 + 49(12) = 10 + 588 = 598$$

Whenever we count, we are sounding off the terms of an arithmetic progression with $a = 1$, and $d = 1$, so the nth term is $1 + (n - 1) \cdot 1$, which is, of course, n. We saw earlier that the sum of the first n counting numbers is given by $\dfrac{n(n + 1)}{2}$. It is possible to develop a formula for the sum of the first n terms of any arithmetic progression. We state the formula as a definition.

DEFINITION 2   Sum of an Arithmetic Progression

The sum of the first n terms of an arithmetic progression with first term a and common difference d is

$$n[a + \frac{n - 1}{2} \cdot d]$$

EXAMPLE 6

In the progression 10, 22, 34, 46, ... what is the sum of the first five terms? Here we have $n = 5$, $a = 10$, and $d = 12$. So the required sum is

$$5\,[10 + \frac{5 - 1}{2} \cdot 12] = 5[10 + 2 \cdot 12] = 5(34) = 170$$

You should verify that $10 + 22 + 34 + 46 + 58 = 170$.

Now we turn to progressions in which each term after the first is multiplied by some number to get the next number. An example is 1, 2, 4, 8, 16, ... Clearly, we have a first term 1, multiplied by 2 to get the second term; the second term is multiplied by 2 to get the third term, and so on. Such a progression is called a *geometric progression*. The general appearance is

$$a,\ ar,\ ar^2,\ ar^3,\ ar^4,\ ar^5,\ \ldots$$

The a is the first number in the progression; the ar is the second number, and so on. Note that the second number is r times the first, then ar is multiplied by r to get the third number ar^2, and so on. The r can be called the *term-by-term multiplier*. Definition 3 gives formulas for the nth term and for the sum of the first n terms of a geometric progression.

DEFINITION 3   Geometric Progression Formulas

In a geometric progression with first term a, and term-by-term multiplier r (called the *common ratio*), the formula for the nth term is

$$ar^{n-1}$$

And the sum of the first n terms is

$$\frac{a(1 - r^n)}{1 - r}$$

The common ratio cannot be 1.

EXAMPLE 7

Let $a = 1$, $r = 3$ in a geometric progression. Write five terms of the progression, and find the tenth term and the sum of the first 10 terms. The progression is

$$1, 1 \cdot 3, 1 \cdot 3 \cdot 3, 1 \cdot 3 \cdot 3 \cdot 3, \ldots$$

So the first five terms are

$$1, 3, 9, 27, 81$$

Using the formula ar^{n-1}, we get the tenth term by letting $n = 10$:

$$1 \cdot 3^{10-1} = 1 \cdot 3^9 = 3^9 = 19683$$

Using the formula for the sum, where $n = 10$, we get

$$\frac{1 \cdot (1 - 3^{10})}{1 - 3} = \frac{1 - 3^{10}}{-2} = \frac{1 - 59049}{-2} = \frac{-59048}{-2} = 29524$$

EXAMPLE 8

Given the geometric progression 1, 1/2, 1/4, 1/8, ... find the sixth term and the sum of the first six terms. The sixth term, where $a = 1$, $r = 1/2$, and $n = 6$ is

$$1 \cdot (1/2)^{6-1} = (1/2)^5 = 1/32$$

The sum of six terms is

$$\frac{1 \cdot [1 - (1/2)^6]}{1 - 1/2} = \frac{1 - 1/64}{1/2} = \frac{63/64}{1/2} = \frac{63}{32} = 1.96875$$

In Example 8, we see that the sum $1 + 1/2 + 1/4 + 1/8 + 1/16 + \ldots$ appears to approach 2 (by thinking of successive sums as measurements on a number line as in Figure 1).

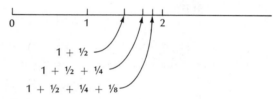

Figure 1   A sum approaching 2.

To achieve 2, we would have to add up *all* the terms in the progression. That is, we need to add up an infinite number of terms. The sum of an infinite geometric progression can be found by a formula, provided that the common ratio is between 0 and 1.

DEFINITION 4   Sum of Infinite Geometric Progression

The sum of an infinite geometric progression, where the common ratio is between 0 and 1, is given by

$$\frac{a}{1-r}$$

EXAMPLE 9

Show that the formula $\dfrac{a}{1-r}$ applies to the sum $1 + 1/2 + 1/4 + 1/8 +$ .... This is the sum of an infinite geometric progression, where $a = 1$ and $r = 1/2$. Since $1/2$ is between 0 and 1, we can use the formula $\dfrac{a}{1-r}$. It gives

$$\frac{1}{1-1/2} = \frac{1}{1/2} = 2$$

Repeating decimals can be thought of as a sum of an infinite geometric progression. Thus $0.333\ldots = 3/10 + 3/100 + 3/1000 + \ldots$ Since the common ratio is $1/10$, we can get this infinite sum by using the formula $\dfrac{a}{1-r}$, with $a = 3/10$ and $r = 1/10$. Thus

$$0.3333\ldots = \frac{3/10}{1-1/10} = \frac{3/10}{9/10} = 3/9 = 1/3$$

The result, that $0.3333\ldots = 1/3$, was certainly known to you. This is another method for converting a repeating decimal to its rational form. You should compare this method to the algebraic method discussed in Chapter 4.

EXAMPLE 10

Show that 0.121212 . . . can be represented as a fraction. The repeating decimal 0.121212 . . . can be written as 12/100 + 12/10000 + 12/1000000 + . . . , which is the sum of an infinite geometric progression, where $a = 12/100$ and $r = 1/10$. So the sum is given by $\dfrac{a}{1-r}$, which is

$$\frac{12/100}{1-1/10} = \frac{12}{100-1} = \frac{12}{99} = \frac{4}{33}$$

Exercise Set 7.1

A desk calculator may be used in the problems that follow.

1 (a) Divide 1001 by 13.
   (b) Divide 123,123 by 13.
   (c) Divide each of these numbers by 13: 465,465; 998,998; 103,103; 725,725
   (d) Divide each of the numbers in (c) by 1001. Do you notice the pattern of each of these divisions? The reason 13 divides numbers of the form *abc,abc* should be clear intuitively. Try to explain it.
   (e) Divide 465,465,465 by 13. Why does 13 *not* divide such nine-digit numbers exactly?

2 (a) Divide each of the following numbers by 3.

       162; 261; 621; 1002; 2001; 123,456; 654,321

   (b) Divide 163 by 3.
   (c) Try to discover a rule for divisibility by 3.

3 (a) If a number is divisible by 3, is it divisible by 9?
   (b) Write the converse of "If a number is divisible by 3, then it is divisible by 9."
   (c) Is the converse true?
   (d) Divide 135, 531, 1089, 27, and 123,453 by 9.
   (e) What is the rule for divisibility by 9?

4 Any number divisible by 6 must also be divisible by 2 and 3. Conversely, if a number is divisible by 2 and 3, it is divisible by 6.
   (a) What is the rule for divisibility by 2?
   (b) Be sure you know the rule of divisibility by 3. (See exercise 2.)
   (c) Without performing the division, test each of the following numbers for divisibility by 6.
   (i)   342   (ii) 243   (iii) 111,132
   (iv) 1,472,004   (v)   3201   (vi)   3312

5 (a) Show that the sum of the first ten counting numbers is $10 + \dfrac{9}{2} \cdot 10$ and that it is also $\dfrac{10(10+1)}{2}$ by calculating each of these expressions and comparing the results to the direct sum $1 + 2 + 3 + \ldots + 10$.

(b) What is the sum of the natural numbers from 1 to
  (i) 50?   (ii) 100?   (iii) 1000?

6 Given the sequence of numbers 1, 4, 9, 16, 25, . . .
   (a) What seems to be the most probable next term?
   (b) A formula for the sum of the squares of the first n counting numbers is

$$\frac{n(n + 1)(2n + 1)}{6}$$

   Find the sum $1 + 4 + 9 + 16 + 25 + 36 + 49$ by column addition
   and by using this formula.
   (c) What is the sum $1 + 4 + 9 + 16 + \ldots + 144$?

7 A formula for the sum of the cubes of the first n counting numbers is

$$\frac{[n^2(n + 1)^2]}{4}$$

   (a) Find $1 + 8 + 27 + 64 + 125$, using this formula.
   (b) How much is $1 + 8 + 27 + 64 + \ldots + 1000$?

8 Consider the progression 6, 14, 22, 30, 38, . . .
   (a) Is this an arithmetic or a geometric progression?
   (b) What is the common difference or common ratio, whichever applies?
   (c) What is the first term?
   (d) What is the seventh term?
   (e) What is the sum of the first seven terms?

9 Write out the first six terms of the arithmetic progression in which $a = 3$ and
   $d = 10$.

10 Write out the first six terms of the geometric progression in which $a = 3$ and
    $r = 10$.

11 A boy is paid 1¢ the first day, 2¢ the second day, 4¢ the third day, 8¢ the
    fourth day, and so on.
    (a) How much is he paid on the tenth day?
    (b) How much is he paid on the thirtieth day?
    (c) How much does he make altogether in 30 days? Express your answer in
        dollars.

12 A chess board is eight squares wide and eight squares long. If you put one
    penny on the first square, two pennies on the second, four pennies on the
    third, and so on, how many would you put on the sixty-fourth?

13 Consider a geometric progression in which $a = 1/2$ and $r = 1/10$.
    (a) Write out the first six terms in fractional form.
    (b) Convert each fraction to a decimal.
    (c) What is the sum of *all* the terms in the progression? Express your answer
        first in decimal form, and then as a single fraction.

14 Show that each of the following is true, by using the ideas of geometric
    progressions:
    (a) $0.1111 \ldots = 1/9$ (b) $0.454545 \ldots = 5/11$
    (c) $0.6666 \ldots = 2/3$ (d) $0.123123123 \ldots = 41/333$
    (e) $0.9999 \ldots = 1$ (f) $0.4999 \ldots = 1/2$ (g) $1.49999 \ldots = 1.5$

°15 (a) Write the first six terms of the sequence in which the general term is in the form $(n + 1)^2$. (That is, let $n = 1, 2, 3, 4, 5,$ and 6 and calculate.)

(b) Directly under your answer to (a) write down the same number of terms of the form n^2.

(c) Draw a line under both lists and subtract each number from the one above it.

(d) What kind of progression is the set of subtractions?

(e) Write a formula for the nth term in this progression, and simplify the result.

(f) In algebra, what is $(n + 1)^2 - n^2$?

(g) True or false: Any two consecutive squares of numbers differ by an odd amount.

16 Would you rather be paid $2000 a day for 30 days, or be paid a penny the first day, double that the second day, and so on, always doubling the previous day's pay, for 30 days?

7.2 PRIMES AND COMPOSITES

Perhaps the most fascinating subset of the natural numbers is the set of primes. While we will delay the definition of "prime number" briefly, we can informally say that it is a natural number that has no exact divisors except itself and 1. Thus 7 is prime since no number divides 7 exactly except 7 and 1. No other set of numbers, it seems, has been such a constant source of theorems and conjectures throughout history. Euclid proved that the number of primes is infinite, yet no one has been able to find a general formula yielding primes. In fact only recently a computer was used to show that $2^{19937} - 1$ is prime. (This number has 6002 digits.) In 1974 this was the largest known prime, yet since there are infinitely many primes, there must be larger primes.

We will investigate certain facts about primes that have been known for years and will state some conjectures that have not yet been proved. The primes will provide endless hours of research in the future for professional mathematicians and amateurs.

DEFINITION 5   Prime Number

A *prime number* is a natural number other than 1 that has no exact divisors in the set N except itself and 1. If a number other than 1 has some divisor other than itself and 1, it is called *composite*.

EXAMPLE 1

The number 12 is not prime since it has exact divisors other than 12 and 1. They are 3, 4, and 6. Thus 12 is a composite number.

In several instances we have used the term "exact divisor." You should be able to see intuitively that an exact divisor is one that leaves no remainder. If a number a, divides another number b, without leaving a remainder, then a is called an exact divisor of b, or a is a *factor* of b. We will write

$$a \mid b$$

which can be read "a divides b." If a is a factor of b, then we will call b a *multiple* of a.

EXAMPLE 2

The factors of 12 are 1, 2, 3, 4, 6, and 12 since each of those numbers divides 12 exactly.

EXAMPLE 3

Since $20 = 4(5)$, both 4 and 5 are factors of 20. Then 20 is a multiple of 4 and a multiple of 5.

The multiples of 5 are elements of a set, namely $\{$ 5, 10, 15, 20, 25, 30, ... $\}$, and so 20 is the fourth multiple of 5. In the same way 18 is the sixth multiple of 3, and 30 is the fifth multiple of 6.

A prime number has no factors other than itself and 1, and a prime is a multiple of no smaller number except 1.

Each of the numbers in the following list is a prime: 2, 3, 5, 7, 11. We know that the set of primes, P, is infinite, but when we write

$$P = \{ 2, 3, 5, 7, 11, 13, 17, 19, 23, \ldots \}$$

the dots are *not* intended to indicate what prime number is next in the list.

Is a number like 571 prime? To find out we must determine whether any smaller number (except 1) divides 571. That is, does any number in the set

$$\{ 2, 3, 4, 5, 6, \ldots, 570 \}$$

divide 571? If we write this set of "candidate" divisors as

$$\{ 2, 3, 4, \ldots, 285, 286, 287, \ldots, 570 \}$$

all the numbers from 286 to 570 can be discarded as possible divisors of 571. This is because 2(286) exceeds 571, 2(287) exceeds 571, 2(288) exceeds 571, and so on. Our set of candidate divisors has been cut in half:

$$\{ 2, 3, 4, 5, \ldots, 285 \}$$

Since no even number divides 571, we can compress the set of possible divisors even more:

$$\{ 3, 5, 7, 9, 11, \ldots, 285 \}$$

Even this set contains 142 numbers, but it will not be necessary to perform all these divisions. The method of "*the sieve of Eratosthenes*" leads us to a technique for shortening the work considerably. The "sieve method" (known since 200 B.C.) yields all the primes *up to* a certain number.

Suppose we want to find all the primes up to 65. (We will come back to 571 later.) Observe the following stepwise procedure, with reference to Figure 2.

Step 1 Write down all the numbers from 2 to 65.

Step 2 Since 2 is prime, circle it. Then cross out all multiples of 2, since they contain 2 as a factor and thus are not primes.

Step 3 Since 3 is prime, circle it. Then cross out all multiples of 3 that are not already crossed out.

Step 4 Circle 5 and cross out all multiples of 5.

Step 5 Circle 7 and cross out all multiples of 7. (The only one remaining is 49, the 7th multiple of 7.)

Step 6 The next possibility is 11, and it is prime. Circle 11. The multiples of 11 are 1(11), 2(11), 3(11), 4(11), 5(11), 6(11), 7(11), and so on. We are not interested in 7(11) since that exceeds 65. And clearly we would not cross out 1(11). All the other multiples of eleven have already been considered; 2(11) was taken care of in step 2, 3(11) was taken care of in step 3, 4(11) was taken care of in step 2, 5(11) was taken care of in step 4, and 6(11) was taken care of in step 2. Thus there are no multiples of 11 to cross out.

Step 7 The second, third, fourth, . . . multiples of 13, 17, 19, 23, 29, 31, 37, 41, 43, 47, 53, 59, and 61, have been crossed out because of reasoning similiar to step 6. Thus these numbers are also primes. All the primes up to 65 are now circled.

The process really ends at step 5. If we circle 7 and cross out all its multiples, all the numbers circled or not crossed out are the primes up to 65.

(2)	(3)	~~4~~	(5)	~~6~~	(7)	~~8~~	~~9~~	~~10~~	
(11)	~~12~~	(13)	~~14~~	~~15~~	~~16~~	(17)	~~18~~	(19)	
~~20~~	~~21~~	~~22~~	(23)	~~24~~	~~25~~	~~26~~	~~27~~	~~28~~	
(29)	~~30~~	(31)	~~32~~	~~33~~	~~34~~	~~35~~	~~36~~	(37)	
~~38~~	~~39~~	~~40~~	(41)	~~42~~	(43)	~~44~~	~~45~~	~~46~~	
(47)	~~48~~	~~49~~	~~50~~	~~51~~	~~52~~	(53)	~~54~~	~~55~~	
~~56~~	~~57~~	~~58~~	(59)	~~60~~	(61)	~~62~~	~~63~~	~~64~~	~~65~~

Figure 2 Finding primes up to 65

The essence of steps 5, 6, and 7 gives us a method for greatly reducing the work in determining whether or not a number is prime.

A method for testing whether or not a number n is prime:

1. Take the square root of n. If the square root is a natural number, n is not prime. So the test is complete.

2. If the square root of n is not a natural number, locate the square root of n between two consecutive natural numbers. Call the smaller of these two numbers x.

3. If x is prime, then the set of possible divisors is $\{\,2, 3, 4, \ldots, x\,\}$.

4. If x is not prime, then consider the next smaller number that *is* prime. Call it y. The set of possible divisors of n is $\{\,2, 3, 4, 5, \ldots, y\,\}$.

5. Clearly we need only investigate the primes in the lists $\{\,2, 3, 4, \ldots, x\,\}$ and $\{\,2, 3, 4, \ldots, y\,\}$ because the nonprimes would be multiples of smaller numbers.

6. The trial divisor list, then, always consists of the numbers $\{\,2, 3, 5, 7, 11, 13, \ldots\,\}$ up to either x or y as defined above.

EXAMPLE 4

Now we return to 571. Is 571 prime? We can calculate that $\sqrt{571} = 23.8^{+}$, so the square root of 571 is between 23 and 24. We take x to be 23. Since 23 is prime, we consider only the set $\{\,2, 3, 5, 7, 11, 13, 17, 19, 23\,\}$ as possible divisors of 571. After trying each of the nine divisions, we conclude that none of them is exact, so 571 is prime.

EXAMPLE 5

Is 581 prime? The square root of 581 is between 24 and 25. The x of our method is 24, but 24 is not prime. So we back up to 23 and consider all the primes $\{\,2, 3, 5, \ldots, 23\,\}$ as possible divisors. It doesn't take long to discover that $7 \mid 581$, so 581 is composite.

Is 4,294,967,297 prime? The process is impractical here, since the square root is between two 5-digit numbers. No one would want to divide each of the primes up to 65,536 into this large number to see if one of them is a factor. You might like to see if 641 is a factor, though.

Consideration of very large numbers should convince you that our method for determining primeness has its limitations. Mathematicians have tried to find a better way to decide whether a number is prime or composite, but to date none has been found. You can see why large prime numbers are fascinating. How are they discovered? We stated that $2^{19937} - 1$ is a prime. How can we find the next higher prime?

One way would be to try to develop formulas that would yield primes. Given such a formula, the search for primes is reduced to substituting in the formula. While the search for a formula is beyond the scope of this chapter, we will list two formulas that were thought to give only primes.

The first is $n^2 - n + 41$, where $n = 1, 2, 3, \ldots$. Some results are listed in Table 1. We leave it to the reader to try to find a factor of 6683, and perhaps to try to discover how we know that $82^2 - 82 + 41$ is not prime. In fact, there is a value of n smaller than 82 that yields a composite number.

Another formula that was thought to yield primes is $2^{(2^n)} + 1$. Table 2 shows the first few values of this expression. Leonard Euler (1707–1783), a Swiss mathematician, showed that the value of $2^{(2^n)} + 1$ when $n = 5$, namely 4,294,967,297, is not prime, since 641 is a factor. You might wonder what happens when n is greater than 5. No one has ever found another prime of the form $2^{(2^n)} + 1$, and it is now conjectured that none exist for n greater than 4.

Table 1

n	$n^2 - n + 41$	Prime or not?
1	41	prime
2	43	prime
3	47	prime
4	53	prime
5	61	prime
6	71	prime
7	83	prime
82	6683	not a prime

Table 2

n	$2^{(2^n)} + 1$	Value	Prime or not?
1	$2^2 + 1$	5	prime
2	$2^4 + 1$	17	prime
3	$2^8 + 1$	257	prime
4	$2^{16} + 1$	65,537	prime

Exercise Set 7.2

1 Write all the primes less than 100.

2 Is it true that the set of primes and the set of composites have no elements in common? Is it true that any natural number is either a prime or a composite?

3 In terms of the definition of a prime number, tell why 16 is not prime.

4 True or false:
   (a) $15 \mid 30$ (b) $30 \mid 15$ (c) $17 \mid 51$
   (d) $3 \mid 17$ (e) $19 \mid 1007$

5 Determine whether or not each of the following is a prime:
   (a) 1007   (b) 655   (c) 477
   (d) 101   (e) 677

6 Use the "sieve" method to find all the primes up to
   (a) 103   (b) 78

7 In determining whether or not a number is prime by the square root method, a certain set of primes are tested as divisors. What is that set of primes for
   (a) 91   (b) 171
   (c) 1523   (d) 487

8 (a) Does $n^2 - n + 41$ generate all the primes?
   (b) In that same formula, let $n = 41$. Is the result a prime?
   (c) Is it possible to choose a natural number n so that $n^2 - n + 41 = 571$?

9 (a) Evaluate the formula $2^{(2^n)} + 1$ for $n = 6$, using a desk calculator.
   (b) Solve $641x = 4{,}294{,}967{,}297$.

10 In one of the examples of this section we asked whether or not 4,294,967,297 is a prime, and mentioned in that connection the number 65,536. What does 65,536 have to do with the question?

11 Look at the list of all primes up to 100:

$$2, 3, 5, 7, 9, \ldots, 97$$
$$1 \ 2 \ 2 \ 2$$

Beneath and between consecutive primes write their difference. Do you think the differences (gaps) keep getting larger?

12 Determine whether 2081 is a prime. Do the same for 2083. Reconsider exercise 11. Mathematicians have proved that we can find two consecutive primes that differ by more than any number we can mention. But strangely enough, every so often there are two consecutive primes that differ by only 2. Such primes, like 3 and 5, 5 and 7, 59 and 61, are called *prime twins*. It has been conjectured, but not proved, that there are infinitely many pairs of prime twins. Is 89 and 91 such a pair?

13 (a) It is provable that the only trial divisors necessary, in checking whether or not a number less than 100 is prime, are 2, 3, 5, and 7. Why is this so?
    (b) What are the only primes needed as trial divisors for numbers less than 200? Less than 1000?

7.3 PRIME FACTORIZATIONS

When 12 is written as $2 \cdot 6$ we understand that 2 and 6 are factors of 12. Another factorization of 12 is $4 \cdot 3$. Of interest in this section is the *prime factorization* of composites, the factorization consisting of only prime numbers. The prime factorization of 12 is $2 \cdot 2 \cdot 3$. We can prove that it is always possible to factor a composite into primes: so many 2's, so many 3's, so many 5's, so many 7's and so on. While we will not go through the details of the proof, we state the following theorem:

Theorem 1   The Unique Prime Factorization Theorem   Every composite number can be written as a product of primes. The set of primes used in the factorization of a particular composite, as well as the number of repetitions of any prime, is unique.

This is the first time we have used the term "unique"; in mathematics it means "only one." So the proof of Theorem 1 would consist of two parts. First we would have to prove that a composite has a set of primes for factors, and second, we would have to prove that no other different set of primes, with possible repetitions, factors that composite.

EXAMPLE 1

Since $12 = 2 \cdot 2 \cdot 3$, the following is true: $\{\,2,3\,\}$ is the set of prime factors of 12 and $2 \cdot 2 \cdot 3$ is the prime factorization of 12.

The prime factorization of a composite number is written in *standard order* when the prime factors, including any repetitions, are written in increasing numerical order from left to right. Thus $2 \cdot 2 \cdot 3$ is in standard order but $2 \cdot 3 \cdot 2$ is not.

EXAMPLE 2

The following prime factorizations are written first in standard order, and then in more compact form, using "power" notation:

$$100 = 2(2)(5)(5) \qquad\qquad = 2^2 \cdot 5^2$$
$$84 = 2(2)(3)(7) \qquad\qquad = 2^2 \cdot 3 \cdot 7$$
$$32 = 2(2)(2)(2)(2) \qquad\qquad = 2^5$$
$$51 = 3(17)$$
$$24{,}750 = 2 \cdot 3 \cdot 3 \cdot 5 \cdot 5 \cdot 5 \cdot 11 \;= 2 \cdot 3^2 \cdot 5^3 \cdot 11$$

When a number is identified as a composite, it is usually easy to obtain its prime factorization. Examine the following procedure for finding the prime factorization of 24,750.

EXAMPLE 3

$$24{,}750 = 2(12375)$$
$$= 2(3)(4125)$$
$$= 2(3)(3)(1375)$$
$$= 2(3)(3)(5)(275)$$
$$= 2(3)(3)(5)(5)(55)$$
$$= 2 \cdot 3 \cdot 3 \cdot 5 \cdot 5 \cdot 5 \cdot 11$$

In the exercises we ask you to write a rule for systematically factoring a composite as in Example 3.

Another interesting way to factor a composite is to construct a "tree diagram" in which branches generate successive prime factors.

EXAMPLE 4

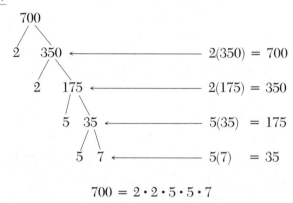

$$2(350) = 700$$
$$2(175) = 350$$
$$5(35) = 175$$
$$5(7) = 35$$

$$700 = 2 \cdot 2 \cdot 5 \cdot 5 \cdot 7$$

In mathematics we often want to deal with the prime factorization of numbers. For example, ordinary reduction of fractions to lowest terms is explained easily in terms of prime factorization. While we all know that 8/16 reduces to 1/2, how do we know 1/2 is the lowest term reduction of 1/2? To put it more simply, how do we know 1/2 cannot be reduced any further? The next examples show a systematic way to reduce fractions that guarantees that the answer is completely reduced.

EXAMPLE 5

Reduce 20/36 completely.
First factor 20 and 36 into primes.

$$20 = 2 \cdot 2 \cdot 5$$
$$36 = 2 \cdot 2 \cdot 3 \cdot 3$$

We write $\dfrac{20}{36} = \dfrac{2 \cdot 2 \cdot 5}{2 \cdot 2 \cdot 3 \cdot 3}$ and then cancel factors repeated in both the numerator and denominator:

$$\frac{20}{36} = \frac{5}{3 \cdot 3} = \frac{5}{9}$$

EXAMPLE 6

Completely reduce $\dfrac{14}{42}$.

$$\frac{14}{42} = \frac{2 \cdot 7}{2 \cdot 3 \cdot 7} = \frac{2 \cdot 7 \cdot 1}{2 \cdot 3 \cdot 7} = \frac{1}{3}$$

Note that when all the factors of the numerator cancel, it is convenient to introduce a factor of 1 to guard against losing the numerator. Also, if both numerator and denominator are prime, the fraction is completely reduced.

In reducing a number like 20/36 we can divide the numerator and denominator by 2, getting 10/18. This can be further reduced to 5/9. This suggests that a more efficient divisor of numerator and denominator would have been 4. It turns out that 4 is the *greatest common divisor* (gcd) of 20 and 36. A fraction can be completely reduced in one step by dividing numerator and denominator by the gcd. We find this special common divisor by looking at the prime factorizations.

EXAMPLE 7

Find the gcd of 60 and 36.
Again, we factor each number into primes:

$$60 = 2 \cdot 2 \cdot 3 \cdot 5$$
$$36 = 2 \cdot 2 \cdot 3 \cdot 3$$

Clearly the greatest common divisor is $2 \cdot 2 \cdot 3$ or 12. Thus a fraction like 36/60 is reduced completely to 3/5 merely by dividing numerator and denominator by 12:

$$36/60 = \frac{36/12}{60/12} = \frac{3}{5}$$

Here is a rule for finding the gcd of two (or more) given numbers: First, write the prime factorizations of each number. Then, the gcd is the largest number that can be formed by taking factors common to each of the factorizations.

EXAMPLE 8

Find the gcd of 180 and 252.
The prime factorizations of 180 and 252 are these:

$$180 = 2 \cdot 2 \cdot 3 \cdot 3 \cdot 5$$
$$252 = 2 \cdot 2 \cdot 3 \cdot 3 \cdot 7$$

Each of the numbers has 2, 2, 3, and 3 as factors. Thus the product 2(2)(3)(3) is the largest number that can be formed from common factors of 180 and 252. So the gcd of 180 and 252 is 36.

EXAMPLE 9

Find the gcd of 350, 490, and 130.

The prime factorizations are as follows:

$$350 = 2(5)(5)(7)$$
$$490 = 2(5)(7)(7)$$
$$130 = 2(5)(13)$$

It should be apparent that 2 and 5 are the only common factors of all three numbers. So the gcd of 350, 490, and 130 is $2(5) = 10$.

Another use of prime factorizations is in finding the lowest common denominator when adding fractions. For example, we know that 24 is the lowest common denominator when the fractions 1/3 and 3/8 are added. That is, 24 is the smallest number that 3 and 8 both divide exactly.

All the numbers 24, 48, 72, 96, . . . are divisible by both 3 and 8. The set { 24, 48, 72, 96, . . . } is the set of *common multiples* of 3 and 8, and 24 is the *least common multiple* (lcm). (The lcm is often called the lowest common denominator when fractions are added.)

It is evident that any number that 8 divides must have $2 \cdot 2 \cdot 2$ as a factor, and any number that 3 divides must have 3 as a factor. So the number $2(2)(2)(3) = 24$ is the smallest number divisible by 3 and 8. Thus we can find the lcm of two numbers from an inspection of prime factorizations.

EXAMPLE 10

Find the lcm of 15 and 35.
Factoring each into primes, we get $15 = 3(5)$ and $35 = 5(7)$.
A multiple of 15 must have a 3 and a 5 factor so that 15 divides the multiple, and a multiple of 35 must have a 5 and a 7 factor so that 35 divides the multiple. The number $3(5)(7)$ has the smallest number of 3's, 5's, and 7's so that 15 and 35 both divide a common multiple. Thus $3(5)(7) = 105$ is the lcm of 15 and 35.

EXAMPLE 11

Find the lcm of 8, 30, and 28. Factoring each number, we get

$$8 = 2(2)(2)$$
$$30 = 2(3)(5)$$
$$28 = 2(2)(7)$$

The lcm is $2(2)(2)(3)(5)(7) = 840$.
Notice that for 8 to divide a number there must be the factor $2(2)(2)$, for 30 to divide a number there must be the factor $2(3)(5)$, and for 28 to divide there must be the factor $2(2)(7)$. If any of the six factors of 840 were dropped, then one of the numbers 8, 30, and 28 would no longer divide the result.

Exercise Set 7.3

1 (a) What is the set of primes that factors 630?
   (b) What is the prime factorization of 630?

2 A certain composite is factored by primes in the set { 3, 7, 11 }. There are two
   3's, two 7's, and one 11 in the prime factorization. What is the number?

3 Find the prime factorization (by any method) of each of the following:
   (a) 162   (b) 288
   (c) 1000   (d) 243
   (e) 1024   (f) 89
   (g) 1,000,000   (h) 363

4 The answer to one part of exercise 3 is "none exists." Which part?

5 True or false: 64 | 1,000,000 [See exercise 3(g).]

6 True or false: $15625 = 5^5$ and 15625 | 1,000,000

7 All natural numbers other than 1 either are prime or can be factored uniquely
   into primes. What does this statement say about
   (a) 31?   (b) 32?
   (c) 1?   (d) 2?

8 Show the factorization of each number below by the tree diagram method.
   (a) 210   (b) 630
   (c) 875   (d) 4236
   (e) 512   (f) 1080

9 Completely reduce the following fractions:
   (a) $\dfrac{60}{420}$ (b) $\dfrac{57}{330}$

   (c) $\dfrac{89}{123}$ (d) $\dfrac{7}{126}$

   (e) $\dfrac{121}{165}$ (f) $\dfrac{54}{74}$

10 How do we know that 5/16 is in lowest terms?

11 Find the gcd of each of the following:
    (a) 60 and 210   (b) 32 and 48
    (c) 100 and 144   (d) 13 and 15
    (e) 23 and 69   (f) 17, and 289, and 85
    (g) 252 and 396   (h) 112 and 616

12 (a) Write down the first six common multiples of 8 and 3.
    (b) What is the least common multiple of 8 and 3?

°13 Two numbers are said to be *relatively prime* if they have no common divisor
     except 1. Thus 4 and 7 are relatively prime since they have no common factor
     other than 1. Find out whether or not each pair of numbers below is relatively
     prime.
     (a) 7, 84   (b) 15, 13
     (c) 19, 23   (d) 51, 17
     (e) 105, 300   (f) 15,625, 1,000,000

*14   True or false: A fraction is completely reduced when numerator and denominator are relatively prime.

15   Since the prime factorization of 12 is 2(2)(3), we can write *all* the divisors of 12 by selecting the factors in all possible ways, one at a time, two at a time, and three at a time. Of course, 1 is also a divisor. Thus the divisors of 12 are { 1, 2, 3, 4, 6, 12 }. Find all the divisors of

(a) 15   (b) 18   (c) 32
(d) 144   (e) 100   (f) $2^2 \cdot 3^3$

16   Suppose you know that 12 and 15 both divide a number x. What other divisors can you be sure that x has?

17   Find the lcm of each pair below:

(a) 14, 18   (b) 7, 9   (c) 12, 18
(d) 20, 42   (e) 24, 60   (f) 14, 21

18   True or false: Given two numbers, the greatest common divisor is no larger than the smallest of the given numbers.

19   True or false: The least common multiple of two numbers is no smaller than the larger of the two numbers.

20   True or false: If two numbers are prime, their least common multiple is never equal to their product.

7.4 NUMBERS WITH INTERESTING PROPERTIES

The primes and composites provide a rich source of theorems and conjectures. A conjecture is an idea suspected to be true, though no proof has been found. We might call conjectures "tentative theorems." In this section a few theorems and conjectures related to primes are investigated.

We have already mentioned *prime twins* (Exercise Set 7.2)—two consecutive primes whose difference is 2. If we write prime twins as ordered pairs, then (3, 5) would be the first pair of prime twins. A few others are

$$\{ (5, 7), (11, 13), (17, 19), (29, 31), (41, 43), \ldots \}$$

The dots would suggest that this is an infinite set, since no one has yet proved that there is no largest pair of prime twins. It is conjectured that the set of prime twins is infinite. Some larger pairs of the twin primes follow:

$$(599, 601), (2081, 2083), (2711, 2713),$$
$$(1{,}000{,}000{,}009{,}649, \ 1{,}000{,}000{,}009{,}651)$$

Many problems in number theory have to do with primes and addition. Fermat proved that "Every prime that is one more than $4n$, where n is a natural number, can be written uniquely as the sum of two squares."

EXAMPLE 1

The notation $4n$ is simply a way to write the multiples of 4. For various values of n we get $4, 8, 12, 16, 20, 24, 28, 32, 36, \ldots$. Adding one to each of these, we get $5, 9, 13, 17, 21, 25, 29, 33, 37, \ldots$, and the primes in this list are $5, 13, 17, 29, 37, \ldots$. Fermat's theorem says that each of these numbers can be written as a sum of squares:

$$5 = 1^2 + 2^2$$
$$13 = 2^2 + 3^2$$
$$17 = 1^2 + 4^2$$
$$29 = 2^2 + 5^2$$

The theorem further states that these sums are unique. For example, we find that the sum of the squares of 2 and 5 gives 29; if the sum is unique, 2 and 5 are the only numbers that, when squared and added, yield 29.

Some time ago there was a conjecture that every odd number greater than 3 might be written as the sum of a prime and some power of 2. Thus $5 = 3 + 2^1$, $7 = 3 + 2^2$, $9 = 5 + 2^2$, $11 = 3 + 2^3$, and so on. Notice that this conjecture does not say that the sum expressions are unique. In fact, $9 = 7 + 2^1$. No proof was found for this conjecture, and today it is known that the conjecture is false since the number 127 *cannot* be written as a sum of a prime and a power of 2. The easy way to check this is to investigate these subtractions:

$$127 - 2^1 = 125$$
$$127 - 2^2 = 123$$
$$127 - 2^3 = 119$$
$$127 - 2^4 = 111$$
$$127 - 2^5 = 95$$
$$127 - 2^6 = 63$$

None of the numbers on the right is a prime. Would the conjecture have worked for 129?

The Goldbach conjecture states that every even number greater than 4 can be expressed as a sum of two odd primes. Thus $6 = 3 + 3, 8 = 3 + 5$, $10 = 3 + 7, 12 = 5 + 7, 14 = 7 + 7$, and $14 = 11 + 3$. This remains a conjecture until an exception or a proof is found.

A similar conjecture (also due to Goldbach) for the sum of three primes states that "Every odd number greater than 5 can be written as the sum of three prime numbers." Thus $7 = 2 + 2 + 3, 9 = 3 + 3 + 3, 11 = 2 + 7 + 2, 13 = 5 + 5 + 3$, and so on. We emphasize that no proof has yet been found.

Many hundreds of hours have been spent trying to find a formula that will generate the primes. Although it is believed that no such formula exists, the search for primes and the test of a certain number for primeness might be simpler if we had a formula. A related question is, "How frequently do primes occur?" That is, as we progress through the natural numbers one by one (in order), how often can we expect to stumble across a prime?

The mathematician Gauss conjectured that the number of primes less than a given number, say x, is roughly

$$\frac{x}{1/1 + 1/2 + 1/3 + \ldots + 1/x}$$

Thus the number of primes less than 10 is approximately

$$\frac{10}{1/1 + 1/2 + 1/3 + 1/4 + 1/5 + 1/6 + 1/7 + 1/8 + 1/9 + 1/10}$$

$$= \frac{10}{2.928968} = 3.4$$

(The arithmetic was performed with a desk calculator.) The actual number of primes less than 10 is 4. Gauss's formula gives 5.56 as the approximate number of primes less than 20. The actual number is 8. if $x = 100$, the formula gives 19.28 and there are actually 25. You can see the importance of the word "roughly," but even though the formula does not give an exact number, it is useful to have an approximation. In 1896, about a hundred years after Gauss made this conjecture, a proof was published, and the conjecture become known as the Prime Number Theorem.

A related conjectural formula gives the number of *twin primes* less than x as

$$\frac{1.32x}{(1 + 1/2 + 1/3 + \ldots + 1/x)^2}$$

For $x = 10$, this gives 1.54. We know that $(3, 5)$ and $(5, 7)$ are the only prime twins less than 10.

It would seem that the primes are plentiful, but it is fairly easy to prove that we can have a "run" of consecutive composites as long as we wish. First, we explain !, the *factorial* symbol. In general $n!$ means to multiply together the first n counting numbers. Thus 3! means 1(2)(3), 4! means 1(2)(3)(4), 5! means 1(2)(3)(4)(5), and so on. Note that $n!$ has n factors.

Suppose we want to prove the existence of ten consecutive composites. All we have to do is write down the numbers 11! + 2, 11! + 3, 11! + 4, 11! + 5, 11! + 6, 11! + 7, 11! + 8, 11! + 9, 11! + 10, 11! + 11. The first one, 11! + 2, is divisible by 2 since 11! contains 2 as a factor and since 2 is divisible by 2. (See examples 2 and 3 for further explanation.) Likewise, 3 divides 11! + 3, 4 divides 11! + 4, 5 divides 11! + 5, and so on. We have created ten consecutive numbers which are composite.

EXAMPLE 2

Show that 2 divides 3! + 2.
3! + 2 can be written 1(2)(3) + 2, which is the same as 2(1 · 3 + 1), which is a multiple of 2 and therefore divisible by 2.

EXAMPLE 3

We know that 4 divides 6! + 4 since we can write 6! + 4 = 1(2)(3)(4)(5)(6) + 4 = 4(1 · 2 · 3 · 5 · 6 + 1), which is a multiple of 4.

EXAMPLE 4

Find seven consecutive composites. The answer can be 8! + 2, 8! + 3, 8! + 4, 8! + 5, 8! + 6, 8! + 7, 8! + 8 or 40,322; 40,323; 40,324; 40,325; 40,326; 40,327; 40,328. You should be able to find a factor of each of these numbers. Note that these seven consecutive composites are not unique since the numbers 90, 91, 92, 93, 94, 95, 96 also work. Another set of seven consecutive composites is 5042, 5043, 5044, 5045, 5046, 5047, 5048.

In general we can create n consecutive composite numbers by formulating all the numbers from $(n + 1)! + 2$ to $(n + 1)! + (n + 1)$.

The last set of numbers we shall mention are the so-called *perfect numbers*. A number is perfect if it is equal to the sum of all its divisors except itself. Thus the divisors of 6 are 1, 2, 3, and 6. Excluding 6 itself, we have 6 = 1 + 2 + 3. Another perfect number is 28, since 28 = 1 + 2 + 4 + 7 + 14. The first four perfect numbers are 6, 28, 496, and 8128. The following formula, known since the time of Euclid, generates perfect numbers:

$$2^{P-1}(2^P - 1)$$

provided that both P and $2^P - 1$ are prime. (A prime P, for which $2^P - 1$ is also prime, is called a *Mersenne prime*.)

EXAMPLE 5

Let $P = 7$. Then $2^7 - 1 = 127$, which is prime. So a perfect number is generated by $2^{7-1} \cdot (2^7 - 1) = 2^6 \cdot (127) = 64 \cdot (127) = 8128$.

No one has ever found an odd perfect number, and it is not known whether or not the set of perfect numbers is infinite. We close this section with the remark that if the sum of the divisors of a number n (excluding n itself) is less than n, the number is called *deficient*, and if the sum of the divisors is greater than the number, the number is called *abundant*. Thus all composite numbers are either perfect, deficient, or abundant. We do not include prime numbers in this statement because a prime's only divisor other than itself is 1.

EXAMPLE 6

The number 12 has divisors 1, 2, 3, 4, and 6. Furthermore, $1 + 2 + 3 + 4 + 6 = 16$, so 12 is abundant. 15 has divisors 1, 3, and 5, and $1 + 3 + 5 = 9$, so 15 is deficient.

Exercise Set 7.4

1 (a) Find two pairs of prime twins between 51 and 100.
    (b) Are 127 and 129 prime twins?
    (c) Are 137 and 139 prime twins?

2 (a) Show how 37 can be written as a sum of squares.
    (b) Do the same for 41, 53, 61, 73, and 89.
    (c) What are the primes less than 100 of the form $4n + 1$?

3 (a) Is it a theorem or a conjecture that every odd number greater than 3 can be written as a sum of a prime and some power of 2?
    (b) Review how we showed that 127 cannot be written as such a sum.
    (c) Investigate 129, 131, 133, 135, and 137 with regard to this same theorem or conjecture.

4 (a) Apply the first Goldbach conjecture we described to the following numbers: 16, 18, 20, 22, 144, 100.
    (b) In the statement of the Goldbach conjecture in the text, why do we use the phrase "greater than 4"?
    (c) Can an even number greater than 4 be expressed *uniquely* as a sum of two odd primes?

5 (a) Apply the second Goldbach conjecture to 15.
    (b) Are the primes you used in (a) necessarily unique?

6 (a) Write down a list of five consecutive composites of the form $6! + n$.
    (b) Next to each, write an obvious divisor.
    (c) Find two other lists of five consecutive composites less than 50.

7 (a) Indicate the procedure for finding 1000 consecutive composites.
    (b) What is a divisor of $1001! + 7$?
    (c) Is it possible to have a gap between consecutive primes as large as 1,000,000? As large as 1,000,000,000?

8 Can consecutive primes be infinitely far apart?

*9 In a library, find a proof due to Euclid that the set of primes is infinite. What assumption is made?

10 If consecutive primes can differ by more than a million, does this mean that the *next* prime is even "farther away"? (See questions 11 and 12, Exercise Set 7.2)

11 (a) Find all the divisors of 496. Hint: First find the prime factorization of 496.
    (b) Show that 496 is perfect.

12 Given that $8128 = 2(2)(2)(2)(2)(2)(127)$,
   (a) Find all the divisors of 8128. (There are 14.)
   (b) Show that 8128 is perfect.

°13 The first 17 Mersenne primes are

$$2, 3, 5, 7, 13, 17, 19, 31, 61, 89, 107, 127,$$
$$521, 607, 1279, 2203, \text{ and } 2281.$$

Using a desk calculator, find the perfect number $2^{P-1}(2^P - 1)$ generated by the Mersenne prime $P = 13$.

14 Determine whether each of these numbers is abundant or deficient:

$$10, 30, 39, 100, 144$$

°15 Write a brief report related to number theory on one of the following persons or topics:
   (a) Christian Goldbach   (b) Fermat's last theorem
   (c) Marin Mersenne   (d) Pythagorean triples
   (e) Carl F. Gauss

SUMMARY

Number theory, as a branch of mathematical study, is concerned with ideas interesting enough to intrigue everyone from the novice to the advanced research mathematician. Many of these ideas describe properties of arithmetic and geometric progressions.

Probably the most interesting part of number theory has to do with the prime numbers. Many properties of the primes continue to baffle mathematicians. We saw how to find all the primes up to a certain number and stated a method for testing whether or not a number is prime. But neither method is practical for testing the primeness of large numbers; to date no one has found a formula that gives all the primes.

Even if we know a very large prime number, we have no way of finding the next one. Yet we have known for about 2000 years that the set of primes is infinite. Beyond any large prime there must be another. This kind of thinking leads to questions about how frequently the primes appear in the natural numbers; that is, how many primes can we expect per hundred thousand natural numbers? We saw how to create a string of consecutive composites of any length, so there are gaps between consecutive primes that are as great as we care to have them. Yet the prime numbers 1,000,000,009,649 and 1,000,000,009,651 differ by two!

Aside from the interesting properties of the primes, we saw that they are useful in arithmetic. Reducing fractions and finding the lowest common denominator are better understood when we know about prime factorizations. The greatest common divisor (gcd) and the least common multiple (lcm) are also found by prime factorization. All natural numbers other than 1 either are primes or can be factored uniquely into primes. This is the essence of the prime factorization theorem.

Before attempting the Review Test, you should be familiar with the following:

1. Arithmetic progressions; the use of the formulas

$$a + (n-1)d \quad \text{and} \quad n[a + \frac{n-1}{2} \cdot d]$$

2. Geometric progressions; the use of the formulas

$$ar^{n-1}, \quad \frac{a(1-r^n)}{(1-r)}, \quad \text{and} \quad \frac{a}{(1-r)}$$

3. Expressing nonterminating but repeating decimals as sums of the terms of geometric progressions and finding the sum by formula.

4. The first ten or twelve primes in order.

5. Finding primes up to a given number by the sieve method.

6. The square root method for testing for primeness.

7. The divisibility notation $a \mid b$.

8. How to find the prime factorization of a composite number.

9. Reducing fractions by prime factorization.

10. Finding the gcd of two numbers.

11. Finding the lcm of two numbers.

12. Given the first Goldbach conjecture, applying it to some even number greater than 4.

13. Using the factorial idea to write a set of, say, eight consecutive composites.

REVIEW TEST

1 Consider the arithmetic progression

$$2, 7, 12, 17, 22, \ldots$$

(a) What is the sixth term?
(b) What is the common difference?
(c) What is the fiftieth term?
(d) What is the sum of the first ten terms?

2 Consider the geometric progression

$$3, 6, 12, 24, 48, \ldots$$

(a) What is the common ratio?
(b) What is the ninth term?
(c) What is the sum of the first nine terms?

3 Find the sum of all the terms in the geometric progression.

$$2, 2/3, 2/9, 2/27, \ldots$$

4 (a) Express 0.2222 . . . as the sum of the terms of a geometric progression.
(b) Find the sum, thus determining the fractional form of this repeating nonterminating decimal.

5 (a) The first prime is 2; the second is 3; the third is 5. What is the tenth prime?
(b) Is a formula known that gives the thousandth prime?

6 (a) What is the greatest prime trial divisor used in determining whether or not 673 is prime (by the square-root method)?
(b) What is the list of prime trial divisors used in showing whether or not 673 is prime?
(c) Is 673 prime?

7 (a) Show the prime factorization of 180.
*(b) Find all 18 exact divisors of 180.

8 (a) What is the gcd of 7350 and 147?
(b) What is the lcm of 7350 and 147?
(c) True or false: 147 | 7350.

9 Completely reduce 28/148, showing first the prime factorizations of 28 and 148.

10 Consider 3/8 + 5/42.
(a) Is 336 a common denominator?
(b) What is the least common denominator?
(c) Find the sum, in completely reduced form.

11 One of the Goldbach conjectures states that every even number greater than 4 can be expressed as a sum of two odd primes. Apply this conjecture to 28.

12 Using the factorial idea, write a set of six consecutive composites.

A mathematician may deal with numbers in a seemingly unemotional way. Yet many of us react to the *qualities* of certain numbers even though we would be quick to call our behavior irrational. Hidden in our psyche is a preference for one number over another. (Think of a number from one to four—for some mysterious reason most people pick three). Throughout history, beliefs and superstitions have often involved numbers. (Why do hotels rarely have a 13th floor?) In this article Evelyn Sharp informs us of some intriguing ancient beliefs about the first few counting numbers.

SOME NUMBER THEORY— ANCIENT SUPERSTITIONS AND UNSOLVED PROBLEMS

EVELYN SHARP

A good thing about the study of number theory is that it is concerned only with whole numbers. There are no decimals, no fractions, not even any negative numbers or zero, since the subject is based on the natural, or counting, numbers (1,2,3, etc.) which were used before any of these sophisticated adjuncts were invented.

Number theory goes back to very ancient times, and is rooted in numerology, just as astronomy began first as astrology. Though the Greeks called it "arithmetica," from their word "arithmos," meaning number, it is not the same branch of mathematics that we today know as arithmetic. *That* the Greeks called "logistica," meaning calculation, and they rather looked down their aristocratic noses at it.

What interested the Greeks was philosophy and the exercise of reason—let the slaves tend to the commerce, the trade, and all the necessary attendant computations. The surprising part is that their philosophy—especially that of the Pythagorean Society—was so imbued with beliefs about numbers.

Pythagoras, who had traveled in Egypt and Babylon absorbing both mathematics and mysticism, thought that numbers were the elements out of which everything else in the world is made, in much the same way that to other thinkers fire and water were the elements. He taught that number is the essence of reality and lies at the base of the real world.

About 530 B.C. his followers banded together at Croton, a Greek colony in southern Italy, into a sort of religious brotherhood devoted to the study of philosophy and mathematics, which to them were intertwined. They lived a rather ascetic life, seeking by rites and abstinences to purify the soul and free it from its fleshly prison, the body. Only men belonged, they took vows for life, and many practiced celibacy (though according to legend, Pythagoras himself married a young wife when he was past 60). Believing in reincarnation and the transmigration of souls, even into the body of an animal, they ate meat only on the occasion of a religious sacrifice.

The members of the order bound themselves with an oath not to reveal to outsiders the mathematical secrets they learned. Down through the centuries men have sworn by their gods and the stars, the bones of the saints and the beard of the prophet, but the Pythagoreans swore by a number. That number was ten—the holy tetractys.

Why did they pick ten? Because $1 + 2 + 3 + 4 = 10$. More clearly, the tetractys was represented by this triangular arrangement of ten dots, showing that ten is composed of 1, 2, 3 and 4.

$$\begin{matrix} & & \bullet & & \\ & \bullet & & \bullet & \\ \bullet & & \bullet & & \bullet \\ \bullet & \bullet & & \bullet & \bullet \end{matrix}$$

To them the first four numbers had a special meaning—they associated them with fire, water, air, and earth. Their sum therefore encompassed everything and stood for the ideal.

So persistent were they in this belief that they thought the whole universe must embody this principle. Their conviction that there *must* be ten heavenly bodies—although they could find only nine—led them to invent another one which they named the counterearth, explaining that it was always in the wrong part of the sky to be seen.

Their picture of the universe was of a central fire, or guiding force, around which revolved ten moving bodies. From the center outward, these were: the counterearth, the earth, moon, sun, the five planets known at the time (Mercury, Venus, Mars, Jupiter, Saturn), and lastly the stars, which they thought were all fixed to a single sphere and counted as one.

This conception was far advanced for the times and is much closer to the truth than the usual ancient (and medieval) idea of an earth which stood still, with all the other objects in the sky moving around it.

They identified the numbers with human characteristics, including a good bit of sex. The odd numbers were thought to be masculine, good, and celestial—the even numbers were feminine, evil, and earthly. There was a correspondence here with the Chinese, who also took the odd numbers to be bright, male and beneficent and the even numbers to be dark, female, and evil. (You notice that the women got the worst of it in all this ancient symbolism. Does that prove that women then were more malevolent than men? Of course not—it just proves that it was the men who invented the symbols.)

In addition, individual numbers had their own distinguishing traits. *One* stood for reason, and was not considered a true number, but rather as the source from which all the numbers were generated, by adding ones together. It was not classified as either even or odd.

Two represented opinion—wavering, indecisive, as today we say, "I'm of two minds about the matter."

Four was identified with justice, because it was the first number that is the product of equals, i.e., two times two. We still use "square" with this meaning (or did, before bebop changed the connotation)—"a square deal," "fair and square," "a square shooter."

Five was the marriage number, made from the union of *two*, the first even (feminine) number, and *three*, the first odd (masculine) number, since *one* didn't count.

Seven was a virgin number, because, of the first ten, it alone is neither factor nor product—that is, none of the others is divisible° by seven, nor can seven be divided evenly by any of them, without a remainder.

Don't laugh at all this. Underneath the numerology, the fundamental idea in Pythagorean philosophy was that only through number and form can man grasp the nature of the universe. Sir William Cecil Dampier, in *A History of Science* (1949), said that "Moseley with his atomic numbers, Planck with his quantum theory, and Einstein with his claim that physical facts such as gravitation are exhibitions of local space-time properties, are reviving ideas that in older, cruder forms, appear in Pythagorean philosophy." Alfred North Whitehead in *Mathematics as an Element in the History of Thought*, said, " . . . we have in the end come back to a version of the doctrine of old Pythagoras, from whom mathematics, and mathematical physics, took their rise."

°Divisible means exactly divisible, with no remainder.

8 AN INTRODUCTION TO ALGEBRA

Expressing mathematical problems in equations made up of letters, numbers, and the familiar equal sign, has been commonplace to mathematicians and scientists since the early sixteenth century. Historically the study of "algebra" is much older, having been introduced to the Arabs from the Hindus by the astronomer Alkarismi (or al-Khwarizmi) about A.D. 830. We know that the Italians learned of Alkarismi's algebra in the year 1202 and that the English first heard of it in 1557.

René Descartes (1596–1650) and François Vieta (1540–1603) introduced much of our present-day algebraic notation and usage. Power, or exponential, notation (like 3^2 or 10^3) was used freely by Descartes but was not seen in the literature much before his time. Descartes made many improvements in algebraic problem solving; he published his new ideas in a book called *Geometry*.

In this chapter we introduce some basic algebra and show a corresponding geometrical, or graphical, interpretation of algebra. The geometry of Descartes was quite different from Euclid's. In Descartes' geometry (called "analytic" or "coordinate" geometry) numbers are associated with points, and algebraic equations correspond to lines, circles, and other geometric figures. Given an equation, figures could be constructed by plotting points whose "coordinates" satisfied the equation. Euclidean geometry deals with properties of the figures themselves, without corresponding equations. Its constructions are restricted to those possible by ruler and compass only.

In coordinate geometry the graph serves as a "visual aid" for algebra. For example, a certain equation may have a graph that is a circle. Any geometric property of the circle is instantly apparent from the graph. Such properties may be more difficult to ascertain from the algebra alone.

In this introduction to algebra, our first concern will be for equations. We will see what is meant by "solving an equation" and will show how to get the geometric interpretation of equations. We will see why mathematicians so often sketch graphs in order to have a clearer picture of algebraic problems. Although Sections 8.6, 8.7, and 8.8 are optional, they deal with the interesting topic of inequalities. Again we will emphasize the graphical approach.

8.1 THE SYMBOLS OF ALGEBRA

Before the late sixteenth and early seventeenth centuries, the notation of algebra was awkward because of the lack of convenient symbolism. Problems were frequently stated in verbal form (as they are today), but there was no way to write them succinctly and concisely. Consider the problem in Example 1.

EXAMPLE 1

When I am 125 years old you will be 100. If you are now 20, how old am I?
Verbal solution. The first sentence tells you I am 25 years older than you.
The second says that you are now 20, so I must now be 20 plus 25, or 45
years old.
Algebraic solution. Let x stand for my age, and y stand for your age. The
first sentence implies that $x - y = 25$. The second sentence asks for x when
y is 20. So we simply solve $x - 20 = 25$. This equation has the solution 45.

The strength of algebra is that once we have reduced a problem to an
equation, we can deal with the equation by a system of rules and not worry
about the words in the problem.

We ordinarily use letters from the alphabet to stand for unknowns in
equations; we call these letters *variables*. In the equations $x + y = 4$ and
$3a^2 + 2a - 1 = 0$, the variables are x, y, and a.

An equation containing unknowns is a complete sentence containing
variables and certain *constants* or numbers, and an "$=$" sign. The equal
sign is the verb of the sentence. Thus $2x = 6$ is an equation with variable x
that might be read "twice a certain number is six." In the examples below
we see the algebraic translation of some English sentences.

EXAMPLE 2

"Seven less than twice a number is five" is written algebraically as

$$2n - 7 = 5$$

Here the variable is chosen to be n.

EXAMPLE 3

"The square of the length of the hypotenuse of a right triangle is equal to
the sum of the squares of the lengths of the other two sides." Given the
sketch below, the algebra is easy:

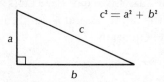

This might become an equation to be solved if, say, a is given as 3 and b is
given as 4, and we want to find c.

Solving an equation means finding numbers that, when substituted for a
variable (or for the variables) make the equation into a true statement. In
Sections 8.2 and 8.3 we will formalize this idea. For now we will use
inspection or organized guesswork in order to solve equations.

EXAMPLE 4

Solve $x - 2 = 7$. Because we are familiar with subtraction, we know by quick mental work that $9 - 2 = 7$, so x must stand for 9 here. We say that the solution to $x - 2 = 7$ is 9.

EXAMPLE 5

Solve $2n - 7 = 5$. It is easy to see that $2n$ has to be 12, since $12 - 7 = 5$. If $2n$ is 12, then n is 6, since $2(6) = 12$. Since $2(6) - 7 = 5$ is a true statement, the solution is 6.

The equation $2n - 7 = 5$ can also be solved by trial and error, where we simply try possible values for n. We could make a table (Table 1) to organize the work. Thus n is 6, since that value makes $2n - 7 = 5$ a true statement. If the tabulation were to continue, other choices for n *might* be found to make $2n - 7 = 5$ a true statement. At least we know that 6 is *one* solution.

Table 1   Solving an equation by trial and error

n	$2n - 7 = ?$	Comment
1	$2(1) - 7 = -5$	too low
2	$2(2) - 7 = -3$	too low
3	$2(3) - 7 = -1$	too low
4	$2(4) - 7 = 1$	too low
5	$2(5) - 7 = 3$	too low
6	$2(6) - 7 = 5$	correct

EXAMPLE 6

Two numbers have a sum that is 12 and a difference that is 2. What are the numbers? Here we have two facts about two numbers. If we let x stand for one of the numbers and y stand for the other, we can write each fact as a separate equation:

$$\begin{cases} x + y = 12 \\ x - y = 2 \end{cases}$$

Experimentation would show that the numbers are 7 and 5. Observe that we were looking for a pair of numbers (x, y) that, when substituted in the equations, make both of them true statements. The solution can be written as $(7, 5)$, using ordered-pair notation.

Obviously certain equations may be too difficult to solve by guesswork, or it may take too long to use a process of organized experimentation.

Formal algebra allows us to apply rules that make the search for solutions straightforward and mechanical. We will discuss such rules in the next several sections.

Exercise Set 8.1

1 Express the following equations in words:
   (a) $3x + 2 = 11$
   (b) $\frac{x}{7} = 2$
   (c) $2x + 3 = x + 4$
   (d) $7 - x = 4$
   (e) $x - 7 = 4$

2 Translate the following sentences into their algebraic equivalents:
   (a) Twice a certain number is three.
   (b) Five less than three times a number is ten.
   (c) One more than twice a number decreased by the number is nine.
   (d) One more than a number is nine.
   (e) Twice a number decreased by half the number is six.

3 (a) Solve the equation in exercise 1(a) by thinking about the "three times table" (multiples of three).
   (b) In 1(b) we can find the solution because we know the multiples of 7. What is the solution?
   (c) Is 7 the solution of 1(c)? Is 6? Is 5? Substitute 7, 6, 5, 4, 3, ... into the equation until you find the solution; tabulate your results.
   (d) Solve 1(d) and 1(e).

4 (a) If you must draw numbers from $N = \{1, 2, 3, 4, \ldots\}$, can you find a solution to the equation you got in exercise 2(a)?
   (b) Find all the numbers referred to in question 2.

5 What numbers in N satisfy $2n - n = n$?

6 (a) If you multiply any number by 3 and then divide it by 3, what do you get?
   (b) Express the idea of 6(a) algebraically.

7 John's age is three years more than twice Ed's. If John is 37, how old is Ed? First, solve this problem by trial and error, tabulating the results as in Table 1. Then express the problem as an equation to be solved, letting x be Ed's age.

8 You have $10.00 to pay for a meal in a restaurant. If you leave $1.00 tip and have $3.50 left, how much did the meal cost? Solve this problem mentally, and also write an equation whose solution will give the result. Check your mental solution in the equation.

9 You remember that your average grade for a class is 88 and that two of your grades were 80 and 94. If there were three grades, what is the grade you forgot?

8.2 THE SOLUTION SET AND THE REFERENCE SET

If a simple equation like $2x = 6$ is to be solved, we are looking for a number such that twice the number is 6. Clearly the number is 3. We will call 3 the "solution" and will invent a set called the "solution set" to contain the 3.

DEFINITION 1   Solution Set

The *solution* to an equation is a set of numbers that, when substituted for the variables in the equation, make the equation a true statement. The set of all solutions is called the *solution set* and is designated by the letter S.

The only number that makes $2x = 6$ a true statement when substituted for x is 3, so the solution set for $2x = 6$ is $S = \{3\}$.

Apparently a solution set *excludes* all numbers that do not make the equation a true statement. The set of all excluded numbers would be the complement of S. If you look up the definition of "complement set" in Chapter 2, you will find that we first need the notion of the universal set, which contains all the elements under consideration. Thus if we are to conceive of the set of excluded numbers when solving equations, we have to know what the universal set is for the problem. In algebra, the universal set is usually called the *reference set*.

The instructions "Solve $2x = 6$" are therefore incomplete. What are the allowable numbers that might belong to the solution set? Suppose we have only the set of even numbers to work with; then there can be no solution to $2x = 6$ since there is no even number that doubled is 6.

DEFINITION 2   Reference Set

The *reference set*, R, is the set of numbers available to test as possible solutions of a particular equation.

EXAMPLE 1

Solve $2x = 6$ with $R = \{2, 4, 6, 8\}$. Since $2(2) = 6$, $2(4) = 6$, $2(6) = 6$, and $2(8) = 6$ are all false statements, we conclude that the solution set is empty. We can write, $S = \{\ \ \}$ or $S = \emptyset$.

EXAMPLE 2

Solve $2x = 6$ with $R = \{ \ldots, -3, -2, -1, 0, +1, +2, \ldots \}$, that is, with $R = I$, the set of integers. Here a choice of the integer $+3$ makes a true statement. By intuition we know that no other integer is a solution, so $S = \{ +3 \}$.

EXAMPLE 3

Solve $2x - x = x$ with $R = \{ 1, 2, 3, 4, \ldots \}$. We try the elements of R in turn:

$$2(1) - 1 = 1$$
$$2(2) - 2 = 2$$
$$2(3) - 3 = 3, \text{ and so on}$$

Since every such statement is true, we conclude that $S = \{ 1, 2, 3, 4, \ldots \}$; that is, $S = R$. This equation has infinitely many solutions.

EXAMPLE 4

Solve $x^2 = 4$ with reference set equal to all the integers (positive, negative, and 0). Is it clear that $S = \{ +2, -2 \}$?

Sometimes we are given equations in more than one variable. An example might be $2x + y = 5$, where x and y can be any real numbers. The choice of x as 1 and y as 3 makes a true statement. We write such a solution as an ordered pair, where we choose the first element to be the x value and the second element to be the y value. Thus $(1, 3)$ is a solution, since $2(1) + 3 = 5$ is a true statement. The solution set is an infinite set of ordered pairs:

$$S = \{ (1, 3),(0, 5),(1/2, 4),(1/3, 13/3),(5, -5), \ldots \}$$

In the example of the last paragraph, $2x + y = 5$, the reference set is the set of reals. The set of reals is used so frequently as a reference set that we make the following agreement:

> *Unless otherwise stated, the reference set will be the set of reals.*
> (The reals include the integers, rationals, and irrationals.)

EXAMPLE 5

Solve $5x - 2 = 28$. After a bit of experimentation you should see that 6 is the only number that you can multiply by 5, then subtract 2 to get 28. Thus, $S = \{ 6 \}$. Note that the set of real numbers is the assumed reference set.

Note that we have defined solutions to be numbers and solution sets to be sets of numbers. Thus in Example 5 it would be incorrect to write the solution as the equation $x = 6$. Equations are not numbers.

Exercise Set 8.2

1 Solve $3x = 9$ with reference set
(a) $R = \{1, 2, 3, 4, 5, \ldots\}$.   (b) $R = \{2, 4, 6, 8, \ldots\}$.
(c) $R = \{-1, -2, -3, \ldots\}$.   (d) $R = \{x \mid x \text{ is a multiple of 5}\}$.
(e) $R = \{x \mid x \text{ is a multiple of 3}\}$.

2 If we say that $S = \{3\}$ is the solution set corresponding to $4x = 12$, what reference set is implied?

3 (a) Find six ordered pairs satisfying $2x + 3y = 12$.
(b) Can you write *all* the ordered pairs satisfying $2x + 3y = 12$?

4 (a) Solve $7x - 8 = 27$ if $R = \{1, 2, 3, 4, 5\}$.
(b) What subset of R is the solution set?
(c) What subset of R is excluded by the solution set?

5 $R = \{-4, -2, 0, +2, +4\}$. Find S if $S = \{x \mid x^2 = 16\}$.

6 (a) Solve $2x^2 = 32$.
(b) Solve $2x^2 - 4 = 46$. (What must $2x^2$ be first?)

7 Given $\{(x, y) \mid x - y = 3, \text{ and } x \text{ and } y \text{ are counting numbers}\}$.
(a) What equation is to be solved here?
(b) What is the reference set?
(c) Write three more ordered pairs belonging to the solution set

$$S = \{(4, 1), (5, 2), (6, 3), (7, 4), \ldots\}.$$

(d) Does $(3,0)$ belong to the solution set?

8 Solve $3w - 4z = 5$, listing a few of the ordered pairs of the form (w, z) in the solution set.

9 How might we list solutions to $x + y + z = 12$?

10 Solve $x = 7$.

8.3   RULES FOR SIMPLIFYING EQUATIONS

Often inspection or guesswork techniques prove too difficult for solving equations. Historically these techniques *were* employed; numbers were successively tried as solutions to verbal problems until the answer was found. When modern efficient notation was invented, systematic techniques for solving equations were also developed. Consider the following sequence of steps:

$$
\begin{align}
2x - 6 &= 4 \tag{1}\\
2x &= 10 \tag{2}\\
x &= 5 \tag{3}\\
S &= \{5\} \tag{4}
\end{align}
$$

What we have shown is a transformation of $2x - 6 = 4$ into succeedingly simpler equations until we arrive at the equation $x = 5$, which has the simple solution set $S = \{5\}$. But what does $x = 5$ have to do with $2x - 6 = 4$? It turns out that $2x - 6 = 4$ has the same solution set as $x = 5$.

DEFINITION 3   Equivalent Equations

Two equations are *equivalent* if they have the same solution set.

Notice that $2x - 6 = 4$, $2x = 10$, and $x = 5$ all have the solution set $S = \{5\}$. They are therefore equivalent equations. If we replace an equation by one equivalent to it, the new equation may be easier to solve. While $2x - 6 = 4$ is not difficult to solve, $2x = 10$ is easier, and in fact, $x = 5$ is easier yet. We list below rules for transforming an equation into an equivalent equation.

RULE 1

If the same quantity is added to, or subtracted from, both sides of an equation, the resulting equation is equivalent to the first. In symbols, if $a = b$, then $a + c = b + c$ and $a - c = b - c$.

RULE 2

If both sides of an equation are multiplied by the same quantity, the resulting equation is equivalent to the first. In symbols, if $a = b$, then $ac = bc$.

RULE 3

If both sides of an equation are divided by the same number, the resulting equation is equivalent to the first. Division by zero is not allowed. In symbols, if $a = b$, then $a/c = b/c$, provided that $c \neq 0$.

EXAMPLE 1

Solve $2x - 3 = 5$.

$$
\begin{array}{llr}
2x - 3 = 5 & & (1) \\
2x = 8 & \text{(Adding 3 to both sides; Rule 1)} & (2) \\
x = 4 & \text{(Dividing both sides by 2; Rule 3)} & (3) \\
S = \{5\} & & (4)
\end{array}
$$

EXAMPLE 2

Solve $\frac{1}{2}x + 4 = 7$.

$$\frac{1}{2}x = 3 \qquad \text{(Subtracting 4 from both sides; Rule 1)} \qquad (1)$$
$$x = 6 \qquad \text{(Multiplying both sides by 2; Rule 2)} \qquad (2)$$
$$S = \{6\} \qquad (3)$$

An equation like $ax + b = c$, where the a, b, and c are considered to be any given real numbers (except $a \neq 0$) and x is the variable, can be solved by applying the basic rules. The result is a *formula solution*:

$$ax + b = c \qquad (1)$$

$$ax = c - b \qquad \text{(Subtracting } b \text{ from both sides)} \qquad (2)$$

$$x = \frac{c - b}{a} \qquad \text{(Dividing both sides by } a) \qquad (3)$$

$$S = \left\{ \frac{c - b}{a} \right\} \qquad (4)$$

We can use the result to get the solution to any equation of the form $ax + b = c$ simply by identifying a, b, and c and substituting in the formula

$$S = \left\{ \frac{c - b}{a} \right\}$$

EXAMPLE 3

Solve $3x + 4 = 7$ by using the formula just developed. Here $a = 3, b = 4$,

and $c = 7$. So $S = \left\{ \frac{(7 - 4)}{3} \right\} = \{1\}$.

We don't wish to overemphasize this formula solution, since it is so easy to solve an equation like $3x + 4 = 7$ by applying the basic rules. However, mathematicians do look for formula solutions to general equations. For more complicated equations the formula may be easier to use than any other technique. Furthermore, the formula provides a way to study all equations to which it applies. For example, given $ax + b = c$, we know a cannot be 0, since it is in the denominator of the fraction given by the formula solution. We cannot divide by 0.

Each time we solve an equation, it is possible to graph the solution set. If the reference set is the set of all real numbers, then the solution set is a subset of the reals. We can sketch a real number line and "plot" the points on it corresponding to the solution. Thus solving an equation can be interpreted geometrically.

EXAMPLE 4

Solve $2x + 3 = 7$ and graph the solution set. We have

$$2x + 3 = 7$$

Then

$$2x = 4 \qquad \text{(Subtracting 3 from both sides)}$$

and

$$x = 2 \qquad \text{(Dividing both sides by 2)}$$

So

$$S = \{2\}$$

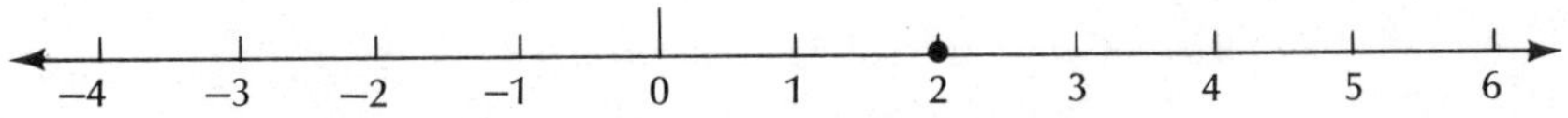

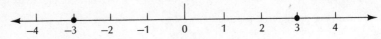

Figure 1   The graph of the solution to $2x + 3 = 7$.

The real number line is sketched in part (Figure 1) and then the solution set $S = \{2\}$ is shown; here it is simply a dot on the line. The graph makes it easy to visualize the numbers that satisfy the equation and those that do not. This technique will assume more importance when we analyze more difficult equations.

Another kind of equation we can solve now is one that has the square of the variable equal to a constant. An example is $x^2 = 9$. There are two numbers that satisfy this equation, namely $+3$ and -3. There were other examples like this in Section 8.2. Another rule will make the procedure a bit more formal:

RULE 4

If $a^2 = b^2$, then $a = b$ or $a = -b$.

EXAMPLE 5

Solve $3x^2 - 5 = 22$ and show a graph of the solution set. We have

$$3x^2 - 5 = 22$$

Then

$$3x^2 = 27 \qquad \text{(Adding 5 to both sides)}$$

and

$$x^2 = 9 \qquad \text{(Dividing both sides by 3)}$$

$$x = 3 \quad \text{or} \quad x = -3 \quad \text{(Rule 4)}$$

$$S = \{3, -3\}$$

The graph of S is shown in Figure 2 as two dots on the real number line.

Figure 2   The graph of the solution to $3x^2 - 5 = 22$.

Exercise Set 8.3

(In each of the following, assume the reference set to be the real numbers, unless otherwise stated.)

1 Solve each equation below, justifying each step in the solving process, and graph the solution set.

(a) $12x - 55 = 5$　　　(b) $7x = 0$　　　(c) $4x - 5 = -3$

(d) $3x - \pi = 2\pi$ (π is an irrational number approximated by 3.14159)

(e) $8x + 8 = -24$　　　(f) $-3x + 5 = 4$　　　(g) $2x - 2 = 5$

2 (a) Solve $2x^2 - 23 = 2$, and graph the solution set.

(b) Do the same for $2x^2 + 1/2 = 1$.　　　(c) Solve $3y^2 + 100 = 400$.

3 (a) Graph the solution set for $2x - x = x$.

(b) Do the same for $5 + x = x + 5$.

(c) Can we graph the entire solution set for these equations?

8.4 GRAPHING SOLUTION SETS IN A PLANE

When an equation involves two variables it is convenient to represent solutions as ordered pairs. Although we cannot graph ordered pairs on a real number line, we can plot them on a plane. We now set up a system of addresses on a plane so that we can find a unique point for each ordered pair (Figure 3).

First we draw a horizontal real number line; this is labeled X in Figure 3 and is called the X *axis*. Then at 0 we construct a perpendicular and set up another real number line; this is labeled Y and called the Y *axis*. The intersection of the X axis and the Y axis at 0 is called the *origin*. The plane is now separated into four regions called *quadrants*. The regions are referred to as quadrants I, II, III, and IV as labeled in Figure 3.

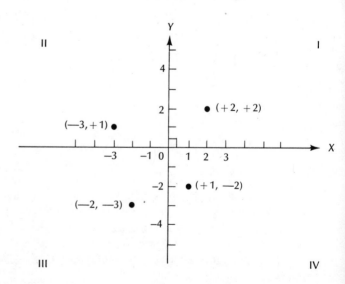

Figure 3
Locating points in a plane.

Ordered pairs of the form (x, y) can be graphed or plotted on the plane. When plotting such ordered pairs we call x the *abscissa* and y the *ordinate*. Together the abscissa and ordinate are the *coordinates* of a point.

The coordinates of a point are, in effect, a unique "address" for that point. Thus, $(+2, +2)$ refers to only one location in quadrant I; it is 2 units to the right of the Y axis and 2 units above the X axis. Also, $(-3, +1)$ is the "address" of only one point in quadrant II, namely, 3 units to the left of the Y and 1 unit above the X axis. Likewise, $(-2, -3)$ refers to a point 2 units to the left of the Y axis and 3 units below the X axis (quadrant III). And $(+1, -2)$ is a point in quadrant IV, 1 unit to the right of the Y axis and 2 units below the X axis. These points are all plotted in Figure 3.

The scheme for locating points is simple. If (a, b) is the point to be plotted, we start at the origin and move a units to the right if a is positive, to the left if a is negative. We then move up b units if b is positive, or down b units if b is negative.

The abscissa is measured right and left from the Y axis and the ordinate is measured up and down from the X axis. Thus, points on the X axis itself have an ordinate that is 0; they are of the form $(c, 0)$, where c is the distance from the origin on the X axis. Points on the Y axis have an abscissa that is 0; they are of the form $(0, a)$, where a is the distance from the origin on the Y axis.

What are the coordinates of the origin itself?

This system, called the rectangular coordinate system, (or cartesian coordinate system, after René Descartes), establishes a one-to-one correspondence between the set of all possible ordered pairs (x, y), where x and y are real numbers, and the set of points in a plane. Thus to every ordered pair of real numbers there corresponds a point, and to every point there corresponds an ordered pair. Now we can plot the graph of the solution sets to an equation in two variables.

EXAMPLE 1

Find and graph the solution set for $x + y = 4$, where the reference set for each of x and y is $R = \{1, 2, 3, 4\}$. We might tabulate the possibilities for x and y as follows:

		y			
		1	2	3	4
	1	(1, 1)	(1, 2)	(1, 3)	(1, 4)
x	2	(2, 1)	(2, 2)	(2, 3)	(2, 4)
	3	(3, 1)	(3, 2)	(3, 3)	(3, 4)
	4	(4, 1)	(4, 2)	(4, 3)	(4, 4)

From these 16 ordered pairs, we must choose those that make $x + y = 4$ a true statement. We substitute each pair, in turn, into the equation.

For $(1,1)$ we get $1 + 1 = 4$, which is not a true statement.
For $(1, 2)$ we get $1 + 2 = 4$, which is not a true statement.
For $(1, 3)$ we get $1 + 3 = 4$, which is a true statement.

Continuing through all 16 possibilities, we arrive at the solution set $S = \{ (1, 3), (2, 2), (3, 1) \}$. The graph is shown in Figure 4:

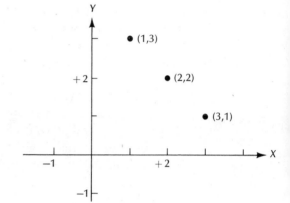

Figure 4
Graph of the solution set
for Example 1.

If the reference set were the set of all real numbers, we would have to consider all possible ordered pairs of reals; this infinite set cannot be tabulated. A slightly different graphing procedure can be used, as we show in the following example.

EXAMPLE 2

Graph the solution set for $x + y = 4$, where each of x and y can be real numbers. The method will be to allow x to take on a few values in turn, and solve for y in each case. We present this work in Table 2.

The graph of the seven ordered pairs calculated so far is shown in Figure 5. The points appear to lie on a straight line.

Table 2   Calculating ordered pairs

$x =$	$x + y = 4$	$y =$
-1	$-1 + y = 4$	5
0	$0 + y = 4$	4
½	$½ + y = 4$	3½
1	$1 + y = 4$	3
1.2	$1.2 + y = 4$	2.8
2	$2 + y = 4$	2
3	$3 + y = 4$	1

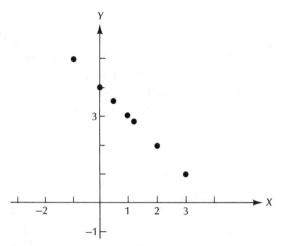

Figure 5   Graphing points whose coordinates satisfy $x + y = 4$.

If we had taken all real possibilities for x (impossible), and solved for y we would have found that all the plotted points fall along the same straight line. It is provable that the graph in a plane of any equation of the form $Ax + By = C$ (where A, B, and C are any real numbers, not all zero) is a straight line. It is also true that every straight line is represented algebraically by $Ax + By = C$. Thus the algebraic counterpart of a geometric line is $Ax + By = C$. For this reason we call this kind of equation a *linear equation*. Since two points determine a line, we need only plot two ordered pairs of the solution set to get the graph of the solution.

EXAMPLE 3

Graph the line $x + y = 4$. (See Example 2 and Table 2.) The points $(2, 2)$ and $(0, 4)$ are convenient points to plot. The line they determine is graphed in Figure 6.

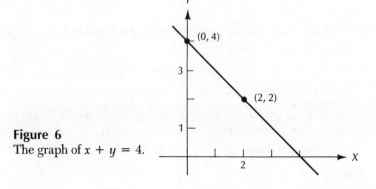

Figure 6
The graph of $x + y = 4$.

EXAMPLE 4

Graph the line $2x - y = 4$. It is easy to see that if $x = 3$, then $y = 2$, and if $x = 2$, then $y = 0$. Thus the solution set contains $(3, 2)$ and $(2, 0)$, as well as many other points. These two points are sufficient to get the graph, which we show in Figure 7.

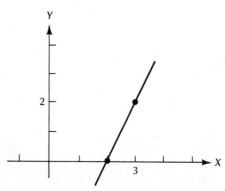

Figure 7   The graph of $2x - y = 4$.

The procedure for finding the ordered pairs satisfying a given equation is a simple one. Choose a first element and solve algebraically for the second element. Thus in $2x - y = 4$ we might let x be 4, getting the following:

$$2(4) - y = 4 \qquad \text{Substituting 4 for } x \tag{1}$$
$$8 - y = 4 \tag{2}$$
$$- y = -4 \qquad \text{Why?} \tag{3}$$
$$y = 4 \qquad \text{Why?} \tag{4}$$

Thus $(4, 4)$ is in the solution set to $2x - y = 4$ and is a point of the graph. We can repeat the above procedure (choosing x and solving for y) for any choice of x in the Real numbers. The solution set for $2x - y = 4$ is infinite; a few ordered pairs are given below:

$$(2, 0), (3, 2), (4, 4), (5, 6), (0, -4), (1, -2)$$

All these points can be found on the graph in Figure 7. (It may need to be extended.)

The procedure for finding points also applies to equations whose graph is not a straight line.

EXAMPLE 5

Sketch the graph of $x + y^2 = 9$. Again we choose x and solve for y. Note our use of the rule "If $a^2 = b^2$, then $a = b$ or $a = -b$" in the calculations that follow (Table 3).

Table 3   Finding points for $x + y^2 = 9$

x	$x + y^2 = 9$	y	Plot		
0	$0 + y^2 = 9,\ y^2 = 9$	± 3	$(0, 3)$	and	$(0, -3)$
9	$9 + y^2 = 9,\ y^2 = 0$	0	$(9, 0)$		
5	$5 + y^2 = 9,\ y^2 = 4$	± 2	$(5, 2)$	and	$(5, -2)$
-7	$-7 + y^2 = 9,\ y^2 = 16$	± 4	$(-7, 4)$	and	$(-7, -4)$

We plot the seven pairs found to belong to the solution set and draw a smooth curve through them (Figure 8). This kind of curve is called a *parabola*.

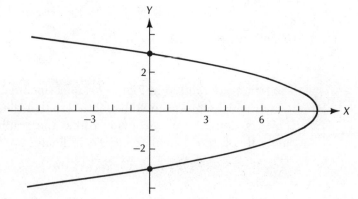

Figure 8   The graph of the parabola $x + y^2 = 9$.

In Example 5 we chose x carefully so that y would be an integer and easily solved for. Of course, we could choose y first, substitute its value in the equation, and then solve for x. You might like to let y be $+1$ and -1; what is x? Are these points on the graph?

EXAMPLE 6

Sketch the graph of $x^2 + y^2 = 25$. Again we tabulate the calculations (Table 4).

Table 4   Finding points for $x^2 + y^2 = 25$

x	$x^2 + y^2 = 25$	y	Plot		
0	$0 + y^2 = 25;\ y^2 = 25$	± 5	$(0, +5)$	and	$(0, -5)$
± 5	$25 + y^2 = 25;\ y^2 = 0$	0	$(+5, 0)$	and	$(-5, 0)$
3	$9 + y^2 = 25;\ y^2 = 16$	± 4	$(3, +4)$	and	$(3, -4)$
-3	$9 + y^2 = 25;\ y^2 = 16$	± 4	$(-3, +4)$	and	$(-3, -4)$
4	$16 + y^2 = 25;\ y^2 = 9$	± 3	$(4, +3)$	and	$(4, -3)$,
-4	$16 + y^2 = 25;\ y^2 = 9$	± 3	$(-4, +3)$	and	$(-4, -3)$

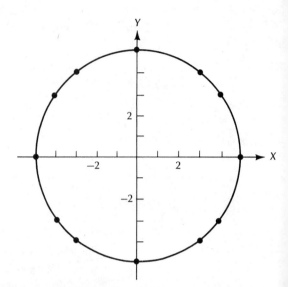

Figure 9
Graphing $x^2 + y^2 = 25$.

The points are plotted in Figure 9, and a smooth curve is drawn through them. Enough points should be plotted so that the shape of the curve becomes obvious. Are you convinced that the graph of $x^2 + y^2 = 25$ is a *circle*? If not, then calculate more coordinates satisfying the equation.

Exercise Set 8.4

1 Plot the following points on one set of coordinate axes:
   (a) $(1, 1)$ (b) $(1, 5)$ (c) $(-1, 4)$ (d) $(-3, -2)$
   (e) $(-2, 5)$ (f) $(6, 3)$ (g) $(+3, -5)$ (h) $(0, 0)$
   (i) $(2, -3/2)$ (j) $(-5/2, -7/3)$

2 The Fundamental Principle of Counting (see page 40) assures us that there are only four quadrants in the plane. Explain. Hint: How many choices for signs are there for an abscissa?

3 Where, in the plane, is a point
   (a) whose abscissa is positive and whose ordinate is negative?
   (b) whose abscissa is negative and whose ordinate is negative?
   (c) whose abscissa is negative and whose ordinate is positive?
   (d) whose abscissa is zero?
   (e) whose ordinate is zero?

4 Sketch lines that meet the following conditions:
   (a) The graph is the set of points all of whose abscissas are 3.
   (b) The ordinate always equals the abscissa.
   (c) The line contains all points where the ordinate is 2 more than the abscissa.
   (d) The sum of the abscissa and ordinate is 3.
   (e) The ordinate is always zero.
   (f) The abscissa is always zero.

5 Write reasonable equations for each of the lines in exercise 4.

6 Quadrant I contains all points whose abscissas and ordinates are positive. Does quadrant I contain the part of the axes that are boundaries?

7 Sketch the lines whose equations are

   (a) $3x + y = 3$ 　　　　　　　　　(b) $x - 2y = 4$

   (c) $y = 2x - 3$ 　　　　　　　　　(d) $\dfrac{x}{3} + \dfrac{y}{4} = 1$

   (e) $0.5x - 0.3y = 1$ 　　　　　　　(f) $5x - 3y = 10$
   (g) $x = y$ 　　　　　　　　　　　　(h) $x = 2$
   (i) $y = 3$ 　　　　　　　　　　　　(j) $3x = 2y$

8 (a) We stated that $Ax + By = C$ is always a line if A, B, and C are constants. Suppose $B = 0$. Where are the lines of the form $Ax = C$?
   (b) Sketch the lines $2x = 3$, $x = 4$, and $x = -5$.
   (c) Sketch the lines $y = 2$, $-2y = 6$, and $y = 4$.
   (d) Describe the graphs of $y = k$, where k is any constant.

9 (a) Graph $2x + 4y = 6$, given that the reference set is $R = \{ 1, 2, 3 \}$.
   (b) Is the graph in (a) a straight line?

10 The graph of $2x + 3y = 6$ is a line if there are no restrictions on x and y. But suppose x and y are each chosen only from the set $\{ 1, 2, 3 \}$. What happens? What is the solution set, given this restricted x and y?

11 Sketch each of the following:

   (a) $x^2 + y = 9$ 　　　　　　　　　(b) $y = x^2$
   (c) $x^2 - y + 2 = 0$ 　　　　　　　(d) $x^2 + y^2 = 4$
   (e) $2x^2 + 2y^2 = 50$ 　　　　　　　(f) $(x - 2)^2 = 4y$
   (g) $y^2 = x + 8$ 　　　　　　　　　(h) $y^2 = -8x + 16$
   (i) $x^2 - y^2 = 9$

8.5 SOLVING PAIRS OF EQUATIONS

To solve an equation like $2x = 10$ we hunt for a number such that twice that number is 10. In set-generator notation the solution set is $S = \{ x \mid 2x = 10 \}$.

To solve an equation like $2x + 3y = 10$ we find ordered pairs such that the sum of twice the first component and three times the second component is 10. In set-generator notation the solution set is $S = \{ (x, y) \mid 2x + 3y = 10 \}$. Remember that the vertical line in set-generator notation is followed by the rule that the variable (or variables) must follow. Also recall that, unless otherwise stated, the number replacements for the variables are chosen from the set of real numbers.

The solution set in the above paragraph could be written this way:

$$S = \{ (x, y) \mid 2x + 3y = 10 \text{ } and \text{ } x \text{ and } y \text{ are reals} \}$$

We can force the variables to obey as many rules as we wish in set-generator notation. Each separate condition on the variables is connected to the others by the word "and." Now consider this example:

EXAMPLE 1

Suppose $S = \{\,(x, y)\,|\,x - y = 1 \text{ and } x + y = 7\,\}$. In words, x and y are two numbers whose difference is 1 and whose sum is 7. After a bit of mental experimentation, we find the numbers to be 4 and 3. Thus $S = \{\,(4, 3)\,\}$.

Now let us examine the solution sets, S_1 and S_2, where the ordered pairs in S_1 obey the first rule in Example 1, and the ordered pairs in S_2 obey the second rule.

$$S_1 = \{\,(x, y)\,|\,x - y = 1\,\}$$
$$S_2 = \{\,(x, y)\,|\,x + y = 7\,\}$$

Certainly S_1 and S_2 are infinite sets of ordered pairs; we can write a few of the ordered pairs in each:

$$S_1 = \{\ldots,(5, 4),(4, 3),(3, 2),(2, 1),(0, -1), \ldots\}$$
$$S_2 = \{\ldots,(6, 1),(5, 2),(4, 3),(3, 4), \ldots\}$$

To find where *both* conditions on (x, y) hold at once, that is, to find where $x - y = 1$ *and* $x + y = 7$, we merely have to find the intersection of these two solution sets. Since $(4, 3)$ is in both (and no other ordered pairs are in both), we call $(4, 3)$ the *simultaneous solution* for $x - y = 1$ and $x + y = 7$.

DEFINITION 4   Simultaneous Solution

Let two equations have the individual solution sets S_1 and S_2. Then the *simultaneous solution* of the two equations is the intersection of S_1 and S_2.

Sometimes a set of equations to be solved is called a *simultaneous system* or just a system. Mathematicians usually put a brace at the left of a pair of equations to be solved simultaneously.

EXAMPLE 2

Find the simultaneous solution to this system:

$$\begin{cases} 2x - y = 5 \\ x + 2y = 0 \end{cases}$$

Here

$$S = \{ (x, y) \mid 2x - y = 5 \text{ and } x + 2y = 0 \}$$

The solution set for $2x - y = 5$ is

$$S_1 = \{ (0, -5),(1, -3),(2, -1),(3, 1), \ldots \}$$

The solution set for $x + 2y = 0$ is

$$S_2 = \{ (0, 0),(1, -1/2),(2, -1),(3, -3/2), \ldots \}$$

The intersection of S_1 and S_2 is $(2, -1)$ so that ordered pair is the simultaneous solution.

The simultaneous solution of two equations is those coordinates (points) that satisfy both of the equations. Thus the solution is on the graph of each of the equations. If a point is on each of two graphs, it can only be on the intersection of the graphs. So we can also solve simultaneous systems graphically.

EXAMPLE 3

Solve the system below graphically.

$$\begin{cases} 2x + y = 6 \\ x - y = 3 \end{cases}$$

Letting $x = 0$ and $x = 2$ in the first equation, we find that $(0, 6)$ and $(2, 2)$ are points on the graph of $2x + y = 6$. Similarly, $(0, -3)$ and $(4, 1)$ are points on the graph of $x - y = 3$. The two lines are graphed in Figure 10; they intersect at $(3, 0)$. You can check that $(3, 0)$ is in both solution sets.

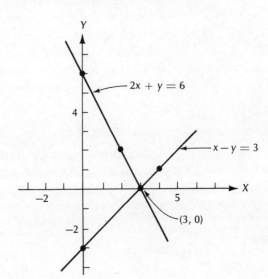

Figure 10
The graphs of $x - y = 3$
and $2x + y = 6$.

EXAMPLE 4

Solve the system below graphically.

$$\begin{cases} x - y = -1 \\ 2x - 2y = 4 \end{cases}$$

The first equation is satisfied by $(0, 1)$ and $(-1, 0)$; the second is satisfied by $(2, 0)$ and $(0, -2)$. Plotting these points, we get the graphs of the lines (Figure 11). Since the lines do not intersect, there is no common point and the solution set is empty.

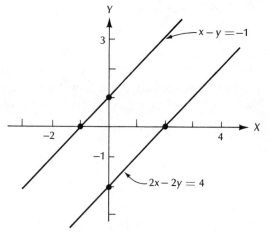

Figure 11   The graphs of $x - y = -1$ and $2x - 2y = 4$.

The lines in Example 4 are, in fact, parallel. You may wish to try to *prove* this by using Euclidean geometry. Hint: If two lines form equal alternate interior angles with a transversal, then they are parallel.

We have just seen how to solve pairs of equations geometrically. This can also be done algebraically, as we shall now see. The first algebraic method for solving simultaneous systems is called the *substitution method*. Suppose we want to solve

$$\begin{cases} 2x + y = 6 & \text{(1)} \\ x - y = 3 & \text{(2)} \end{cases}$$

An equation equivalent to (2) is $x = y + 3$ (adding y to both sides). This says x *is* $y + 3$, so we may replace x in (1) by $y + 3$. Substituting, we get

$$2(y + 3) + y = 6$$

Then

$$2y + 6 + y = 6 \qquad \text{(Distributive law)}$$

and

$$2y + y = 0 \qquad \text{(Subtracting 6, both sides)}$$

and

$$3y = 0 \qquad \text{(Adding } 2y \text{ and } y\text{)}$$

So

$$y = 0 \qquad \text{(Dividing by 3, both sides)}$$

Now we can substitute $y = 0$ into equation (2) to get $x = 3$. Thus the solution is $S = \{(3, 0)\}$. Compare this method to the geometrical solution of the same set of equations in Example 3.

EXAMPLE 5

Solve the system below using the substitution method.

$$\begin{cases} y + 3x = 33 & \text{(1)} \\ 3x + 5y = 45 & \text{(2)} \end{cases}$$

From (1) we get $y = 33 - 3x$; why? Substituting in (2), we get $3x + 5(33 - 3x) = 45$. Removing parenthesis by the distributive law, we get $3x + 165 - 15x = 45$. Combining $3x$ and $-15x$ we get $-12x + 165 = 45$. Subtracting 165 from both sides, we get $-12x = -120$. Dividing both sides by -12, we get $x = 10$. Then $y = 3$; why? Thus the solution set is $S = \{(10, 3)\}$.

Another algebraic method for solving systems can be called the *addition method* because at one point in finding the solution the equations are added together. A rule involved here is "if $a = b$ and $c = d$, then $a + c = b + d$."

EXAMPLE 6

Solve the system

$$\begin{cases} 2x - y = 7 & \text{(1)} \\ 5x + y = 21 & \text{(2)} \end{cases}$$

Equation (1) can be thought of as $a = b$ and Equation (2) as $c = d$. Therefore we can add (1) to (2), left side to left side and right side to right side. We get $(2x - y) + (5x + y) = 7 + 21$, which simplifies to $2x + 5x = 28$. Adding $2x$ and $5x$, we get $7x = 28$, which is equivalent to $x = 4$. Now substitute 4 for x in either (1) or (2) to find y:

$$2(4) - y = 7 \qquad \text{(1)}$$
$$8 - y = 7$$
$$y = 1 \qquad \text{(By inspection)}$$

Thus the solution set is $S = \{(4, 1)\}$.

Once you examine Example 6 and understand each step in the solution, the work can be compressed as follows:

$$\begin{cases} 2x - y = 7 \\ 5x + y = 21 \end{cases} \qquad (1) \\ (2)$$

Adding (1) and (2) we got

$$7x = 28$$

Therefore

$$x = 4$$

Substituting in (1), we got

$$y = 1$$

Therefore

$$S = \{\,(4, 1)\,\}$$

The crucial step in the addition method is adding Equation (1) to Equation (2) to get another equation *free of one of the variables*. It should be evident that Example 6 is particularly suited to the method.

Sometimes both sides of one or both of the equations can be multiplied first by appropriate constants so that when the equations are added the result is free of one of the variables. Study the following example to see how this works.

EXAMPLE 7

Solve the system

$$\begin{cases} 3x + 4y = 25 \\ x + 2y = 11 \end{cases} \qquad (1) \\ (2)$$

Multiply both sides of (2) by -2, getting an equivalent system.

$$\begin{cases} 3x + 4y = 25 \\ -2x - 4y = -22 \end{cases} \qquad (1) \\ (2)$$

Adding (1) and (2), we get

$$x = 3$$

Substituting into the original Equation (2), we have

$$3 + 2y = 11$$

which implies that $2y$ must be 8, or $y = 4$. Thus

$$S = \{\,(3, 4)\,\}$$

Notice that we eliminated the variable y after the addition step in Example 7. We could multiply Equation (2) of the given system by -3, then add, thus eliminating the variable x.

Exercise Set 8.5

1 Let
$$S_1 = \{ (1, 1),(3, 5),(-2, 7),(-6, 4),(5, 4), \ldots \}$$
$$S_2 = \{ (6, 2),(5, -3),(7, 4),(1/2, 6),(3, 5), \ldots \}$$

These two sets of ordered pairs might be solution sets. What is their intersection? (In set notation we are asking for $S_1 \cap S_2$.)

2 (a) Find several ordered pairs in the solution set of $3x + y = 9$.
  (b) Do the same for $2x - y = 1$.
  (c) What is the intersection of the two solution sets?

3 (a) Express the following problem algebraically, using two equations: "The sum of two numbers is 10 and their difference is 2. What are the numbers?"
  (b) See if you can guess at the solution to this problem. Then solve the two equations by a method from this chapter.

4 Express the following in words:
$$\{ (x, y) \mid x + y = 12 \text{ and } y - 8 = x \}$$

5 Solve each of the following simultaneous systems graphically:

(a) $\begin{cases} y - 2x = 6 \\ x + y = 3 \end{cases}$
(b) $\begin{cases} 2x + y = 8 \\ x - y = -2 \end{cases}$

(c) $\begin{cases} x - y = 0 \\ 3x - y = 2 \end{cases}$
(d) $\begin{cases} x + y = 4 \\ y = 2x - 8 \end{cases}$

(e) $\begin{cases} 2x + 4y = 3 \\ 3x - y = 1 \end{cases}$
(f) $\begin{cases} 6x - 3y = -5 \\ 3x + 3y = -4 \end{cases}$

6 (a) Draw the graph of each of the lines for this system: $\begin{cases} x + 2y = 4 \\ 6y = 12 - 3x \end{cases}$
  (b) What is the solution set for the system?

7 (a) Draw the graph of each of the lines for this system: $\begin{cases} x + 2y = 4 \\ x + 2y = 7 \end{cases}$
  (b) What is the solution set for this system?

8 Solve the following simultaneous systems algebraically, using either the substitution or the addition method:

(a) $\begin{cases} x - y = -2 \\ 3x - y = 4 \end{cases}$
(b) $\begin{cases} 2y = x + 4 \\ y = 1/3(x + 4) \end{cases}$
(c) $\begin{cases} 3x + y = 7 \\ 2x - y = 8 \end{cases}$

(d) $\begin{cases} 5x + 4y = 12 \\ -x + 3y = 2 \end{cases}$
(e) $\begin{cases} 25x - 20y = 50 \\ 3x + 4y = 70 \end{cases}$

9 Many problems result in two equations that can be solved simultaneously. In the exercises below, first set up two appropriate equations and then find the solutions.
  (a) I have a certain number of quarters and a certain number of nickels. Altogether I have 25 coins worth $2.25. How many quarters and how many nickels do I have?
  (b) If you invest part of $5000 at 4% and the other part at 5% and receive a total of $230 interest for a year, how much did you have invested at each rate?

(Hints: What equation results in the $230 figure? Also recall that to find, say, 3% of a number x, we multiply x by 0.03)

(c) In a certain course you want your average of three grades to be 85. On the first test you got 70, on the second 90. What must you get on the third test? Hint: Let x be your third test grade. The first equation might be $70 + 90 + x = y$, where y is the sum of the three grades. The second equation expresses that the sum divided by 3 is the desired average.

(d) Solve the same problem as (c), where grade one is 92, grade two is 78, grade three is 82, and you want an average of 85. What should grade four be?

(e) Suppose you want an average for part (d) of 90. What would grade four have to be? What is the highest average you can get, given 100 as a maximum grade on test 4?

(f) Suppose x pounds of coffee A selling at 90¢ a pound are to be mixed with y pounds of coffee B selling at $1.00 a pound. The yield is to be 100 pounds of coffee selling at 98¢ a pound. How many pounds of each coffee would be used? (Be sure to work either entirely in cents or entirely in dollars.)

(g) A wire 30 inches long is to be cut in two pieces. Twice the length of the first piece must be equal to three times the length of the second piece. How long should the pieces be?

*8.6 INEQUALITIES AND THEIR GRAPHS

In the twentieth century, mathematicians deal as much with quantities that are unequal as they do with quantities that are equal. An *inequality* is an algebraic expression comparing quantities that are not equal. Here are some examples:

$$x \neq 5, \text{ read "}x \text{ is not equal to 5"}$$
$$x > 5, \text{ read "}x \text{ is greater than 5"}$$
$$x < 5, \text{ read "}x \text{ is less than 5"}$$

The so-called "*trichotomy rule*" tells us that for any two real numbers exactly one of the following statements is true:

$$a = b$$
$$a > b$$
$$a < b$$

It is helpful to think of the signs $>$ and $<$ as pointing to the smaller quantity. Occasionally these inequality signs are combined with the sign "$=$" to form $\geq$ and $\leq$ (sometimes printed as $\geqslant$, $\leqslant$ or $\geqq$, $\leqq$). Then we would read $x \geq 4$ as "x is greater than or equal to 4." Similarly, $x \leq 5$ would read "x is less than or equal to 5." Because of the trichotomy rule, if we write $x \ngtr 2$ (read "x is not greater than 2"), then x must be smaller than 2 or equal to 2; that is, $x \leq 2$. Likewise, $x \nless 2$ means $x \geq 2$.

All of the signs below are called inequality signs:

"Greater than" and "less than" are generally referred to as *order relations* on the set of real numbers. Assuming that we know how to recognize when one real number is greater than or less than another, it is easy to see that the real numbers are ordered. The real number line (Figure 12) shows the order; larger numbers are always farther to the right.

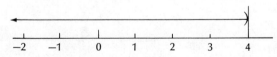

Figure 12   The real number line

The relation $>$ (greater than) on the set of reals is not reflexive; that is, $a > a$ is not true for any real number a. "Greater than" is also *not symmetric* on the reals, since if $a > b$, then b is not greater than a. However, "greater than" is *transitive*, since for any reals if $a > b$, and $b > c$, then $a > c$. (See Section 3.4.)

Similarly, $<$ (less than) is not reflexive and not symmetric, but is transitive. Transitivity of $<$ implies that if $a < b$, and $b < c$, then $a < c$. Transitivity of the order relations $>$ and $<$ is what we mean when we say the real numbers are ordered.

A simple inequality like $x < 4$ can be graphed as part of the real number line. Clearly $x < 4$ is satisfied by any real smaller than 4. (That includes zero and all the negatives.) The solution set to $x < 4$ can be written

$$S = \{\, x \mid x \text{ is a real and } x \text{ is smaller than } 4 \,\}$$

using set-generator notation, but we cannot list all the real numbers in S. (There is no *first* number smaller than 4.) We graph the solution set in Figure 13. Compare the graph of $x < 4$ with the graph of $x \le 4$, shown in Figure 14.

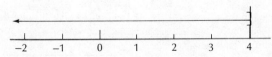

Figure 13   The graph of $x < 4$.

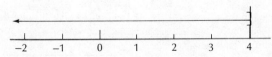

Figure 14   The graph of $x \le 4$.

Several things should be noticed about Figures 13 and 14. First, it is convenient to draw a short vertical line at 4 and plot the solutions above the real number line. This avoids sketching on top of the real line and makes the graph clearer. Second, the arrowhead indicates that the solution set continues indefinitely far in the direction indicated. Third, the parenthesis excludes the endpoint 4, while the bracket includes the endpoint 4.

We sometimes have so-called *double inequalities*, where there are two conditions on the variable. The notation $-1 < x < 2$ means x is less than 2 and greater than -1. This would be sketched as shown in Figure 15.

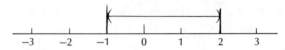

Figure 15   The graph of $-1 < x < 2$.

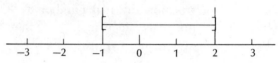

Figure 16   The graph of $-1 \leq x \leq 2$.

The notation $-1 < x < 2$ shows x *between* -1 and 2, and the graph does indicate the points between -1 and 2. Once again, the parentheses exclude -1 and 2. For $-1 \leq x \leq 2$ we would use brackets to indicate the inclusion of the endpoints; the graph of this double condition on x is shown in Figure 16.

The statement $x > 1$ or $x < -2$ is also a double condition on x, but x would not be between 1 and -2 because of the word "or." (We could not have $x > 1$ *and* $x < -2$ as a double condition on x because no number is greater than 1 *and* less than -2.) The graph of $x > 1$ or $x < -2$ is shown in Figure 17.

As a final example of the graphs of double inequalities we sketch in Figure 18 the graph of $x \geq 2$ or $x < -1/2$. Notice the $\geq$ sign and the use of a bracket, and the $<$ sign and the use of the parenthesis.

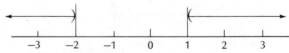

Figure 17   The graph of $x > 1$ or $x < -2$.

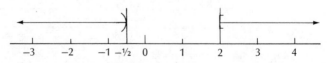

Figure 18   The graph of $x \geq 2$ or $x < -1/2$.

We emphasize that where the reference set is considered to be made up only of reals, these graphed intervals include *all* the real numbers under the graph. Thus in the solution set to $x > 1$ we have a continuous set of numbers greater than 1, including decimals like 1.01, 1.05, 2.345, and 100.67, as well as irrationals like $\sqrt{2}$, $\sqrt{3}$, and π and rationals like $3\frac{1}{2}$, $1\frac{7}{8}$, and $5\frac{3}{4}$.

Exercise Set 8.6

1 Translate each of the following into words:
(a) $2 > 0$ (b) $7 < x$ (c) $9 \geq y$
(d) $8 \leq 10$ (e) $x \neq 4$ (f) $x \not> 2$
(g) $y \not< 15$ (h) $x + y \geq 2$

2 Express each of the following verbally:
(a) $-1 < x < 1$ (b) $0 \leq x \leq 1$
(c) $x > 2$ or $x < 0$ (d) $x \geq 5$ or $x \leq -2$
(e) $0 < x \leq 3$ (f) $-2 \leq x < 5$
(g) $x > 1$ or $x \leq 0$ (h) $x \geq 1$ or $x < 0$

3 How can we express each of the following in symbols?
(a) x is positive   (b) x is negative
(c) x is not positive   (d) x is not negative

4 True or false:
(a) If a number is not positive, then it is negative.
(b) If a number is negative, then it is not positive.
(c) A nonnegative number is zero or positive.
(d) If a number is not negative, then it must be positive.

5 True or false:
(a) $7 > 5$ (b) $5 < 7$ (c) $7 \geq 5$
(d) $-1 > 0$ (e) $-1 > -2$ (f) $-1 < 2$
(g) $-7 < 5$ (h) $-7 < -5$

6 Identify each of these graphs with corresponding equalities or inequalities.

(a) 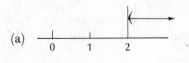   (d)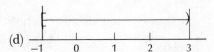

(b) 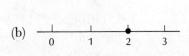   (e)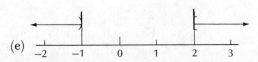

(c) 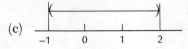   (f)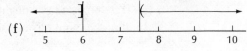

7 Sketch a graph of each of these situations:
   (a) x is between -4 and $+2$
   (b) x is outside the interval from -2 to $+3$ and x is not -2 or $+3$
   (c) x is positive   (d) x is negative or zero
   (e) $x > 2$ or $x < 2$ (f) $x \neq 2$ (g) $x > 2$ and $x < 2$

8 Sketch a graph of each of the inequalities in exercise 2.

9 Is 3.4 in the solution set to $x > -2$?

10 (a) Is there a smallest number in the solution set to $x > 1$?
   (b) Is there a smallest number in the solution set to $x \geq 1$?

*8.7 SOLVING INEQUALITIES

Algebraic procedures for solving inequalities are very much like those used for solving equations. There are, however, some important differences. We begin by stating rules that can simplify a given inequality. Application of any of these rules results in an inequality that is equivalent to the original inequality.

RULE 1
Given an inequality, the same quantity may be added to or subtracted from each side. In symbols this says that

(a) if $a > b$, then $a + c > b + c$;
(b) if $a < b$, then $a + c < b + c$;
(c) if $a > b$, then $a - c > b - c$;
(d) if $a < b$, then $a - c < b - c$.

RULE 2
Given an inequality, each side may be multiplied or divided by the same *positive* quantity. In symbols,

(a) if $a > b$, and c is positive, then $ac > bc$ and $a/c > b/c$;
(b) if $a < b$, and c is positive, then $ac < bc$ and $a/c < b/c$.

RULE 3
Given an inequality, each side may be multiplied or divided by the same *negative* quantity, but the *sense* of the inequality is changed. That is, $>$ is changed to $<$ and $<$ is changed to $>$. In symbols,

(a) if $a > b$, and c is negative, then $ac < bc$ and $a/c < b/c$;
(b) if $a < b$, and c is negative, then $ac > bc$ and $a/c > b/c$.

Rule 3 embodies what is probably the most important difference between the rules for transforming equations and those for transforming
inequalities. A numerical example will clarify why the sense is changed in
Rule 3.

EXAMPLE 1

Suppose we have $7 > 5$ and decide to multiply both sides by -1. This
changes 7 to -7 and 5 to -5. But considering that -7 is to the left of -5 on
a real number line (Figure 19), we must write $-7 < -5$. Likewise, if we
observe that $-8 < -2$ on the real number line, and divide both sides by
-2, then the result is $4 > 1$.

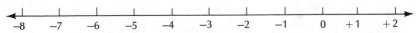

Figure 19   The real number line, emphasizing order of negative numbers.

When multiplying or dividing by a negative quantity, it is the change
from negative to positive or vice versa, that forces us to change the sense
of the inequality. Of course, the same thing is true when inequalities
involving variables are dealt with.

EXAMPLE 2

Solve $-\frac{1}{2}x > 4$. Multiply both sides by -2 and change $>$ to $<$:

$$x < -8$$

Then

$$S = \{\, x \mid x < -8 \,\}$$

Note that -10 is in S, for example. Substituting -10 for x in the given
inequality, we get $(-\frac{1}{2})(-10) > 4$, which is the same as $5 > 4$, which is
true.

Suppose we did not change the sense of the inequality when multiplying or dividing by a negative quantity. Then in Example 2 we would have
had $x > -8$. One number greater than -8 is -6, for example. But -6
substituted for x in $-\frac{1}{2}x > 4$ yields $-\frac{1}{2}(-6) > 4$ or $3 > 4$, which is false.

Now let us solve some inequalities, using the basic rules, and graph the
results.

EXAMPLE 3

Solve $2x - 4 > 5$.

$2x - 4 > 5$	(The given inequality)	(1)
$2x > 9$	(Adding 4 to both sides)	(2)
$x > 9/2$	(Dividing both sides by 2)	(3)

Therefore

$$S = \{\, x \mid x > 9/2 \,\}$$

The graph is shown in Figure 20.

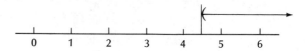

Figure 20   The graph of the solution to $2x - 4 > 5$.

EXAMPLE 4

Solve $-2x + 3 < 11$. First, we can subtract 3 from both sides, getting $-2x < 8$. Then we can divide both sides by -2, getting $x > -4$. (Note the change of $<$ to $>$.) Then the solution set is $S = \{\, x \mid x > -4 \,\}$, and the graph is shown in Figure 21.

Figure 21   Graphing $x > -4$.

Sometimes we want to change the order of a given inequality from, say, $a < b$ to $b > a$. Note that these two inequalities mean the same thing; in both a is the smaller quantity.

EXAMPLE 5

Solve $x + 2 < 2x + 3$. This is the same as $2x + 3 > x + 2$. Then, subtracting x from both sides, we get $x + 3 > 2$; and subtracting 3 from both sides, we get $x > -1$. So

$$S = \{\, x \mid x > -1 \,\}$$

and the graph is shown in Figure 22.

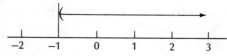

Figure 22   Graphing $x > -1$.

For equalities we had a fourth rule stating that if $a^2 = b^2$, then $a = b$ or $a = -b$. There are similar rules for inequalities.

RULE 4

If $a^2 < b^2$, then $a < +b$ and $a > -b$. In other words, if $a^2 < b^2$, then a is *between* b and $-b$, so we might have written this rule thus: If $a^2 < b^2$, then $-b < a < +b$.

Restriction   In this rule b must be positive.

RULE 5

If $a^2 > b^2$, then $a > b$ or $a < -b$.
Restriction   In this rule b must be positive.

EXAMPLE 6

Solve $x^2 < 4$. By Rule 4 this becomes $x < 2$ *and* $x > -2$, or x is between -2 and 2. This can be written as the *double inequality*, $-2 < x < 2$. So

$$S = \{\, x \mid -2 < x < 2 \,\}$$

and the graph is shown in Figure 23.

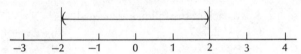

Figure 23   Graphing $-2 < x < 2$.

EXAMPLE 7

Solve $3x^2 - 9 > 6$. Then

$$3x^2 > 15 \qquad\qquad \text{(Adding 9 to both sides)}$$

Then

$$x^2 > 5 \qquad\qquad \text{(Dividing both sides by 3)}$$

So

$$x > \sqrt{5} \qquad \text{or} \qquad x < -\sqrt{5} \quad \text{(Rule 5)}$$

Then

$$S = \{\, x \mid x > \sqrt{5} \quad \text{or} \quad x < -\sqrt{5} \,\}$$

Note that $\sqrt{5}$ is approximated by 2.236. The graph is shown in Figure 24.

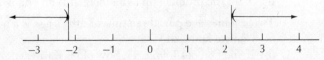

Figure 24   Graphing $x > \sqrt{5}$ or $x < -\sqrt{5}$.

Exercise Set 8.7

1 Give the justification for each numbered step in the following:

(a) $+2x < 7$
$$x < 7/2 \qquad (1)$$

(b) $2x - 7 > 0$
$$2x > 7 \qquad (1)$$
$$x > 7/2 \qquad (2)$$

(c) $3x - 1 \le 5$
$$3x \le 6 \qquad (1)$$
$$x \le 2 \qquad (2)$$

(d) $2x^2 - 5 < 3$
$$2x^2 < 8 \qquad (1)$$
$$x^2 < 4 \qquad (2)$$
$$-2 < x < 2 \qquad (3)$$

(e) $3y^2 + 5 > 32$
$$3y^2 > 27 \qquad (1)$$
$$y^2 > 9 \qquad (2)$$
$$y > 3 \text{ or } y < -3 \qquad (3)$$

(f) $-4x^2 - 3 \le 5x^2 - 12$
$$5x^2 - 12 \ge -4x^2 - 3 \qquad (1)$$
$$9x^2 - 12 \ge -3 \qquad (2)$$
$$9x^2 \ge 9 \qquad (3)$$
$$x^2 \ge 1 \qquad (4)$$
$$x \ge 1 \text{ or } x \le -1 \qquad (5)$$

(g) $-6x + 4 < 2x + 12$
$$-8x + 4 < 12 \qquad (1)$$
$$-8x < 8 \qquad (2)$$
$$x > -1 \qquad (3)$$

(h) $-5x^2 + 3 > 2x^2 + 24$
$$-7x^2 + 3 > 24 \qquad (1)$$
$$-7x^2 > 21 \qquad (2)$$
$$x^2 < -3 \qquad (3)$$
$$S = \emptyset \text{ (empty)} \qquad (4)$$

2 Solve each of the following inequalities:

(a) $x - 2 > 7$

(b) $2x + 6 > 8$

(c) $-2x + 3 < 4$

(d) $-7y + 3y < 12$

(e) $-\dfrac{7y}{3} > 7$

(f) $\dfrac{x + 3}{2} \le -5$

(g) $\dfrac{x}{2} + \dfrac{1}{2} > \dfrac{1}{2}$

(h) $3x^2 < 75$

(i) $3y^2 - 2 > 16$

(j) $4x^2 + 4 > x^2 + 20$

(k) $\dfrac{1}{2}x^2 + \dfrac{1}{3}x^2 > \dfrac{5}{6}$

(l) $-3x^2 - 4 > -2x^2 - 5$

(m) $-3x^2 - 4 > -2x^2 + 5$

3 Rule 4 states that if $a^2 < b^2$, then $-b < a < b$. What happens when b is negative? Hint: Try to apply this rule to the solution of $x^2 < (-3)^2$

4 By $a < x < b$ we mean "$x < b$ *and* $x > a$." Suppose we write $4 < x < -4$. Does such an x exist?

5 Graph $x > 5$ or $x < 3$.
   (a) Can this be written $3 > x > 5$?
   (b) If $a < x < b$, how is a related to b?

6 Certain integers increased by 6 always are greater than 15. What integers are they? Set the problem up as an inequality to be solved.

7 Twice John's age plus three times Kathy's age is a number bigger than 82.
   (a) Can we find Kathy's age?
   (b) If you find out John is 17, how old is Kathy?
   (c) If John is 17 and Kathy is 20, is the given sentence true?

*8.8   GRAPHING INEQUALITIES IN TWO VARIABLES

Recall that an equation of the form $Ax + By = C$ is a linear equation whose graph is a line in the coordinate plane. In this section we discuss the graphs of *linear inequalities* of the form

$$Ax + By > C$$

or

$$Ax + By < C$$

Sometimes they are combined with equalities. For example,

$$Ax + By \geq C$$

means $Ax + By > C$ or $Ax + By = C$.

The graph of a linear inequality will, of course, contain those points whose coordinates satisfy the inequality. The points will not lie on a line, but will be associated with a line. The graph of $x > 1$ is associated with the line $x = 1$; the graph of $y > x + 1$ is associated with the line $y = x + 1$. The next two examples will show just what the association is for these two cases.

EXAMPLE 1

Sketch a graph for $x > 1$. This inequality states that "x is greater than 1," so we want all the points whose abscissas (x coordinates) are greater than 1. If we sketch the line $x = 1$ in a plane, then the graph of $x > 1$ (Figure 25) contains all those points to the right of that line.

Clearly any point in the shaded portion of Figure 25 has an abscissa greater than 1. The shaded portion is understood to continue indefinitely far to the right, and indefinitely up and down. Such a region is termed a

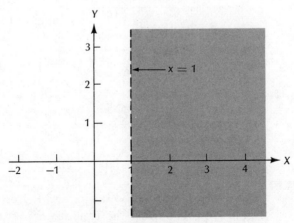

Figure 25   The shaded portion is the graph of $x > 1$.

half-plane. Note that the line $x = 1$ is broken (dashed) to indicate that it is *not* part of the graph of $x > 1$. Yet the line does serve as the *boundary* of the half-plane $x > 1$.

EXAMPLE 2

Sketch the graph of $y > x + 1$. The associated line is $y = x + 1$; the line is the set of points whose ordinate (y) is one more than the abscissa. Similarly, $y > x + 1$ is the set of points whose ordinates are greater than one more than the abscissa. Thus the half-plane determined by $y > x + 1$ is *above* the line $y = x + 1$ (Figure 26).

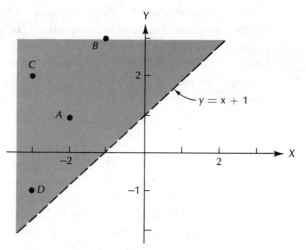

Figure 26   The shaded portion is the graph of $y > x + 1$.

Table 5   Checking that four points satisfy $y > x + 1$.

Point	Coordinates	Substituting in $y > x + 1$	Simplifying	Point satisfies?
A	$(-2, 1)$	$1 > -2 + 1$	$1 > -1$	yes
B	$(-1, 3)$	$3 > -1 + 1$	$3 > 0$	yes
C	$(-3, 2)$	$2 > -3 + 1$	$2 > -2$	yes
D	$(-3, -1)$	$-1 > -3 + 1$	$-1 > -2$	yes

In Figure 26, we have labeled four points out of the infinitely many in the shaded region. Table 5 is a check to see that they satisfy $y > x + 1$. Once again, points on the line $y = x + 1$ do not satisfy $y > x + 1$. [For example, $(0, 1)$ satisfies the equation. Does it satisfy the inequality?] In general, points on the boundary of a half-plane *are* contained in the solution set only when the inequality is of the $\geq$ or $\leq$ type.

It will always be convenient to convert the associated linear equation $Ax + By = C$ to the form $y = Dx + E$. Then the linear inequality $Ax + By > C$, or $Ax + By < C$, can be converted to one of $y > Dx + E$ or $y < Dx + E$.

EXAMPLE 3

Convert $2x + 3y = 6$ to the form $y = Dx + E$. Subtracting $2x$ from both sides, we get $3y = -2x + 6$. Dividing both sides by 3, we get $y = -\frac{2}{3}x + 2$, which is of the form $y = Dx + E$. (What is D? What is E?)

EXAMPLE 4

Convert $2x + 3y > 6$ to one of the forms $y > Dx + E$ or $y < Dx + E$. By steps similar to those in Example 3, we get $y > -\frac{2}{3}x + 2$.

If a linear inequality is of the form $y > Dx + E$, then the graph is the half-plane *above* the associated line $y = Dx + E$. If a linear inequality is of the form $y < Dx + E$, then the graph is the half-plane *below* the associated line $y = Dx + E$.

EXAMPLE 5

Sketch the half-plane $2x + 3y > 6$. This is equivalent to $3y > -2x + 6$, which is equivalent to $y > -\frac{2}{3}x + 2$. Thus the half-plane is the set of points above the line $y = -\frac{2}{3}x + 2$. Since the points $(0, 2)$ and $(3, 0)$ are on the line, the graph is as shown in Figure 27.

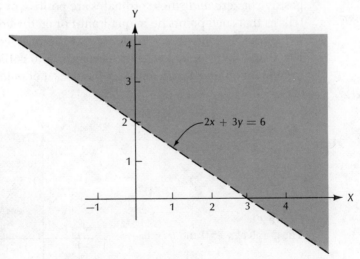

Figure 27   The shaded portion is the graph of $2x + 3y > 6$.

The next example shows the graph of an inequality, where the boundary is included. The line is not broken, indicating that it is part of the graph.

EXAMPLE 6

Graph $y - 2x \leq 1$. First, it is convenient to change this to the equivalent inequality $y \leq 2x + 1$. Then we draw the line $y = 2x + 1$, which is part of the graph (Figure 28). The half-plane $y < 2x + 1$ is the region below the line.

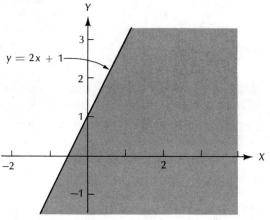

Figure 28   The graph of $y - 2x \leq 1$

Interesting graphs result if two or more inequalities hold at once. For example, if $x \geq 0$ and $y \geq 0$, we want those points whose abscissas are positive or zero *and* whose ordinates are positive or zero. A bit of thought tells us that such points lie in quadrant I or on the boundaries to quadrant I (Figure 29). Do you see Figure 29 as the intersection of the graph of $x \geq 0$ and the graph of $y \geq 0$? Clearly, $x \geq 0$ is the set of points to the right of (or on) the Y axis, and $y \geq 0$ is the set of points above (or on) the X axis.

Figure 29
The graph of $x \geq 0$ and $y \geq 0$.

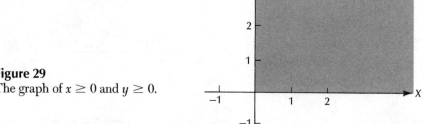

Figure 30 is the graph of the set of points that satisfy all the following conditions at the same time.

$$x \geq 0$$
$$y \geq 0$$
$$y \leq 2$$
$$x \leq 3$$

If you are not convinced that the shaded portion of Figure 30 is correct, construct the graphs of $x \geq 0$, $y \geq 0$, $y \leq 2$, and $x \leq 3$ all on the same plane. Then find their intersection.

Figure 30
A graph satisfying four conditions.

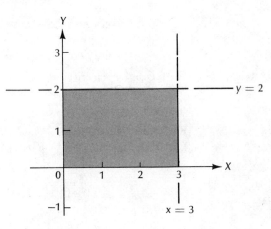

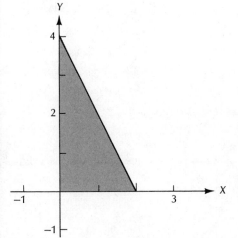

Figure 31
A graph satisfying three conditions.

The interior of a certain triangle (see Figure 31), along with the triangle itself, is defined by these conditions

$$x \geq 0$$
$$y \geq 0$$
$$y \leq -2x + 4$$

If we think of the procedure in reverse, we can say that the triangle of Figure 31 (including sides and interior) is given by

$$S\{(x, y) \mid x \geq 0 \text{ and } y \geq 0 \text{ and } y \leq -2x + 4\}$$

Exercise Set 8.8

1 Sketch the graph of each of the following:
(a) $x > 0$ (b) $x > 3$
(c) $y < 1$ (d) $y > 2$
(e) $y < 5$ (f) $x < -1$

2 (a) Sketch the line $y = 2x - 1$.
   (b) Shade the half-plane corresponding to $y > 2x - 1$, on the same graph.
   (c) Taken together, do (a) and (b) make up the graph of $y > 2x - 1$ or $y \geq 2x - 1$?

3 True or false:
   (a) The graph of $2x - 3y > 3$ is above the line $2x - 3y = 3$.
   (b) The graph of $x - 7y < 4$ is above the line $x - 7y = 4$.
   (c) The graph of $2x + y > -3$ is below the line $y = -2x - 3$.

4 Graph each of the following:
(a) $y + 2x > 1$ (b) $2x - y < 3$
(c) $3x - 4y \leq 12$ (d) $y \leq \frac{2}{3}x + 3$
(e) $y > \frac{1}{3}x$ (f) $x \geq 2y + 4$

5 Sketch a graph in which each of the following conditions is true:
   (a) $y > 1$ and $x < 2$
   (b) $x + y < 3$ and $y > 0$
   (c) $x - y < 3$ and $2y < 0$ and $x > 0$
   (d) $y \leq 3$ and $0 \leq x \leq 2$
   (e) $0 \leq y \leq 3$ and $0 \leq x \leq 2$ and $x + y < 4$

6 Sketch each of the following:
   (a) $\{ (x, y) \mid x \geq 1 \text{ and } y \geq 0 \text{ and } x + y \leq 5 \text{ and } y \leq \frac{2}{3}x + 2 \}$
   (b) $\{ (x, y) \mid x \geq -3 \text{ and } 0 \leq y \leq -x \}$

7 What is the set of linear inequalities that describes
   (a) quadrant III?
   (b) quadrant IV?
   (c) A square 2 units on a side in quadrant I with one vertex at the origin, including its interior?
   (d) A rectangle passing through $(-1, 0)$, $(-1, 3)$, $(4, 0)$, and $(4, 3)$, including its interior?
   (e) The interior of the square in 7(c)?
   (f) The interior of the rectangle in 7(d)?

SUMMARY

It is possible to study algebra without drawing graphs; it is also possible to study geometry without writing equations. But in our approach to algebra, we sketched a graph almost every time we solved an algebraic problem. The graph helps us to visualize sets of numbers or sets of ordered pairs belonging to solution sets.

Mathematicians have as much trouble visualizing relationships as anyone else, but they are quick to let a variable stand for an unknown. Often the variable can then be manipulated by the rules of algebra. Once a solution to a problem is found, the graph is useful in portraying the algebraic relationship. We have seen that some problems can be solved by using graphs alone. For example, we can solve a system of equations by graphical means.

We started to solve equations by either trial and error or tabulation methods. Problem-solving by such methods was known to the ancient Babylonian and Egyptian civilizations (3000 to 2000 B.C.). The establishment of convenient symbols, notation, and usage in the sixteenth century made algebra what it is today.

Once we introduced the solution set, the reference set, and a few rules, we saw that it is easy to solve equations in one variable. We extended the techniques to solving systems of equations, where two equations describe conditions on two variables.

Meanwhile, we graphed solution sets to equations in one variable as points on a line. The solution set for an equation in two variables is graphed as a line in a plane. The solution of a system of linear equations is interpreted graphically as the intersection of two lines.

Sections 8.6, 8.7, and 8.8 dealt with inequalities. First we explained inequality notation and graphed inequalities in one variable. Then the rules for solving inequalities were set down, and we solved inequalities in one variable. Section 8.8 showed how to graph inequalities in two variables. Some interesting graphs result: half-planes (with and without their boundaries), squares, and rectangles. An implied conclusion is that the inequality enables us to describe entire bounded and unbounded regions in a plane.

Before attempting the Review Test, be sure you are familiar with the following:

1. Expressing problems algebraically.

2. Solving equations in one variable with a given reference set.

3. How to apply rules to an equation to get an equivalent equation.

4. The use of the solution set.

5. Graphing solution sets for equations in one variable.

6. Setting up the coordinate plane.

7. Using ordered pairs as solutions to an equation in two variables.

8. Graphing the solution set for an equation in two variables.

9. Solving pairs of equations (simultaneous systems) by the substitution method, the addition method, and graphically.

*10. Inequality notation.

*11. Graphing inequalities like $x < 3$, $x \geq 4$, and $1 \leq x < 2$, with reference to the real number line.

*12. Solving inequalities in one variable, algebraically.

*13. Graphing inequalities in two variables.

*14. Graphing simultaneous sets of inequalities.

REVIEW TEST

1 Express the following statement algebraically: Three less than the product of two numbers is seven.

2 Solve $5x - 7 = 2$
   (a) with the real numbers as reference set.
   (b) with $R = \{2, 4, 6, 8, \ldots\}$.

3 Solve $2x^2 - 13 = 85$. Graph the solution set on a real number line.

4 Solve $\frac{3}{2}x + 6 = 30$.

5 In what quadrant do we find
   (a) $(-3, 2)$ (b) $(-7, -35)$
   (c) $(4.1, 7.9)$ (d) $(+7, -2.3)$

6 Sketch a graph of each of the following:
   (a) $y = 3x - 4$
   (b) $y + x^2 = 4$
   (c) $2x + 5y = 10$

7 Solve this system graphically: $\begin{cases} x + 4y = 2 \\ -x + y = 3 \end{cases}$

8 Solve this system algebraically: $\begin{cases} 2x - y = 4 \\ 7x + y = 50 \end{cases}$

9 I am thinking of two numbers. Twice the first plus the second is 25. If the first is subtracted from the second, the result is 4. What are the numbers?

*10 Graph $-2 < x \leq 5$ on a real number line.

*11 Solve $-2x + 4 < 11$.

*12 Solve $3x^2 - 12 > 0$ and graph the solution set.

*13 Graph $y \leq 2$.

*14 Graph those points where $x \geq -2$ and $y < 5$.

*15 Graph $y > 2x - 5$.

*16 Describe, in terms of inequalities, the rectangle and its boundary
     sketched below.

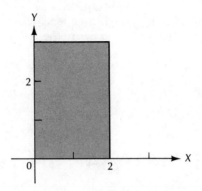

*17 Sketch those points that obey all these conditions:

$$3x + 4y \le 12$$
$$x \le 3$$
$$y \ge 0$$
$$x \ge 0$$

William John Macquorn Rankine (1820–1872), a
Scottish engineer and physicist, is credited with the
founding of the science of thermodynamics. He
made important contributions to the design theory of
masonry dams, mines, steam engines, and ships. In
1859, he coined the term energy as it relates to
thermodynamics and wrote the first discourse on this
subject. Though most of his publications were
technical, the following poem attests to his
lighthearted sense of humor. It was published
posthumously in *Songs and Fables* (1874).

THE MATHEMATICIAN IN LOVE
W. J. M. RANKINE

1

A mathematician fell madly in love
   With a lady, young, handsome, and charming:
By angles and ratios harmonic he strove
Her curves and proportions all faultless to prove.
   As he scrawled hieroglyphics alarming.

2

He measured with care, from the ends of a base,
   The arcs which her features subtended:
Then he framed transcendental equations, to trace
The flowing outlines of her figure and face,
   And thought the result very splendid.

3

He studied (since music has charms for the fair)
   The theory of fiddles and whistles—
Then composed, by acoustic equations, an air,
Which, when 'twas performed, made the lady's long hair
   Stand on end, like a porcupine's bristles.

"The Mathematician in Love" by W. J. M. Rankine, from *Songs and Fables*, 1874.

4

The lady loved dancing—he therefore applied,
　　To the polka and waltz, an equation;
But when to rotate on his axis he tried,
His center of gravity swayed to one side,
　　And he fell, by the earth's gravitation.

5

No doubts of the fate of his suit made him pause,
　　For he proved, to his own satisfaction,
That the fair one returned his affection—"because,
"As every one knows, by mechanical laws,
　　"Reaction is equal to action."

6

"Let x denote beauty—y, manners well-bred—
　　"Z, Fortune—(this last is essential)—
"Let L stand for love"—our philosopher said—
"Then L is a function of x,y, and z,
　　"Of the kind which is known as potential."

7

"Now integrate L with respect to $d\,t$,
　　"(t Standing for time and persuasion);
"Then, between proper limits, 'tis easy to see,
"The definite integral *Marriage* must be—
　　"(A very concise demonstration)."

8

Said he—"If the wandering course of the moon
　　"By Algebra can be predicted,
"The female affections must yield to it soon"—
—But the lady ran off with a dashing dragoon,
　　And left him amazed and afflicted.

9 PROBABILITY

If we were to record everything spoken by an individual during a single day, the chances are good that there would be many references to probability. The following sentences all use the language of probability.

1. What are my chances of getting into Professor Green's course?
2. There is a 4-in-10 chance of rain tomorrow.
3. This road will probably bring us out in the right place.
4. The odds for winning this game are in my favor.
5. In all probability I will go to the carnival.

In fact, the first sentence of this paragraph is a probabilistic statement.

Probability has to do with events that *may* occur, as opposed to those in which logic, facts, or proof *assure* the outcome. "When I am 18 years old I can vote" is a sentence devoid of probability, since it is a factual statement. Factual sentences tend not to be very interesting, since there is no chance that they can be wrong. It is the *probability*, or the *chance*, or the *possibility*, of events occurring that makes life interesting. It would be a dull world if only known cause and effect influenced events and there were no doubt about the probability of something happening in the future.

Gambling owes its appeal to the chances of outcomes, as typified in tossing coins. One never knows for certain that a coin will land heads. Yet we can say that the probability of heads is 1/2 or .5, meaning that in a large number of tosses of a coin we can expect nearly 50% heads. We will clarify such ideas in this chapter.

A gambler of seventeenth century France, Antoine Gombaud (Chevalier de Meré) posed a question to Blaise Pascal (1623–1662) about rolling dice. The question had to do with a game in which the dice were to be rolled, say, 20 times. Bets were placed on the outcome. How should the money be divided between two players if the game is stopped before the game is completed, say after 15 rolls? Pascal corresponded with Pierre de Fermat (1601–1665) about the problem. Together these two mathematicians not only solved Gombaud's problem, but also provided the foundations for probability as a science.

Today the theory of probability is quite advanced. It has applications in the study of heredity, in life insurance, in nuclear physics, in testing, in public opinion polling, and in many other areas.

Our goal here will be to explore probabilities as numbers that measure the likelihood of an occurrence. Before we get to probability itself, we will need to explore "combinations" and "permutations." These two terms refer to methods of counting events that occur without listing or enumerating each and every possibility.

9.1 COMBINATIONS

Out of a set of five possible grades { A, B, C, D, F }, Professor Jones uses only three. How many different subsets of three grades can he choose from the universe of these five grades? Of course, the set { B, C, D }, for example, is the same as the set { C, D, B }.

This is a question about *combinations*. We might ask more generally "How many ways can three objects be chosen from a set of five?" In the language of combinations this question becomes "What is the number of combinations of five things taken three at a time?"

Professor Jones could choose any of the following three-grade sets:

$$
\begin{array}{lllll}
\{ A, B, C \} & \{ A, C, D \} & \{ B, C, D \} & \{ C, D, F \} & \{ A, B, D \} \\
\{ A, C, F \} & \{ B, C, F \} & \{ A, B, F \} & \{ A, D, F \} & \{ B, D, F \}
\end{array}
$$

Can you find a pattern in this listing of the possible sets of three grades? The pattern assures us that we have all the possibilities. Since there are ten possible subsets of three in the given set of five, we will say that the number of combinations of five things taken three at a time is ten. The notation for this is $_5C_3$ or even more simply $\binom{5}{3}$. So $_5C_3 = 10$.

In general, the number of combinations of n things taken k at a time is denoted $_nC_k$ or $\binom{n}{k}$.

Suppose Professor Jones were to choose four grades from the set { A, B, C, D, F } instead of three. The possibilities are as follows:

$$
\begin{array}{lll}
\{ A, B, C, D \} & \{ A, B, D, F \} & \{ B, C, D, F \} \\
\{ A, B, C, F \} & \{ A, C, D, F \}
\end{array}
$$

In general the number of combinations of five things taken four at a time is five. That is, $_5C_4 = 5$, or $\binom{5}{4} = 5$.

Have you discovered a pattern yet for listing subsets containing k elements of a set containing n elements? In any case, there is an indirect way to find the *number* of these subsets in a given problem. It is clear that we need a way to find $_nC_k$ if we consider trying to find the number of ways we could select five different letters of the alphabet, that is, $_{26}C_5$. The answer turns out to be 65,780. Even if we knew *how* to list the subsets containing five elements, we would not want to list that many. But before we get a formula for $_nC_k$, let's review *factorial notation*.

In Chapter 7 we stated that $n!$ means to multiply together all the counting numbers from 1 to n. Thus $5! = 5(4)(3)(2)(1)$. In general,

$$ n! = 1(2)(3)(4) \ldots (n - 2)(n - 1)(n) $$

which is the same as

$$ n! = n(n - 1)(n - 2)(n - 3) \ldots (4)(3)(2)(1) $$

EXAMPLE 1

$$7! = 7(6)(5)(4)(3)(2)(1) = 5040$$
$$6! = 6(5)(4)(3)(2)(1) = 720$$
$$5! = 5(4)(3)(2)(1) = 120$$
$$4! = 4(3)(2)(1) = 24$$
$$3! = 3(2)(1) = 6$$
$$2! = 2(1) = 2$$

Now we state a formula for finding combinations of n things taken k at a time.

$$_nC_k = \binom{n}{k} = \frac{n!}{k!(n-k)!}$$

Each of n and k can be any counting numbers, provided that k is never greater than n. In order to use the formula for any such values of n and k, it will be necessary to define both 1! and 0! as 1.

$$1! = 1 \qquad \text{and} \qquad 0! = 1$$

Although these are strange definitions (given what $n!$ means), they will be consistent with our work.

Study the following examples carefully to see how to use the formula for $\binom{n}{k}$ and to see why our definitions of 0! and 1! make sense.

EXAMPLE 2

Find $\binom{5}{3}$. Here $n = 5$ and $k = 3$. So

$$\binom{5}{3} = \frac{5!}{3!(5-3)!} = \frac{5!}{3!2!} = \frac{5(4)(3)(2)(1)}{3(2)(1)(2)(1)}$$

$$= \frac{5(4)}{2(1)} \qquad \text{(Canceling } 3 \cdot 2 \cdot 1 \text{ in numerator and denominator)}$$

$$= \frac{20}{2} = 10$$

EXAMPLE 3

Find $_5C_4$. Here $n = 5$ and $k = 4$. So

$$_5C_4 = \frac{5!}{4!(5-4)!} = \frac{5!}{4!1!}$$

$$= \frac{120}{24(1)} \qquad \text{(Using factorial values given in Example 1)}$$

$$= 5$$

EXAMPLE 4

Find the number of ways of choosing the days of the week two at a time.

The choice Monday–Tuesday is considered the same as the choice Tuesday–Monday. We can denote this number as $\binom{7}{2}$ or as $_7C_2$:

$$\binom{7}{2} = \frac{7!}{2!5!} = \frac{7 \cdot 6 \cdot 5 \cdot 4 \cdot 3 \cdot 2 \cdot 1}{2 \cdot 1 \cdot 5 \cdot 4 \cdot 3 \cdot 2 \cdot 1}$$

$$= \frac{7 \cdot 6}{2 \cdot 1} \quad \text{(Canceling repeated factors)}$$

$$= 21$$

EXAMPLE 5

How many committees of four can be chosen from a class of thirty? Here $n = 30$ and $k = 4$. Thinking ahead, we see that 30! is needed. But 30! = $30(29)(28)(27) \ldots (3)(2)(1)$, and this can be written as 30! = $30(29!)$ or as $30(29)(28!)$ or as 30! = $30(29)(28)(27!)$, and so on.

$$\binom{30}{4} = \frac{30 \cdot 29 \cdot 28 \cdot 27 \cdot 26!}{4!(26!)}$$

$$= \frac{30 \cdot 29 \cdot 28 \cdot 27}{4 \cdot 3 \cdot 2 \cdot 1} \quad \text{(Canceling 26!)}$$

$$= 15 \cdot 29 \cdot 7 \cdot 9 \quad \begin{array}{l} \text{(Dividing 2 into 30,} \\ \text{3 into 27,} \\ \text{and 4 into 28)} \end{array}$$

$$= 27{,}405$$

EXAMPLE 6

In how many ways can we choose a subset containing four elements from a set containing four elements? Although the answer is obviously 1, the formula can still be used:

$$\binom{4}{4} = \frac{4!}{4!(4-4)!} = \frac{4!}{4!0!}$$

$$= \frac{4!}{4!(1)} \quad \text{(Since } 0! = 1\text{)}$$

$$= 1$$

Exercise Set 9.1

(Answers not easily calculated can be left in factorial form.)

1 In how many ways can 52 cards be selected
   (a) 1 at a time?   (b) 2 at a time?
   (c) 48 at a time?   (d) 13 at a time?

2 Five names are written on slips of paper and dropped in a hat. In how many ways can they be drawn
   (a) none at a time?   (b) one at a time?
   (c) two at a time?   (d) three at a time?
   (e) four at a time?   (f) five at a time?

3 Note that 2(c) and 2(d) result in the same answer. Try to explain why. What other pairs of answers are the same?

4 (a) Given a set with four elements. In how many ways may subsets be chosen containing no elements? One element? Two elements? Three elements? Four elements?

  (b) Given a set with six elements. How many subsets are there containing no elements? One element? Two elements? Three elements? Four elements? Five elements? Six elements?

  (c) How many subsets does a set with four elements have?

  (d) How many subsets does a set with six elements have?

5 Calculate

  (a) $\binom{3}{3}$ (b) $\binom{3}{2}$ (c) $\binom{3}{1}$ (d) $\binom{3}{0}$

6 Calculate $\binom{3}{3} + \binom{3}{2} + \binom{3}{1} + \binom{3}{0}$

7 Calculate

  (a) $_7C_6$ (b) $_9C_8$ (c) $_{11}C_{10}$ °(d) $_nC_{n-1}$

8 (a) Try to apply the formula definition for $\binom{n}{k}$ to $\binom{3}{4}$. What happens?

  (b) In how many ways can we select four books from a set of three?

9 Note in Example 5 that we do not have to write the complete expression for 30! because of the cancellation of 26!. We can also write $8! = 8 \cdot 7!$, $5! = 5 \cdot 4 \cdot 3!$, and $100! = 100 \cdot 99!$. Using the ideas implicit in these statements, calculate

  (a) 9!, knowing that $7! = 5040$.

  (b) $\binom{15}{14}$ (c) $\binom{15}{13}$ (d) $7!/5!$

10 What must n be in the formulas below?

  (a) $n(n-1)! = 8!$

  (b) $n(n-1)(n-2)! = 10!$

11 True or false: $n = \dfrac{n!}{(n-1)!}$

12 True or false: $\binom{n}{k} = \binom{n}{n-k}$

13 (a) How many different outfits can be made from four skirts and three blouses? Hint: Use the Fundamental Principle of Counting.

  (b) In how many ways can two different letters be selected from the set { a, b, c, d, e, f, g }?

14 In a word game a player draws the letters b, d, f, and g.

  (a) In how many ways can a set of three letters be chosen from the four?

  °(b) How many three-letter "words" can be formed? (We mean by "word" a sequence of letters, so bdf would be a different word from dbf.)

15 How many different three-note chords can be played on a piano (88 keys)? (This is equivalent to selecting 3 keys from a set of 88.)

9.2 PERMUTATIONS

When we calculated $\binom{n}{k}$ or $_nC_k$, we were determining the number of ways a subset could be selected from a given set. The order of elements in the subsets was not considered in working with combinations.

EXAMPLE 1

Given the set $\{1, 2, 3, 4\}$, the subsets containing three elements are $\{1, 2, 3\}, \{1, 2, 4\}, \{1, 3, 4\}$, and $\{2, 3, 4\}$. That there are four such subsets is assured by the fact that $\binom{4}{3}$ or $_4C_3$ is 4.

One of the subsets in Example 1 is $\{1, 2, 3\}$. As a set, $\{1, 2, 3\}$ is the same as $\{3, 2, 1\}$ or $\{2, 1, 3\}$, since equal sets contain the same elements. But suppose we are interested in knowing how many different three-digit numbers can be formed from the digits 1, 2, 3, and 4. Then certainly the three-digit numbers 123 and 321 and 213 are all different. In this case the different *orderings* or *arrangements* of elements in a set are important.

EXAMPLE 2

How many three-digit numbers can be formed from the numbers in the set $\{1, 2, 3, 4\}$? We must *select* subsets containing three digits, and then *arrange* them in all possible ways. The selection part is given by $_4C_3 = 4$ (see Example 1). But each of these selections can be rearranged or ordered in six ways. For example, the numbers 1, 2, and 4 can be arranged

$$\begin{array}{ccc} 124 & 214 & 412 \\ 142 & 241 & 421 \end{array}$$

Thus there must be $4 \times 6 = 24$ ways to form three-digit numbers using all four subsets of three digits.

The product 4×6 in Example 2 is a result of the Fundamental Principle of Counting. If we can select three digits in four ways, and if we can order each of these selections in six orders, then there are 4×6 ways to select and order the digits.

Table 1 organizes the 24 three-digit numbers referred to in Example 2.

We call an arrangement of a set of objects a *permutation* of those objects. Thus, AB and BA are the only permutations of the letters A and B. If k objects are selected from a set containing n elements and then arranged, we refer to the permutations of n objects taken k at a time. The number of such permutations is denoted $_nP_k$. In Table 1 we list the permutations of four objects taken three at a time; $_4P_3 = 24$.

Table 1   The 24 three-digit numbers formed from $\{1, 2, 3, 4\}$

Three-digit subsets	Three-digit numbers formed by ordering each subset					
$\{1, 2, 3\}$	123,	132,	213,	231,	312,	321
$\{1, 2, 4\}$	124,	142,	214,	241,	412,	421
$\{1, 3, 4\}$	134,	143,	314,	341,	413,	431
$\{2, 3, 4\}$	234,	243,	324,	342,	423,	432

EXAMPLE 3

Run-off elections are to be held between Smith, Jones, and Green. The top two vote-getters will be president and vice-president. In how many ways can the results occur? Given three candidates, $\binom{3}{2} = 3$ gives the number of ways two persons can be elected: Smith–Jones, Smith–Green, or Jones–Green. But then each of the two people in the winning pair can be president. So there are $3 \cdot 2$ or 6 ways the results may occur. They are shown in Table 2.

Table 2   Permutations of three names taken two at a time

President	Vice-president
Smith	Jones
Jones	Smith
Smith	Green
Green	Smith
Jones	Green
Green	Jones

Once a subset is selected, it is easy to figure out the possible number of permutations of that subset. If we are arranging three elements, we can think of placing them in boxes:

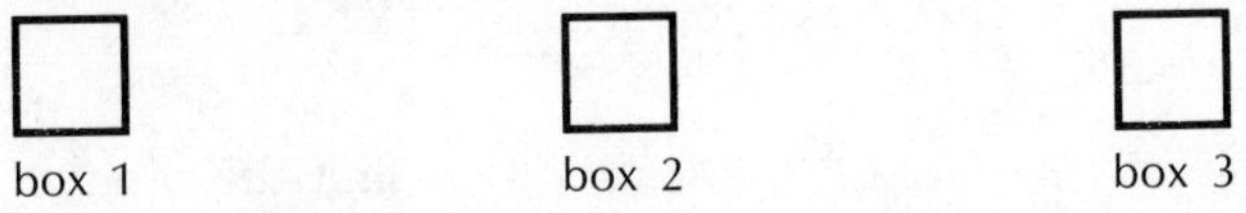

box 1   box 2   box 3

Figure 1   Arranging three elements.

For the three elements, there are three ways of filling box 1. Two elements are left, so then there are two ways to fill box 2. Then there is only one way of filling box 3. Thus there are $3 \cdot 2 \cdot 1$ or 3! ways to arrange 3 elements.

Similarly, there are $4 \cdot 3 \cdot 2 \cdot 1 = 4!$ ways to arrange 4 elements. And there are $5 \cdot 4 \cdot 3 \cdot 2 \cdot 1 = 5!$ ways to arrange 5 elements.

In general, there are $k!$ ways to arrange k elements.

EXAMPLE 4

Let { a, e, i, o, u } be the set of vowels. How many ways can we list the vowels? Since there are 5 vowels, there are 5! or 120 ways to order them.

Now suppose we ask for $_{26}P_5$, which might be the number of permutations of the 26 letters of the alphabet taken 5 at a time. We can select subsets of 5 from 26 elements in $\binom{26}{5}$ ways, and then arrange each subset of 5 in 5! or 120 ways. So 5 letters can be selected *and* arranged in $\binom{26}{5} \cdot 5!$ ways. The calculation of $_{26}P_5$ is as follows:

$$\binom{26}{5} \cdot 5! = \frac{26!}{5!21!} \cdot 5!$$

$$= \frac{26!}{21!} \qquad \text{(Canceling the 5!)}$$

$$= \frac{26 \cdot 25 \cdot 24 \cdot 23 \cdot 22 \cdot 21!}{21!}$$

$$= 26 \cdot 25 \cdot 24 \cdot 23 \cdot 22$$

$$= 7{,}893{,}600$$

(An interesting conclusion of this is that there is a maximum of 7,893,600 five-letter words in the English language, each of the five letters being different.)

EXAMPLE 5

On a small spice shelf there is room for only three spices on the top shelf. If you have ten different spices, in how many different ways can you arrange the top shelf? First, select 3 spices from 10 spices; this can be done in $\binom{10}{3} = 120$ ways. Then arrange each set of 3 spices; this can be done in 3! = 6 ways. Thus you can select and arrange spices for the top shelf in $120 \times 6 = 720$ ways.

To find $_nP_k$ we can always find the number of subsets containing k elements in the set of n elements, and then find the number of ways to arrange each of the subsets. We would multiply these two results. Since the number of subsets is given by $_nC_k$ or $\binom{n}{k}$, and since the arrangement of the subsets is always $k!$, we can find a formula for $_nP_k$ by multiplying $\binom{n}{k}$ by $k!$. That is,

$$_nP_k = \binom{n}{k} \cdot k!$$

$$_nP_k = \frac{n!}{k!(n-k)!} k! \qquad \text{[Definition of } \binom{n}{k} \text{]}$$

Then after canceling the $k!$ in numerator and denominator, we get a simpler formula for the number of permutations of n things taken k at a time:

$$_nP_k = \frac{n!}{(n-k)!}$$

EXAMPLE 6

How many different three-digit numerals are there if no digit is repeated and zeros are not allowed? Here the allowable numerals are $\{1, 2, 3, 4, 5, 6, 7, 8, 9\}$, so $n = 9$. Since we are to select three different digits, $k = 3$. Thus

$$_9P_3 = \frac{9!}{(9-3)!} = \frac{9!}{6!} = \frac{9 \cdot 8 \cdot 7 \cdot 6!}{6!} = 9 \cdot 8 \cdot 7 = 504$$

In Example 6 we required the three digits to be different. If repetitions of digits were allowed, then each of the three digits could be chosen in nine ways, so using the Fundamental Principle of Counting, we find that there would be $9 \cdot 9 \cdot 9 = 729$ three-digit numerals, allowing for repetitions (but no zeros). In problem-solving along these lines, it is important to know when to use combination–permutation ideas and when to simply apply the Fundamental Principle of Counting.

EXAMPLE 7

How many three-note tunes can be played on a piano? A "three-note tune" consists of first one note, then a second note, and then the third note, played successively. This question is not clear until we say whether or not the same note can be repeated. Suppose a note *can* be used repeatedly. Then there are 88 choices for the first note, 88 choices for the second note, and 88 choices for the third note. Thus there are $88 \cdot 88 \cdot 88 = 681,472$ three-note tunes, allowing repetitions. If a note *cannot* be used more than once, then there are $_{88}C_3$ ways to select three notes, and $3!$ ways to arrange them. Thus there are

$$_{88}P_3 = \frac{88!}{85!} = \frac{88 \cdot 87 \cdot 86 \cdot 85!}{85!} = 88 \cdot 87 \cdot 86 = 658,416$$

different three-note tunes.

Exercise Set 9.2

1 (a) Find $_5C_2$ and then find $_5P_2$.
   (b) How many permutations of two objects are there?

2 Calculate each of the following:
   (a) $_{10}P_7$
   (b) $_6P_1$
   (c) $_6P_5$
   (d) $_{26}P_3$
   (e) $_5P_5$
   (f) $_5P_0$

3 It is true that $_{10}C_3 = {_{10}C_7}$. Is it true that $_{10}P_3 = {_{10}P_7}$?

4 (a) In how many ways can five books be selected from a set of eight and arranged on a shelf?
   (b) Given five books, in how many ways can they be arranged?
   (c) In how many ways can five books be chosen from a set of eight?

5 (a) In how many ways can a poker hand of 5 cards be dealt from a deck of 52 cards? (Order is not important.)
   (b) How many different poker hands of five cards are there, where the order in which they are dealt is important?

6 (a) How many different nine-digit social security numbers are possible, where no digit is used more than once?
   (b) How many different nine-digit social security numbers are possible if a digit may be used repeatedly?

7 In how many ways can the letters of the word TEXAS be arranged?

8 List all possible distinguishable arrangements of the letters in each of the following words:
   (a) TWO   (b) TOO   (c) DEAR   (d) DEER

°9 If you have worked exercise 8 correctly, you have discovered that repetitions of letters alter the number of distinguishable permutations. A rule for finding permutation of n elements when there are r repetitions of the same element is to divide by $r!$. Also, if there are r repetitions of one element, s repetitions of another element, t repetitions of a third element, and so on, we divide by $r!$ and by $s!$ and by $t!$, and so on.

EXAMPLE

How many permutations of the letters a a b b b c are there?

$$\frac{6!}{2!\ 3!} = 60$$

The 6! arranges all six letters *as if* they were different letters. The 2! divides out the two repetitions of "a," while the 3! divides out the three repetitions of "b."

In how many ways can we arrange the letters of the word
   (a) BOOK?   (b) ELEVEN?   (c) MISSISSIPPI?

10 Given the set $\{1, 2, 3, 4, 5\}$.
   (a) How many three-digit numerals can be formed from the set if no digit can be repeated?
   °(b) How many three-digit numerals can be formed from the set if digits can be repeated? Hint: Do not use the ideas of exercise 9.
   (c) In how many ways may we select, from the given set, a subset of three digits?
   (d) How many different arrangements of the subsets in (c) are there?

°11 Which would be easier: to write down all the permutations of the digits of 12345 or to do so for 123453? Why?

°12 Generally, which is larger, $_nC_k$ or $_nP_k$? Can you find a situation in which $_nC_k = {_nP_k}$?

9.3 CALCULATING PROBABILITIES

Suppose we were to perform an experiment in which we toss a pair of coins 100 times. In how many tosses out of the 100 can we expect to get double heads? (That is, both coins land heads up.)

If the experiment was actually performed and double heads appeared, say, 28 times, is 28 the number we *expect*? The experiment can be repeated a few times. Suppose there are 24 double heads out of 100 tosses the second time, and 23 double heads the third time. The numbers 28, 24, and 23 are not the same. Apparently the performance of the experiment three times does not give us the expected number of double heads in 100 tosses of two coins.

However, if the experiment were repeated 50 times, the number of double heads would begin to cluster about 25. Probability can be used to tell us in advance that 25 is the number of double heads we can expect.

Before we define probability, let us explain some terms related to such an experiment. In each toss of two coins, either can come up heads or tails. By the Fundamental Principle of Counting, then, there are $2 \times 2 = 4$ possible outcomes for the pair of coins. The set of all possible outcomes in an experiment is called the *sample space*.

The sample space for our experiment is shown in Figure 2, where *HT*, for example, means heads on the first coin and tails on the second. There are four elements in the sample space. An *event* is any subset of the sample space. In our experiment we are interested in the event *HH*, a subset of the sample space.

In some other experiment we might be interested in the event *HH* ∪ *TT*, that is, double heads *or* double tails. (Notice our use of the union symbol for "or.") Then we can denote the event *HH* ∪ *TT* as a subset of the sample space by circling these two outcomes, as in Figure 3.

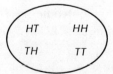

Figure 2 The sample space for tossing two coins.

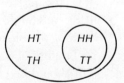

Figure 3 The event *HH* ∪ *TT* as a subset of a sample space.

EXAMPLE 1

Show the sample space for rolling a die (plural: "dice"). Also show the event "rolling an even number on a die." Since dice have spots on them corresponding to the numbers 1, 2, 3, 4, 5, and 6, we can denote the sample space as the set $\{\,1, 2, 3, 4, 5, 6\,\}$ (see Figure 4).

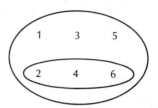

Figure 4   The sample space for rolling a die and the subset event "rolling an even number."

The four possible outcomes of a single toss of two coins are said to be *mutually exclusive*. In general, events (or outcomes) are mutually exclusive if the occurrence of one outcome prevents the occurrence of any other. Only one of *HH, HT, TH, TT* occurs in a single toss of two coins.

Furthermore, each outcome (*HH, HT, TH,* and *TT*) is *equally likely.* That is, any one outcome has as much chance of occurring as any other.

In rolling a die, the outcomes are $\{\,1, 2, 3, 4, 5, 6\,\}$. The six possible outcomes are mutually exclusive and equally likely. In this chapter we will deal only with equally likely events; later, we will see what happens when events are not mutually exclusive. Now we can define the probability of an event.

DEFINITION 1   Probability

Suppose a sample space consists of n equally likely and mutually exclusive outcomes. If an event A contains a of the outcomes, then the *probability* of event A is a/n. We write the probability of event A as follows:

$$P(A) = a/n$$

In tossing two coins, we have four mutually exclusive and equally likely outcomes, so $n = 4$. Only one of the outcomes is *HH*, so the event "tossing double heads" contains 1 outcome, $a = 1$. Then, on each toss, the probability of double heads is $1/4 = 0.25$.

It should be clear that in rolling a die $n = 6$. If event A is "rolling an even number," A contains 3 outcomes, so $a = 3$. Then $P(A) = 3/6 = 1/2$

= 0.5. (It is common to convert probability fractions to their decimal equivalents.)

To say P(double heads) = 0.25 means that we can expect two heads in 25% of a large number of tosses of two coins. This does not mean that we *must* get two heads at least once in four tosses. Nor does it mean that two heads will come up in the fourth toss if they have not come up in the first three. We can, however, expect 25 double heads in 100 tosses, or 250 in 1000 tosses, and so on.

If two coins are tossed 1000 times, we *expect* 250 double heads, but we will not be surprised if 249 or 252 show up. Probability does not tell us what *will* happen, it tells us only what to expect.

EXAMPLE 2

Suppose three cards from a bridge deck are placed upside down on a table. You are told that two of the cards are aces and one is a queen. If you draw one card, what is the probability of getting an ace? Here the number of outcomes (n) is the number of possible draws, so $n = 3$. Since there are two draws that could result in an ace, a is 2. The event A is "getting an ace." The probability of the event, $P(A)$, is then 2/3.

EXAMPLE 3

What is the sample space for tossing three coins? We can write this *HHH*, *HHT*, *HTH*, and so on, or we can tabulate the outcomes. Since each coin can be H or T, there are $2(2)(2) = 8$ outcomes in the sample space (see Table 3).

Table 3   Tabulating the sample space for tossing three coins

Coin 1	Coin 2	Coin 3
H	H	H
H	H	T
H	T	H
H	T	T
T	H	H
T	H	T
T	T	H
T	T	T

EXAMPLE 4

What is the probability of getting at least one tail in a toss of three coins? The sample space contains eight outcomes, so $n = 8$. (See Table 3.) Since seven of the outcomes contain a T, $a = 7$. Hence, P(at least one tail) = 7/8.

Table 4   The sample space for rolling a pair of dice

	1	2	3	4	5	6
1	(1, 1)	(1, 2)	(1, 3)	(1, 4)	(1, 5)	(1, 6)
2	(2, 1)	(2, 2)	(2, 3)	(2, 4)	(2, 5)	(2, 6)
3	(3, 1)	(3, 2)	(3, 3)	(3, 4)	(3, 5)	(3, 6)
4	(4, 1)	(4, 2)	(4, 3)	(4, 4)	(4, 5)	(4, 6)
5	(5, 1)	(5, 2)	(5, 3)	(5, 4)	(5, 5)	(5, 6)
6	(6, 1)	(6, 2)	(6, 3)	(6, 4)	(6, 5)	(6, 6)

When a pair of dice are rolled, each die can show six different numbers. Then there must be $6 \times 6 = 36$ elements in the sample space. If the first die comes up a 3 and the second comes up a five, we can indicate this outcome as the ordered pair (3, 5). Table 4 shows the sample space in tabular form.

Remember that an event is any subset of the sample space. Since there are 36 ordered pairs in the sample space for rolling a pair of dice, there are 2^{36} or 68,719,476,736 different subsets of the sample space. That is, we can ask for the probability of almost 69 billion events when two dice are rolled. No wonder people are so fascinated by dice games!

EXAMPLE 5

What is the probability of tossing doubles with a pair of dice? The event (as a subset of the sample space in Table 4) is $\{ (1, 1),(2, 2),(3, 3),(4, 4), (5, 5),(6, 6) \}$. Since there are 6 ordered pairs in the event and 36 ordered pairs in the sample space,

$$P(\text{tossing doubles}) = 6/36 = 1/6$$

EXAMPLE 6

What is the probability of throwing two dice and getting at least a sum of 9? The event is $\{ (6, 3),(5, 4),(4, 5),(3, 6),(6, 4),(5, 5),(4, 6),(6, 5), (5, 6),(6, 6) \}$; that is, there are 10 successful outcomes. Thus

$$P(\text{sum at least 9}) = 10/36 = 5/18$$

Sometimes the sample space is too large to list all the elements. For example, if a committee of four is chosen at random (by drawing lots) from among five men and three women, then there are $\binom{8}{4} = 70$ such committees. We would not want to list 70 four-person committees. Yet we can still ask probability questions about certain subsets of the 70 possible committees.

EXAMPLE 7

For the 70 committees described in the previous paragraph, what is the probability that the committee chosen will contain all men? Since there are five men, we can select them four at a time in $\binom{5}{4}$ = 5 ways. Certainly these five all-male committees are counted in the 70 committees. Thus P(all men) = 5/70 = 1/14.

Remember that $\binom{8}{4}$ and $\binom{5}{4}$ are symbols standing for numbers of combinations. Thus we used combination ideas to determine the size of the sample space, and the size of a particular subset. Sometimes we will want to ask about ordered subsets, so permutation ideas will be used.

EXAMPLE 8

From a set of paints containing red, orange, yellow, green, blue, and violet, three colors are chosen at random. What is the probability that the result is yellow, red, and green, in that order? The number of ordered subsets of three in a set of six is given by $_6P_3 = \dfrac{6!}{3!} = 120$. Clearly, only one of these ordered sets is referred to, so P (yellow, red, green) *in order* = 1/120.

Sometimes we use the Fundamental Principle of Counting to determine the size of the sample space.

EXAMPLE 9

If a monkey happens to dial a seven-digit number on a telephone, what is the probability that it is your number? There are ten choices for each of seven dials; thus the monkey could dial 10(10)(10)(10)(10)(10)(10) or 10,000,000 different telephone numbers. Since only one of them is your number, the probability that the monkey dials your number is

$$\frac{1}{10,000,000} = .0000001$$

Exercise Set 9.3

1 You and your friend have chances of getting either an A, a B, or a C in this course. (Assume that nothing but chance affects your grades.)
   (a) List the sample space in some convenient form.
   (b) What is the probability that you both get the same grade?
   (c) What is the probability that only one of you gets an A?
   (d) What is the probability that one or both of you get an A?
   (e) What is the probability that your friend gets a higher grade than you do?

2 How many elements are there in the sample space corresponding to tossing
   (a) two coins?   (b) three coins?   (c) four coins?
   (d) five coins?   (e) *n* coins?

3 What is the probability of getting all heads when you toss six coins?

4 (a) If three coins are tossed, what is the probability of getting at least two
       heads? (See Table 3.)
   (b) If three coins are tossed 1000 times, approximately how many tosses are
       expected to result in at least two heads?
   (c) Is it possible to toss *HHH* 1000 times in a row?
   (d) What would you say about the probability of this happening?

5 Gregor Mendel, the great pioneer in genetic theory, found in certain hybrid-
   ization experiments (in which he recorded results of 19,959 crosses of hybrid
   pea plants) that the second generation of plants showed 14,949 plants with
   dominant characteristics and 5010 with recessive characteristics. This is a
   ratio of 2.98 to 1, or approximately 3 to 1.
   (a) If you were to duplicate this experiment, what would be the probability
       that any single plant of the second generation would exhibit dominant
       characteristics?
   (b) How many plants out of 100 would you expect to have dominant char-
       acteristics?

6 If you toss two dice, what is the probability that you get a
   (a) 6 on one of them?   (b) sum that is even?   (c) sum of 5 or less?
   (d) sum of 12 or less?   (e) sum of 7 or 11?

7 Given a bridge deck consisting of 52 cards, A 2 3 4 5 6 7 8 9 10 J Q K in each of
   four suits, hearts (red), diamonds (red), spades (black), and clubs (black). On a
   draw of a single card from the deck, what is the probability of drawing
   (a) a club?   (b) a red card?
   (c) a black card?   (d) an ace?

8 Given a bridge deck, find the probability of each of the following:
   (a) Drawing three red cards in a row, replacing and shuffling after each draw.
   (b) Drawing three red cards at one time (without replacement).
   (c) Drawing a hand (13 cards) in which all cards are red.

9 (a) Toss 3 coins 48 times, keeping track of the number of times at least one
       head comes up. Report your results to class.
   (b) Calculate the probability of getting at least one head when three coins are
       tossed. (See Table 3.)
   (c) In view of your answer to 9(b), how many times out of 48 would you expect
       at least one head?
   (d) Would you bet a dollar that at least 40 times out of 48 tosses of three coins
       you will get at least one head?

10 (a) Would you bet someone a dollar that when three coins are tossed once,
       exactly two heads will come up?
   (b) What is the probability that this will happen?
   (c) What percentage of the time, when tossing three coins, do you expect to
       toss exactly two heads?

9.4   PROBABILITIES OF COMPOUND EVENTS

When two coins are tossed the sample space contains four elements: *HH*, *HT*, *TH*, *TT*. The sample space can be thought of as containing four separate events, each of which is equally likely. Table 5 names the events $A_1, A_2, A_3,$ and A_4 and shows the probability of each event.

Table 5   Probabilities of outcomes when two coins are tossed

Event name	Event or Outcome	Probability of events
A_1	*HH*	$P(A_1) = 1/4$
A_2	*HT*	$P(A_2) = 1/4$
A_3	*TH*	$P(A_3) = 1/4$
A_4	*TT*	$P(A_4) = 1/4$

First we should note from Table 5 that the sum of the probabilities is 1; that is,

$$P(A_1) + P(A_2) + P(A_3) + P(A_4) = 1$$

When two coins are tossed, exactly one of the events must occur, that is, A_1 or A_2 or A_3 or A_4 is certain to happen. It is also true that if A_1 occurs, for example, none of the other events can happen; that is, the events are mutually exclusive.

So when the elements of a sample space are thought of as separate mutually exclusive events, the sum of the probabilities of the events is 1. Since some event in the sample space is certain to occur, this means that the probability of certainty is 1.

What is the probability of getting either two heads or two tails on one toss of two coins? In Table 5, we notice that $P(HH) = 1/4$ and $P(TT) = 1/4$. The probability of *HH* or *TT* is $1/4 + 1/4 = 1/2$. This can be written in set notation as

$$P(HH \cup TT) = 1/2$$

We can add probabilities of separate events only if they are mutually exclusive. What happens when the events are not mutually exclusive?

<u>EXAMPLE 1</u>

Consider the sample space for tossing two coins and these events:

E_1: tossing a head on the first coin.
E_2: tossing a head on the second coin.

Since E_1 is the subset $\{ HH, HT \}$, we have $P(E_1) = 2/4 = 1/2$. Since E_2 is the subset $\{ HH, TH \}$, we have $P(E_2) = 1/2$. The probability of $E_1 \cup E_2$

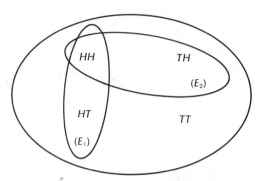

Figure 5   A sample space showing non-mutually exclusive events.

$(E_1$ or $E_2)$ is not $1/2 + 1/2 = 1$, since $(E_1$ or $E_2)$ as a compound event is not certain; that is, the coins *could* land *TT*. This time we cannot add $P(E_1)$ and $P(E_2)$ to get $P(E_1 \cup E_2)$. The reason is that E_1 and E_2 are not mutually exclusive. If *HH* occurs, then both E_1 and E_2 have happened simultaneously. Clearly $E_1 \cup E_2 = \{\, HH, TH, HT \,\}$, so $P(E_1 \cup E_2) = 3/4$.

An examination of Venn diagrams will show how to find the probability of a compound event like $E_1 \cup E_2$ when the events are not mutually exclusive.

Figure 6 is the Venn diagram for $E_1 \cup E_2$.

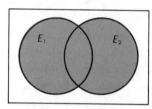

Figure 6   The Venn diagram for $E_1 \cup E_2$.

The number of points in $E_1 \cup E_2$ is the number of points in E_1, plus the number of points in E_2, minus the number of points in $E_1 \cap E_2$ (the intersection of E_1 and E_2). Schematically, we can show this with Venn diagrams (Figure 7). (Recall that $n(A)$ is notation for the cardinality of set A, that is, the number of elements in A.)

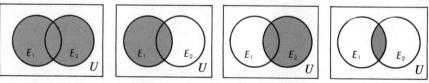

Figure 7 $n(E_1 \cup E_2) = n(E_1) + n(E_2) - n(E_1 \cap E_2)$

Let us now divide both sides of this equation by the number of elements in the sample space. It is appropriate to write this number as $n(U)$ referring to the universal set.

$$\frac{n(E_1 \cup E_2)}{n(U)} = \frac{n(E_1)}{n(U)} + \frac{n(E_2)}{n(U)} - \frac{n(E_1 \cap E_2)}{n(U)}$$

But these fractions are probabilities. For example $n(E_1)/n(U)$ is the number of elements in event E_1 divided by the number of elements in the sample space, which is $P(E_1)$. Thus we have

$$P(E_1 \cup E_2) = P(E_1) + P(E_2) - P(E_1 \cap E_2)$$

The compound event $E_1 \cap E_2$ can be thought of as E_1 and E_2. In Example 1, $E_1 \cap E_2 = \{HH\}$, since that outcome is heads on the first coin and heads on the second coin. Then $P(E_1 \cap E_2) = 1/4$. Applying our formula to Example 1, we get $P(E_1 \cup E_2) = 1/2 + 1/2 - 1/4 = 3/4$, which agrees with our result in Example 1.

EXAMPLE 2

Let E_1 = rolling a sum of at least 11 on a pair of dice, and
E_2 = rolling doubles. (See Table 4.)

Then $E_1 = \{(6, 5),(5, 6),(6, 6)\}$ and $E_2 = \{(1, 1),(2, 2),(3, 3),(4, 4),(5, 5),$ $(6, 6)\}$. We see also that $E_1 \cap E_2 = \{6, 6\}$. Then $P(E_1) = 3/36$, $P(E_2) = 6/36$, $P(E_1 \cap E_2) = 1/36$. So the probability of rolling a sum of at least 11 or rolling doubles is

$$P(E_1 \cup E_2) = 3/36 + 6/36 - 1/36 = 8/36 = 2/9$$

What is the probability of rolling a 7 with one die? Since the sample space is $\{1, 2, 3, 4, 5, 6\}$, this cannot happen. If we call "rolling a 7" event E_1, then E_1 is an empty subset of the sample space; at any rate, $n(E_1) = 0$, so $P(E_1) = 0/6 = 0$. Thus, if an event is impossible, its probability is 0. Remembering that the probability of certainty is 1, we now know that the probability of any event A is 0, 1, or some number in between. That is, for all events A,

$$0 \le P(A) \le 1$$

If the probability of an event A is a number a/n, then the probability that A does not happen is $1 - (a/n)$. Once again, we can explain this in terms of sets. Event A is a subset of U, the sample space. The probability of A is

$$P(A) = \frac{n(A)}{n(U)}$$

The event "A does not happen" would be a subset of U outside the set A, and can be denoted A' (the complement of A). Since $A \cup A' = U$, and A and A' are mutually exclusive (or disjoint sets), then $P(A \cup A') = P(A) + P(A') = 1$. So $P(A') = 1 - P(A)$. Hence, if $P(A) = a/n$, it follows that $P(A') = 1 - a/n$.

EXAMPLE 3

A pair of dice is thrown. The probability that doubles come up with a pair of dice is $6/36 = 1/6$. The probability that doubles do *not* come up is $1 - 1/6 = 5/6$.

EXAMPLE 4

The probability of drawing an ace from a deck of 52 cards is $4/52 = 1/13$. The probability of not drawing an ace is $1 - 1/13 = 12/13$.

Sometimes compound events occur in which the probability of the first event in no way affects the probability of the second event. Such events are called *independent events*. If John first rolls a die and then tosses a coin, the events are independent. We can still ask probability questions about such compound events. For example, what is the probability that John rolls a 3 on the die and then tosses a head on the coin? To find the answer let us first tabulate the sample space. The die can land 1, 2, 3, 4, 5, or 6 and the coin can land H or T. By the Fundamental Principle of Counting there must be $6 \cdot 2 = 12$ mutually exclusive events in the sample space (see Table 6). The event "rolling a 3 and then tossing a head" is $\{(3, H)\}$. Since there is only one success out of 12 possible outcomes, the probability is $1/12$.

Note that if we let $E_1 =$ rolling a 3 and $E_2 =$ tossing a head, then $P(E_1) = 1/6$ and $P(E_2) = 1/2$. Also, $P(\text{rolling a 3 and then tossing a head}) = 1/12$. This time $P(E_1 \text{ and } E_2) = 1/6 \cdot 1/2 = 1/12$.

In general, when two events are independent we find the probability that both of them occur by multiplying their separate probabilities.

Table 6   The sample space for rolling a die and then tossing a coin

		Coin	
		H	T
	1	1, H	1, T
	2	2, H	2, T
Die	3	3, H	3, T
	4	4, H	4, T
	5	5, H	5, T
	6	6, H	6, T

EXAMPLE 5

In rolling a die twice, what is the probability of rolling less than 4 on the first roll (E_1) and more than 4 on the second roll (E_2)? Clearly, $E_1 = \{1, 2, 3, \}$ and $E_2 = \{5, 6\}$. Then $P(E_1) = 3/6$ and $P(E_2) = 2/6$. Thus $P(E_1$ and then $E_2) = 3/6 \cdot 2/6 = 6/36 = 1/6$.

Rolling a die twice is equivalent to rolling a pair of dice once, provided that we can identify the first die and the second die. Thus the two independent events, E_1 and E_2, in Example 5 are equivalent to the event $E_3 =$ "rolling less than 4 on the first die and rolling more than 4 on the second." The event E_3 is a subset of the sample space for rolling a pair of dice, Table 4:

$$E_3 = \{ (1, 5),(1, 6),(2, 5),(2, 6),(3, 5),(3, 6) \},$$
$$\text{and so } P(E_3) = 6/36 = 1/6.$$

EXAMPLE 6

Suppose the probability of a boy being born is 1/2 (It is actually slightly more.) What is the probability that a family of two children contains two boys? Assuming that the sex of the second child in no way depends upon the sex of the first, the events are independent. Thus the probability of two boys being born is

$$1/2 \cdot 1/2 = 1/4$$

Exercise Set 9.4

1 A bag contains a mixture of red and white marbles. One marble is selected at random.
   (a) What is the probability that it is a green marble?
   (b) What is the probability that it is either a red or a white marble?
   (c) We cannot determine the probability that the marble is red. Why not?

2 Consider Figure 2, the sample space for tossing two coins. Event E_1 is tossing a head on the first coin, event E_2 is tossing a head on the second coin, and event E_3 is tossing double heads or double tails.
   (a) List the elements in E_1, E_2, and E_3.
   (b) Are any two of these events mutually exclusive?
   (c) Let event $E_4 = E_1 \cup E_3$. Describe E_4 in words and list the elements of E_4.
   (d) Find $E_1 \cap E_3$.
   (e) What is the probability of E_4? Use the formula

$$P(E_4) = P(E_1) + P(E_3) - P(E_1 \cap E_3)$$

3 One card is drawn from a bridge deck (52 cards). What is the probability that the card is
   (a) an ace?
   (b) an ace or a jack?
   (c) a picture card (king, queen, or jack)?
   (d) a picture card or an ace?
   (e) an ace or a spade? (not mutually exclusive)
   (f) a spade or a diamond?
   (g) a spade or a diamond or an ace?

4 If the probability of event A is .3, what is the probability of A'?

5 Three coins are tossed. Given that the probability of three heads is 1/8, what is the probability of at least one tail?

6 In a given telephone book it is estimated that 300 last names are Smith and that there are 50,000 names in the book.
   (a) If a listed telephone number is dialed at random, what is the probability that the party reached will be named Smith?
   (b) What is the probability that the party will not be named Smith?

7 A bag contains 13 red marbles and 7 white marbles.
   (a) If one marble is drawn at random, what is the probability that it is red? That it is white?
   (b) If two marbles are simultaneously drawn at random, what is the probability that they are both white? That they are both red?
   (c) Using the results of both parts of 7(b), find the probability that one of the marbles drawn is red and the other is white. Hint: The only possibilities for two marbles are both white, or both red, or one red and the other white.

8 A pair of dice is rolled. What is the probability of rolling
   (a) a sum of 7?   (b) a sum of 11?
   (c) first a sum of 7 and then a sum of 11?

9 A pair of dice is rolled. What is the probability of rolling
   (a) doubles?   (b) doubles twice in a row?
   (c) doubles three times in a row?

10 What is the probability of first rolling a die and getting a 3 and then tossing a coin and getting tails?

11 (a) If the probability that John will get an A is 1/3 and the probability that Mary will get an A is 2/3, what is the probability that John and Mary will both get A's?
   (b) What is the probability that Mary will not get an A?
   (c) What is the probability that John will get an A but Mary won't?

12 If your chances of getting a parking space are 1 out of 4 on any given day, what is the probability that you will
   (a) get a parking space on any given day?
   (b) get a parking space five days in a row?
   (c) not get a parking space five days in a row?
   (d) get a parking space on Monday but not on Tuesday?

9.5 ODDS AND MATHEMATICAL EXPECTATION

Many interesting probability problems can be stated in terms of *odds*. Odds are the bettor's way of stating probabilities.

EXAMPLE 1

In a certain automobile race in which there are 8 cars, let us assume that every car has an equal chance of winning. Then the *odds in favor* of car A winning are 1:7 and the *odds against* car A winning are 7:1. The probability that car A will win is $1/8$ and the probability that car A will not win is $7/8$.

DEFINITION 2   Odds

The *odds against* event A happening are given by the ratio $P(A')/P(A)$; that is,

$$\frac{\text{probability that } A \text{ does not occur}}{\text{probability that } A \text{ does occur}}$$

The *odds in favor* of event A are given by the ratio $P(A)/P(A')$.

The ratios are sometimes written as fractions and sometimes written with a colon. For example, the odds in favor of A can be written $P(A):P(A')$.

EXAMPLE 2

At the racetrack the probability that a horse named Jet Fire wins is $3/4$. The probability that Jet Fire loses is therefore $1/4$. The odds against Jet Fire are $\dfrac{1/4}{3/4} = 1/3$. The odds in favor of Jet Fire are 3:1.

At racetracks odds are given instead of probabilities, and are based upon the amount of money bet. If \$10,000 is bet for Jet Fire to win and \$70,000 is bet against the horse, then the odds against Jet Fire are 7:1. The term "odds against" is frequently more descriptive of an event that seems unlikely to happen.

EXAMPLE 3

What are the odds against rolling a pair of dice and getting a sum of 12?

This only happens when the dice land (6, 6). The probability $P(6, 6) =$ 1/36 and the probability of not getting double sixes is 35/36. Thus the odds against rolling a sum of 12 are

$$\frac{35/36}{1/36} = 35/1 \text{ or } 35{:}1$$

When the probabilities for and against an event are equal, that is, both 1/2, the odds are 1:1. Such odds indicate that the event is just as likely to happen as not.

EXAMPLE 4

What are the odds in favor of drawing a red card from a bridge deck? Since there are 52 cards and 26 of them are red, the probability of drawing a red card is $26/52 = 1/2$. The probability of not drawing a red card is also 1/2. So the odds in favor of a red card are $\frac{1/2}{1/2} = 1/1$ or 1:1.

There is an alternative way of defining odds. If an event can happen in x ways and can fail to happen in y ways, then the odds in favor would be the ratio $x{:}y$. The odds against the event would be $y{:}x$.

EXAMPLE 5

There are 6 ways of rolling doubles with a pair of dice and 30 ways of failing to roll doubles. The odds against rolling doubles are 30:6. Note that this reduces to 5:1. Observe that the probability of rolling doubles is 6/36 and of not rolling doubles is 30/36. When 30/36 is divided by 6/36 in calculating odds against, the 36 cancels, giving the same result, 30/6.

If odds for an event A are given, the probability of event A is easily calculated. Suppose the odds for event A are 3:5. This suggests that event A may happen three times and fail to happen five times if the event is repeated eight times. Thus the probability of event A is 3/8. Given odds of $x{:}y$, the denominator of the probabilities involved is always $x + y$. This can be shown algebraically, but for the sake of simplicity we will state it as a rule.

RULE

If the odds in favor of event A are $x{:}y$, then the probability that A happens is $\frac{x}{x + y}$. The probability that A does not happen is $\frac{y}{x + y}$.

EXAMPLE 6

The odds in favor of Jet Fire winning are established as 3:1. The probability that Jet Fire will win is 3/(3 + 1) or 3/4. The probability that Jet Fire will not win is 1/(3 + 1) = 1/4.

This last example should indicate to you that given odds we can calculate probability and given probability we can calculate odds. Probability and odds are mathematically equivalent ways of measuring the degree to which we can expect an event to occur. In use, odds are more closely associated with gambling than probability is.

Often when probabilistic games are played there is a certain "payoff," that is, an amount is paid to the players who win. We can use the term "payoff" more generally to stand for the value of an outcome. Sometimes the payoff is a negative amount; the player who loses must pay a certain penalty. To combine probability and payoff amounts, we use the term *mathematical expectation* or just *expectation*.

DEFINITION 3   Mathematical Expectation

The *mathematical expectation* associated with an event E, denoted $M(E)$, is the product of the probability that E occurs and the payoff value, V, resulting when E happens. In symbols,

$$M(E) = P(E) \cdot V$$

EXAMPLE 7

In a state lottery 1,000,000 tickets are sold and the winning ticket yields $50,000. If you buy one ticket, what is your expectation? You have $\frac{1}{1,000,000}$ probability of winning a $50,000 payoff. Hence your expectation is

$$M(E) = \frac{1}{1,000,000} \times \$50,000 = \$0.05$$

This 5¢ expectation means that if you were to buy lottery tickets over a long period of time, you could expect to average 45¢ loss for each 50¢ spent. The state gains 45¢ for each 50¢ ticket bought, on the average. Thus, if 1,000,000 tickets are sold at 50¢ apiece, the state receives 45/50 of $500,000 or $450,000.

To make a game "fair" the expectation should equal the cost to play. In order for the expectation in Example 7 to be 50¢, the cost of a ticket, the prize must be $500,000:

$$\frac{1}{1,000,000} \times \$500,000 = \$0.50$$

But what would the state gain in that case?

EXAMPLE 8

Is the following a fair game? You pay a dollar to play, and a coin is flipped. If it comes up heads, you win $2.00. The probability of heads is 1/2, so your mathematical expectation is $1/2 \times \$2.00 = \$1.00$. Since this is the same as the cost to play, the game is fair.

For the game in Example 8, if you were the dealer and wanted to make money, you would have to pay less than $2.00 each time a player won. If you paid $1.50, the player's expectation would be 75¢, and the dealer would average 25¢ profit per play.

Often an event E consists of several mutually exclusive outcomes. If the probabilities of the separate outcomes are known, the expectation of E is a sum of several terms. For example, if E is made up of outcomes A_1, A_2, A_3, A_4, and A_5, with corresponding probabilities $P(A_1), P(A_2), P(A_3), P(A_4)$, and $P(A_5)$, and corresponding payoffs V_1, V_2, V_3, V_4, and V_5, then

$$M(E) = P(A_1) \cdot V_1 + P(A_2) \cdot V_2 + P(A_3) \cdot V_3 + P(A_4) \cdot V_4 + P(A_5) \cdot V_5$$

EXAMPLE 9

In a certain game with one die, a player gets 10¢ if six "shows," 5¢ if five shows, 4¢ if four shows, 3¢ if three shows, 2¢ if two shows, and −10¢ if one shows (−10¢ means he loses 10 cents). Find the expectation per play, knowing that the probability of each different show is 1/6. The expectation for 10¢ is $1/6 \cdot 10$¢; for 5¢ is $1/6 \cdot 5$¢, and so on. Thus the total expectation is simply the sum of the individual expectations.

$$\begin{aligned} M(E) &= \tfrac{1}{6}(10) + \tfrac{1}{6}(5) + \tfrac{1}{6}(4) + \tfrac{1}{6}(3) + \tfrac{1}{6}(2) - \tfrac{1}{6}(10) \\ &= 10/6 + 5/6 + 4/6 + 3/6 + 2/6 - 10/6 \\ &= 14/6 = 2\tfrac{1}{3}¢ \end{aligned}$$

In Example 9, the player can expect to gain $2\tfrac{1}{3}$¢ per play. To make it a fair game, the player would have to pay exactly $2\tfrac{1}{3}$ cents to roll the die (or 7¢ for three rolls of the die).

EXAMPLE 10

A careful observer established the following probabilities that birds would be at a certain birdfeeder at any moment of the day.

Number of birds	0	1	2	3	4	5
Probability	.60	.20	.11	.04	.03	.02

Find the expected number of birds at any moment. The payoffs are numbers of birds. Thus the expectation for 0 birds is $.60(0) = 0$, for 1 bird is $.20(1)$, for 2 birds is $.11(2)$, and so on. Thus

$$M(E) = .6(0) + .2(1) + .11(2) + .04(3) + .03(4) + .02(5)$$
$$= 0 + .2 + .22 + .12 + .12 + .10$$
$$= .76$$

While .76 live birds is an impossibility, the number still makes sense. If you were to check the birdfeeder 100 times, you could expect to have counted a total of 76 birds. Note that the probabilities are expressed as decimals and that the sum of the probabilities is 1.00. What is the significance of this sum in this example?

Exercise Set 9.5

1 What are the odds against drawing an ace from a bridge deck?

2 What are the odds in favor of getting at least one head in one toss of three coins?

3 There are three red and five black books on a shelf. If you select a book at random (say in the dark), what are the odds in favor of getting a black book?

4 A pair of dice is rolled. What are the odds against getting
   (a) double ones?
   (b) a sum equal to 7?
   (c) an even sum?
   (d) at least one 6?

5 If the odds in favor of an event are 3:11, what is the probability that the event will occur? What is the probability that the event will not occur?

6 On an assembly line it is found that 40 out of every 1000 cars produced have defective door locks.
   (a) What are the odds in favor of a particular car having defective door locks?
   (d) What are the odds against a defective door lock?
   (c) What is the probability that any particular car has defective door locks?
   (d) How many cars out of every hundred are expected not to have defective locks?

7 Play the game of Example 8 twenty times and keep track of the results in a table labeled as follows:

Game number	Player pays	Result on coin	Pay to Player

   (a) What amount does the player's opponent take in?
   (b) In your 20 plays how much does the player actually receive?
   (c) How much *should* the player have expected?

8 Change the game to yield a payoff of $1.50 instead of $2.00.
   (a) Answer the same questions as in exercise 7.
   (b) In 100 plays how much can the player expect to win or lose?

9 In a certain game, a pair of dice is rolled. The payoff is 50¢ if double sixes result.
   (a) What is the player's expectation?
   (b) If it costs 10¢ to play the game, who wins in the long run?
   (c) After 20 rolls of the dice, how much can the player expect to win or lose?
   (d) Roll a pair of dice 20 times, tabulating the results. Compare the results with those in 9(c).

10 The probability that a man at age 50 will die in the next year is .012, and so the probability that he will live another year is .988. Suppose he buys an insurance policy for $100. If the man dies, his estate receives $10,000, and if he lives he receives nothing.
   (a) What is the expectation of the man (or his estate) for the year in which he paid $100 for insurance?
   (b) What is the expectation of the insurance company?
   (c) Should the insurance company raise its rates?

11 Cars arriving at a certain college campus are observed over a period of time and the following data determined:

Number of occupants per car	1	2	3	4
Probability	.5	.3	.1	.1

(The probability of five or more students per car was considered too small to be of concern.)
   (a) What is the expected number of occupants per car?
   (b) In 100 cars how many students are expected to arrive on campus?
   (c) If the campus was built for 5000, all of whom must arrive on campus by car, how many student parking places should be provided?

12 A die is rolled. If a 1 shows, the player receives $1; if a 2 is rolled the player receives $2; and so on.
   (a) What is the player's mathematical expectation?
   (b) What is a fair amount to pay for playing this game?
   (c) If the player pays $3.00 each time the die is rolled, how much profit or loss can he expect in 10 rolls?

13 A pair of dice is tossed. If an even sum appears, the player wins $1.00; if an odd sum appears, the player loses $1.00. Is this a fair game?

SUMMARY

Probability has to do with predictions of the future, not as a fortune teller predicts the course of one's life, but on the basis of sound mathematical principles. It is strange mathematics, indeed, that can tell how many times out of a hundred we can expect to get heads on a toss of a coin. But though we *expect* 50 heads, there is a chance that the coin will land tails 100 times out of 100. The probability of this happening is very low, but it is not zero.

Furthermore, there is a probability that the sun will not come up tomorrow, or that all four tires of your car go flat simultaneously, or that you will inherit a million dollars, or that a monkey can, at random, type a perfect copy of the Gettysburg Address. Of course, all these probabilities are low. We are disturbed when something we expect to happen does not. Yet if every event in life were known in advance, life would be uninteresting.

As a probability, the number 2/5 can be thought of as a fractional part of 1. The event involved is a subset of the sample space. So an event whose probability is 2/5 contains 2/5 of the elements in the sample space. The sample space contains all possible outcomes, so we might say that the probability of the sample space as an event, is 1. Since we can speak of an event that is impossible (getting three heads on a toss of 2 coins, for example) the event must be the empty subset of the sample space, and it must have probability 0.

So all probabilities fall in the range from 0 to 1, inclusive. We can represent probabilities as fractions, like 2/5, or as decimals (2/5 = .4). Given the decimal form, it is easy to convert a probability to a percentage (.4 = 40%). Thus an event whose probability is 2/5 is expected to happen 40% of the time in a large number of repetitions of the experiment involved.

Once probability is viewed as a measure of what to expect, we can formalize the notion of mathematical expectation. This combines payoff values with probability. Thus, through probability theory, we not only know what percentage of the time to expect a certain outcome, but we can also calculate expected profits or losses.

You should be familiar with each of the following before trying the Review Test:

1. Notation for numbers of combinations and permutations, $\binom{n}{k}$ or $_nC_k$ and $_nP_k$, and how to evaluate these numbers.

2. How to find the number of arrangements of k objects.

3. How to tabulate sample spaces.

4. How to calculate probabilities of events.

5. How to add probabilities for mutually exclusive events.

6. How to use the formula
$$P(E_1 \cup E_2) = P(E_1) + P(E_2) - P(E_1 \cap E_2).$$

7. How to find probabilities for independent events.

8. How to find the probability of event A', given the probability of event A.

9. How to calculate odds against and odds in favor.

10. Given odds in favor of event A, how to find $P(A)$.

11. How to find mathematical expectation, given probabilities and payoff values.

REVIEW TEST

1 Find each of the following:
    (a) $\binom{9}{2}$ (b) $_7C_3$ (c) $_6P_3$

2 In how many ways can a committee of four be chosen from ten students?

3 How many different arrangements of the letters r, s, t, v are there?

4 How many license plates are there consisting of three numerals followed by three letters?

5 A pair of dice is tossed. What is the probability of tossing
    (a) 4?
    (b) more than 4?

6 (a) Construct the sample space for tossing four coins.
    (b) What is the probability of getting exactly two tails when four coins are tossed?
    (c) What is the probability of getting three or more heads when four coins are tossed?
    (d) In 160 tosses of 4 coins, how many times do you expect to get either all heads or all tails?

7 (a) Two coins are selected from a change purse containing a nickel, dime, quarter, and half-dollar. What is the probability that the sum is 75¢?
    (b) If one coin is selected from the purse, what is the probability it is a nickel?
    (c) If one coin is selected from the purse, what is the probability that it is a nickel or a dime?
    (d) If two coins are selected, what is the probability that one is a nickel or that the sum is less than 50¢?

8 First a coin is tossed and then a card is drawn from a bridge deck. What is the probability that we see a head on the coin and an ace from the deck?

9 If the probability of hitting a home run is .28, what is the probability of not hitting a home run?

10 In a certain lottery you pay $1.00 for a ticket and 1000 tickets are sold. The prize is worth $50.
    (a) What is your mathematical expectation?
    (b) How much would the prize have to be worth to make the lottery fair (that is, to give you an expectation equal to the price of the ticket)?
    (c) In a fair lottery, how much money is realized by those running the lottery?

11 (a) If the odds in favor of an event are 4:11, what is the probability that the event will *not* occur?
    (b) What are the odds against this same event?

12 On a true/false test there are five questions. The probability that you get any question right is ½ (you are guessing). What is the probability that you will get
    (a) all five questions right?
    (b) the first question right but all the others wrong?
    (c) at least one question right? Hint: That is, you do not get all the questions wrong.

The most improbable event has a probability! While
the probability of flipping a coin one hundred times
and getting a head *every* time is small, it is not zero.
Russell Maloney in his 1940 *New Yorker* story,
"Inflexible Logic," explores an event which we know
is impossible. (But *is* it?) Behind many a tale lies an
author's fascination with the improbable, or almost
impossible. This story is no exception.

INFLEXIBLE LOGIC
RUSSELL MALONEY

When the six chimpanzees came into his life, Mr. Bainbridge was
thirty-eight years old. He was a bachelor and lived comfortably in a
remote part of Connecticut, in a large old house with a carriage drive, a
conservatory, a tennis court, and a well-selected library. His income was
derived from impeccably situated real estate in New York City, and he
spent it soberly, in a manner which could give offence to nobody. Once a
year, late in April, his tennis court was resurfaced, and after that
anybody in the neighborhood was welcome to use it; his monthly
statement from Brentano's seldom ran below seventy-five dollars; every
third year, in November, he turned in his old Cadillac coupé for a new
one; he ordered his cigars, which were mild and rather moderately
priced, in shipments of one thousand, from a tobacconist in Havana;
because of the international situation he had cancelled arrangements to
travel abroad, and after due thought had decided to spend his travelling
allowance on wines, which seemed likely to get scarcer and more
expensive if the war lasted. On the whole, Mr. Bainbridge's life was
deliberately, and not too unsuccessfully, modelled after that of an English
country gentleman of the late eighteenth century, a gentleman
interested in the arts and in the expansion of science, and so sure of
himself that he didn't care if some people thought him eccentric.

Mr. Bainbridge had many friends in New York, and he spent several
days of the month in the city, staying at his club and looking around.

"Inflexible Logic" by Russell Maloney, *The New Yorker*, February 3, 1940, pp. 19–22.
Reprinted by permission; © 1940, 1968 The New Yorker Magazine, Inc.

Sometimes he called up a girl and took her out to a theatre and a night club. Sometimes he and a couple of classmates got a little tight and went to a prizefight. Mr. Bainbridge also looked in now and then at some of the conservative art galleries, and liked occasionally to go to a concert. And he liked cocktail parties, too, because of the fine footling conversation and the extraordinary number of pretty girls who had nothing else to do with the rest of their evening. It was at a New York cocktail party, however, that Mr. Bainbridge kept his preliminary appointment with doom. At one of the parties given by Hobie Packard, the stockbroker, he learned about the theory of the six chimpanzees.

It was almost six-forty. The people who had intended to have one drink and go had already gone, and the people who intended to stay were fortifying themselves with slightly dried canapés and talking animatedly. A group of stage and radio people had coagulated in one corner, near Packard's Capehart, and were wrangling about various methods of cheating the Collector of Internal Revenue. In another corner was a group of stockbrokers, talking about the greatest stockbroker of them all, Gauguin. Little Marcia Lupton was sitting with a young man, saying earnestly,"Do you really want to know what my greatest ambition is? I want to be myself," and Mr. Bainbridge smiled gently, thinking of the time Marcia had said that to him. Then he heard the voice of Bernard Weiss, the critic, saying, "Of course he wrote one good novel. It's not surprising. After all, we know that if six chimpanzees were set to work pounding six typewriters at random, they would, in a million years, write all the books in the British Museum."

Mr. Bainbridge drifted over to Weiss and was introduced to Weiss's companion, a Mr. Noble. "What's this about a million chimpanzees, Weiss?" he asked.

"Six chimpanzees," Mr. Weiss said. "It's an old cliché of the mathematicians. I thought everybody was told about it in school. Law of averages, you know, or maybe it's permutation and combination. The six chimps, just pounding away at the typewriter keys, would be bound to copy out all the books ever written by man. There are only so many possible combinations of letters and numerals, and they'd produce all of them—see? Of course they'd also turn out a mountain of gibberish, but they'd work the books in, too. All the books in the British Museum."

Mr. Bainbridge was delighted; this was the sort of talk he liked to hear when he came to New York. "Well, but look here," he said, just to keep up his part in the foolish conversation, "what if one of the chimpanzees finally did duplicate a book, right down to the last period, but left that off? Would that count?"

"I suppose not. Probably the chimpanzee would get around to doing the book again, and put the period in."

"What nonsense!" Mr. Noble cried.

"It may be nonsense, but Sir James Jeans believes it," Mr. Weiss said, huffily. "Jeans or Lancelot Hogben. I know I ran across it quite recently."

Mr. Bainbridge was impressed. He read quite a bit of popular science, and both Jeans and Hogben were in his library. "Is that so?" he murmured, no longer feeling frivolous. "Wonder if it has ever actually been tried? I mean, has anybody ever put six chimpanzees in a room with six typewriters and a lot of paper?"

Mr. Weiss glanced at Mr. Bainbridge's empty cocktail glass and said drily, "Probably not."

Nine weeks later, on a winter evening, Mr. Bainbridge was sitting in his study with his friend James Mallard, an assistant professor of mathematics at New Haven. He was plainly nervous as he poured himself a drink and said, "Mallard, I've asked you to come here—Brandy? Cigar?—for a particular reason. You remember that I wrote you some time ago, asking your opinion of . . . of a certain mathematical hypothesis or supposition."

"Yes," Professor Mallard said, briskly. "I remember perfectly. About the six chimpanzees and the British Museum. And I told you it was a perfectly solid popularization of a principle known to every schoolboy who had studied the science of probabilities."

"Precisely," Mr. Bainbridge said. "Well, Mallard, I made up my mind. . . It was not difficult for me, because I have, in spite of that fellow in the White House, been able to give something every year to the Museum of Natural History, and they were naturally glad to oblige me. . . . And after all, the only contribution a layman can make to the progress of science is to assist with the drudgery of experiment . . . . In short, I—"

"I suppose you're trying to tell me that you have procured six chimpanzees and set them to work at typewriters in order to see whether they will eventually write all the books in the British Museum. Is that it?"

"Yes, that's it," Mr. Bainbridge said. "What a mind you have, Mallard. Six fine young males, in perfect condition. I had a—I suppose you'd call it a dormitory—built out in back of the stable. The typewriters are in the conservatory. It's light and airy in there, and I moved most of the plants out. Mr. North, the man who owns the circus, very obligingly let me engage one of his best animal men. Really, it was no trouble at all."

Professor Mallard smiled indulgently. "After all, such a thing is not unheard of," he said. "I seem to remember that a man at some university put his graduate students to work flipping coins, to see if heads and tails came up an equal number of times. Of course they did."

Mr. Bainbridge looked at his friend very queerly. "Then you believe that any such principle of the science of probabilities will stand up under an actual test?"

"Certainly."

"You had better see for yourself." Mr. Bainbridge led Professor Mallard downstairs, along a corridor, through a disused music room, and into a large conservatory. The middle of the floor had been cleared of plants and was occupied by a row of six typewriter tables, each one supporting a hooded machine. At the left of each typewriter was a neat stack of yellow copy paper. Empty wastebaskets were under each table. The chairs were the unpadded, spring-backed kind favored by experienced stenographers. A large bunch of ripe bananas was hanging in one corner, and in another stood a Great Bear water-cooler and a rack of Lily cups. Six piles of typescript, each about a foot high, were ranged along the wall on an improvised shelf. Mr. Bainbridge picked up one of the piles, which he could just conveniently lift, and set it on a table before Professor Mallard. "The output to date of Chimpanzee A, known as Bill," he said simply.

" '*Oliver Twist*, by Charles Dickens,' " Professor Mallard read out. He read the first and second pages of the manuscript, then feverishly leafed through to the end. "You mean to tell me," he said, "that this chimpanzee has written—"

"Word for word and comma for comma," said Mr. Bainbridge. "Young, my butler, and I took turns comparing it with the edition I own. Having finished *Oliver Twist*, Bill is, as you see, starting the sociological works of Vilfredo Pareto, in Italian. At the rate he has been going, it should keep him busy for the rest of the month."

"And all the chimpanzees—" Professor Mallard was pale, and enunciated with difficulty—"they aren't all—"

"Oh, yes, all writing books which I have every reason to believe are in the British Museum. The prose of John Donne, some Anatole France, Conan Doyle, Galen, the collected plays of Somerset Maugham, Marcel Proust, the memoirs of the late Marie of Rumania, and a monograph by a Dr. Wiley on the marsh grasses of Maine and Massachusetts. I can sum it up for you, Mallard, by telling you that since I started this experiment, four weeks and some days ago, none of the chimpanzees has spoiled a single sheet of paper."

Professor Mallard straightened up, passed his handkerchief across his brow, and took a deep breath. "I apologize for my weakness," he said. "It was simply the sudden shock. No, looking at the thing scientifically—and I hope I am at least as capable of that as the next man—there is nothing marvellous about the situation. These chimpanzees, or a succession of similar teams of chimpanzees, would in a million years write all the books in the British Museum. I told you some time ago that I believed that statement. Why should my belief be altered by the fact that they produced some of the books at the very outset? After all, I should not be very much surprised if I tossed a coin a hundred times and it came up heads every time. I know that if I kept at it long enough, the ratio would reduce itself to exactly fifty percent. Rest assured, these chimpanzees will begin to compose gibberish quite soon. It is bound to happen. Science tells us so. Meanwhile, I advise you to keep this experiment secret. Uninformed people might create a sensation if they knew."

"I will, indeed," Mr. Bainbridge said. "And I'm very grateful for your rational analysis. It reassures me. And now, before you go, you must hear the new Schnabel records that arrived today."

During the succeeding three months, Professor Mallard got into the habit of telephoning Mr. Bainbridge every Friday afternoon at five-thirty, immediately after leaving his seminar room. The Professor would say, "Well?" and Mr. Bainbridge would reply, "They're still at it, Mallard. Haven't spoiled a sheet of paper yet." If Mr. Bainbridge had to go out on Friday afternoon he would leave a written message with his butler, who would read it to Professor Mallard: "Mr. Bainbridge says we now have Trevelyan's *Life of Macaulay*, the Confessions of St. Augustine, *Vanity Fair*, part of Irving's *Life of George Washington*, the Book of the Dead, and some speeches delivered in Parliament in opposition to the Corn Laws, sir." Professor Mallard would reply, with a hint of a snarl in his voice, "Tell him to remember what I predicted," and hang up with a clash.

The eleventh Friday that Professor Mallard telephoned, Mr. Bainbridge said, "No change. I have had to store the bulk of the manuscript in the cellar. I would have burned it, except that it probably has some scientific value."

"How dare you talk of scientific value?" The voice from New Haven roared faintly in the receiver. "Scientific value! You—you—chimpanzee!" There were further inarticulate sputterings, and Mr. Bainbridge hung up with a disturbed expression. "I am afraid Mallard is overtaxing himself," he murmured.

Next day however, he was pleasantly surprised. He was leafing through a manuscript that had been completed the previous day by Chimpanzee D, Corky. It was the complete diary of Samuel Pepys, and Mr. Bainbridge was chuckling over the naughty passages, which were omitted in his own edition, when Professor Mallard was shown into the room. "I have come to apologize for my outrageous conduct on the telephone yesterday," the Professor said.

"Please don't think of it any more. I know you have many things on your mind," Mr. Bainbridge said. "Would you like a drink?"

"A large whisky, straight, please," Professor Mallard said. "I got rather cold driving down. No change, I presume?"

"No, none. Chimpanzee F, Dinty, is just finishing John Florio's translation of Montaigne's essays, but there is no other news of interest."

Professor Mallard squared his shoulders and tossed off his drink in one astonishing gulp. "I should like to see them at work," he said. "Would I disturb them, do you think?"

"Not at all. As a matter of fact, I usually look in on them around this time of day. Dinty may have finished his Montaigne by now, and it is always interesting to see them start a new work. I would have thought that they would continue on the same sheet of paper, but they don't, you know. Always a fresh sheet, and the title in capitals."

Professor Mallard, without apology, poured another drink and slugged it down. "Lead on," he said.

It was dusk in the conservatory, and the chimpanzees were typing by the light of student lamps clamped to their desks. The keeper lounged in a corner, eating a banana and reading *Billboard*. "You might as well take an hour or so off," Mr. Bainbridge said. The man left.

Professor Mallard, who had not taken off his overcoat, stood with his hands in his pockets, looking at the busy chimpanzees. "I wonder if you know, Bainbridge, that the science of probabilities takes everything into account," he said, in a queer, tight voice. "It is certainly almost beyond the bounds of credibility that these chimpanzees should write books without a single error, but that abnormality may be corrected by—*these!*" He took his hands from his pockets, and each one held a .38 revolver. "Stand back out of harm's way!" he shouted.

"Mallard! Stop it!" The revolvers barked, first the right hand, then the left, then the right. Two chimpanzees fell, and a third reeled into a corner. Mr. Bainbridge seized his friend's arm and wrested one of the weapons from him.

"Now I am armed, too, Mallard, and I advise you to stop!" he cried. Professor Mallard's answer was to draw a bead on Chimpanzee E and shoot him dead. Mr. Bainbridge made a rush, and Professor Mallard fired at him. Mr. Bainbridge, in his quick death agony, tightened his finger on the trigger of his revolver. It went off, and Professor Mallard went down. On his hands and knees he fired at the two chimpanzees which were still unhurt, and then collapsed.

There was nobody to hear his last words. "The human equation . . . always the enemy of science . . ." he panted. "This time . . . vice versa . . . I, a mere mortal . . . savior of science . . . deserve a Nobel . . ."

When the old butler came running into the conservatory to investigate the noises, his eyes were met by a truly appalling sight. The student lamps were shattered, but a newly risen moon shone in through the conservatory windows on the corpses of the two gentlemen, each clutching a smoking revolver. Five of the chimpanzees were dead. The sixth was Chimpanzee F. His right arm disabled, obviously bleeding to death, he was slumped before his typewriter. Painfully, with his left hand, he took from the machine the completed last page of Florio's Montaigne. Groping for a fresh sheet, he inserted it, and typed with one finger, "UNCLE TOM'S CABIN, by Harriet Beecher Stowe. Chapte . . ." Then he, too, was dead.

10 A LOOK AT STATISTICS

The concepts of statistics allow us to make predictions about a large set of objects (the *population*) by studying the properties of a smaller subset (the *sample*). For example, a light-bulb manufacturer produces 100,000 light bulbs; he would like to know how long each bulb will burn. If he finds that 1000 of the bulbs burn an average of 750 hours each, then he may feel safe in advertising that his brand of light bulbs will burn 750 hours. Obviously, he cannot test every one until it burns out. So from his population of 100,000 bulbs, he tests a 1000-bulb sample to make a prediction about the entire population.

Taking a sample to judge a whole population is commonplace in today's world. We sample a piece of candy to judge the box, spray a bit of perfume to see if we like the brand, test drive a demonstrator to judge all cars of that model. "Pollsters" sample a small subset of our population to make predictions or inferences about the opinions, habits, beliefs, or tastes of 200,000,000 Americans.

In addition to making predictions about populations, statistics is concerned with the organization and presentation of data in ways that emphasize their important aspects. In a college of 3000 students, where each student may take 5 courses, the list of 15,000 grades tells us little about the students as a group. But if these data are organized in certain ways, we may be able to see trends or groupings. It might be interesting to know what the average grade is, how many A's or B's were given, and so on.

Organizing and presenting data is the objective of *descriptive statistics*. Using data from a small sample to make a prediction about a large population is called *inferential statistics*. (An inference is a conclusion about a situation, made by extending available evidence.)

As a branch of mathematics, statistics dates from about the mid-seventeenth century. One of the first to contribute to the field was John Graunt. In 1662 Graunt published a small book (based upon the "Bills of Mortality," published weekly by the London government), in which statistics on births, deaths, accidents, christenings, and diseases were analyzed. In 1693 Edmund Halley (predictor of Halley's comet) published the first mortality tables that could be used to calculate life insurance premiums.

Much of the early work in statistics was advanced by biologists who were trying to determine the laws of heredity. By collecting and organizing data on various crosses (using, for example, pea plants or fruit flies), it became possible to statistically predict the ratio of dominant to recessive characteristics in hybrid offspring. In Exercise Set 9.3 we mentioned Gregor Mendel's work along these lines. Mendel's contributions were made in the mid-nineteenth century. Others who developed statistics while working with the problems of heredity and genetics were Francis Galton, Karl Pearson, and Sir Ronald A. Fisher.

Statistics is of great interest to those whose specialties lie outside mathematics proper. For the biologist interested in genetics, a firm statistical background is imperative. Insurance premiums are determined by actuaries, men who need to be expert statisticians. Much psychological research is conducted by statistical experiments. Sociology and Anthropology also rely heavily on statistical research. In medicine, a certain drug may be found to be 80% effective in the treatment of a disease, implying statistical experimentation with populations. In fact, we can say that medicine has wiped out smallpox and polio, meaning that the frequency of the disease is so low as to be almost negligible. Certainly, this is a statistical remark.

With the development of automatic tabulating equipment and computers, huge masses of data can now be handled. We are in the greatest information-handling period in all history. Everyone seems to use statistics to support his or her opinions (although they may be used misleadingly.) Indeed, we can even buy books of statistics: almanacs, encyclopedias, yearbooks, and books of records. Thus it is almost essential nowadays to know something about statistical methods and procedures.

10.1   ORGANIZING AND PRESENTING DATA

Statistics is the study of the collection, organization, analysis, and interpretation of data obtained in numerical form or capable of being dealt with in numerical form.

To study the collected data, it is often helpful to display them in a table or a graph. Suppose the following are the prices of 30 different canned goods, noted at random, in a supermarket (figures in cents):

73	25	33	37	54	20	48	42	26	57
89	53	99	90	25	29	80	79	99	63
70	66	57	47	17	31	85	16	97	76

With the data written this way, we see only a blur of figures; nothing of significance is apparent about a long list of numbers. We can organize the data in several different ways.

In Table 1 we set up nine price intervals after scanning the data and discovering that the prices vary from 16¢ to 99¢. Then we simply enter the can prices in the second column. In the third column we place the prices in order.

Grouping the data makes the list of prices easier to scan. At least in this form (Table 1), it is clear that no one price interval is very much more common than any other. Furthermore, the can prices do not seem to be massed in any one price interval.

Table 1   Distributing and ordering can prices in intervals

Price interval	Can prices	Can prices in order
10–19	17, 16	16, 17
20–29	25, 20, 26, 25, 29	20, 25, 25, 26, 29
30–39	33, 37, 31	31, 33, 37
40–49	48, 42, 47	42, 47, 48
50–59	54, 57, 53, 57	53, 54, 57, 57
60–69	63, 66	63, 66
70–79	73, 79, 70, 76	70, 73, 76, 79
80–89	89, 80, 85	80, 85, 89
90–99	99, 90, 99, 97	90, 97, 99, 99

We could make a *frequency table* (Table 2) that would show the same thing as Table 1, but without the actual prices. We tally the number of prices in each interval; the total of the tallies in an interval is the *frequency* assigned to that interval. We can also accumulate (add up) the frequencies as we go from price interval to price interval.

(Observe that two can prices fall in the 10–19 interval and five can prices fall in the 20–29 price range. The cumulative frequency at the end of the second price interval is obtained by adding 2 and 5. In Table 2 the cumulative frequency of 17, for example, means that 17 can prices fall in the 10–59 price range.)

We can now take the information from Table 2 and display it in a *histogram* or bar graph (Figure 1). The histogram is constructed of rectangles whose width is the width of the price interval and whose height is a number of units corresponding to the frequency. It is understood that the first interval is 10–19 cents, and that the second interval starts at 20 cents, and so on. Once again, we see no exceptional grouping of prices per

Table 2   Frequency table for 30 can prices

Price interval	Tally	Frequency	Cumulative frequency
10–19	//	2	2
20–29	/////	5	7
30–39	///	3	10
40–49	///	3	13
50–59	////	4	17
60–69	//	2	19
70–79	////	4	23
80–89	///	3	26
90–99	////	4	30

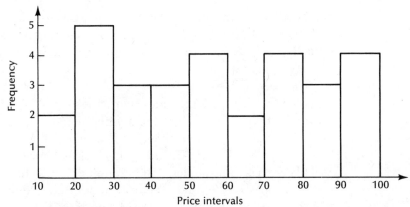

Figure 1   Histogram for 30 can prices.

can. In broad terms, the price of a can appears to fall as often toward the low end of the scale as toward the high end. In observing this, our eye lumps the areas of rectangles together; there is not significantly more area to the left of center than there is to the right of center.

You will see that the "center of the histogram" is emphasized for a reason.

The middle value of a set of data is called the *median*. Half the data values (also referred to as data points) are above the median, and half are below. In the column titled "can prices in order" of Table 1, we count halfway down the 30 data values. "Halfway down" is somewhere between the fifteenth and sixteenth values, that is, between 54 and 57. We take the median to be halfway between these two, or 55½.

In general, the median of an even number of data values is one-half the sum of the two middle values. The median of an odd number of data values is the middle value. In both cases the data values are considered to be arranged in order.

We can also say that the median price interval is 50–59 cents. From Table 2, there are $2 + 5 + 3 + 3 = 13$ can prices below the median price interval, and $2 + 4 + 3 + 4 = 13$ can prices above it. From the histogram we can calculate 13 units of area to the left of the 50–59 interval and 13 units of area to the right. The median tells us what we expected; there is no tendency of the data to be more to the high end of the scale than to the low end.

The *mean* of a set of data is simply the average value, computed by adding up all the data values and dividing by the number of pieces of data. The mean is denoted by $\bar{x}$ (read "x bar").

On a desk calculator we have computed the mean price per can to be 56.1 cents. Notice that both the mean and the median give some estimate of the middle value. We call the mean and the median *measures of central tendency*.

Two other measures are used by statisticians. The first, the *range*, is simply the largest data value minus the smallest data value. In our example, the range is $99 - 16 = 83$, computed by subtracting the lowest price from the highest price.

The second measure is the *mode*, that value that occurs with the greatest frequency. Sometimes there are several modes, indicating that several different data values occur with the same frequency (but no other data value has any higher frequency). In our example, the prices occurring most frequently are 25, 57, and 99. Each of these prices occurs twice. This set of data has three modes. After grouping (see the histogram in Figure 1) we can speak of the interval mode. In our example the interval mode is 20–29, since that interval has the highest frequency, 5.

We have now organized our 30 can prices into intervals, tabulated frequencies and cumulative frequencies, and constructed a histogram. In addition we know that the median can price is 55½ cents, the median price interval is 50–59 cents, the mean is 56.1 cents, the range is 83, the data mode is 25 or 57 or 99, and the interval mode is 20–29.

A range of 83 along with a mean of 56.1 cents indicates a wide distribution of can prices. Thus the mean is not very typical of the 30 data values. Yet the store owner would be more accurate in saying "Our can prices average 56 cents" than in saying "More can prices fall in the 20–29 cent range than in any other." (We hasten to point out that the sample of 30 cans is far too small for any accurate analysis of the entire stock.)

The mean, median, mode and range all tell us something about a set of data. To use one measure without any of the others could distort one's view of the sample.

EXAMPLE 1

Statistically analyze the following set of test grades.

> 62, 72, 75, 67, 84, 89, 83, 83, 80, 91, 96,
> 59, 67, 73, 79, 67, 93, 98, 52, 73, 79

Organizing the grades from lowest to highest gives us

> 52, 59, 62, 67, 67, 67, 72, 73, 73, 75, 79,
> 79, 80, 83, 83, 84, 89, 91, 93, 96, 98

We then construct a frequency table (Table 3), using (arbitrarily) intervals of 10. The corresponding histogram is shown in Figure 2.

From the histogram we can see that the interval mode is 70–79. The mean is the sum of the 21 grades divided by 21, or 77. Since there are 21 data points, the median grade is the eleventh one in the lowest to highest list, or 79. The range is 46, found by subtracting 52 from 98.

Table 3   Frequency table for Example 1

Interval	Frequency	Cumulative frequency
50–59	2	2
60–69	4	6
70–79	6	12
80–89	5	17
90–99	4	21

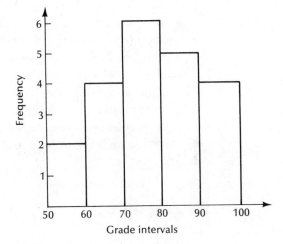

Figure 2
Histogram for Example 1.

An interpretation of Example 1 is as follows: there is a central tendency in the 70's, which is apparent from considerations of histogram areas, and which is confirmed by the mean (77), the median (79), and the interval mode (70–79). Half the class got more than 79 and half got less. We could associate the intervals with letters, so that there would be, perhaps, two F's, four D's, six C's, five B's, and four A's.

The teacher who gave the grades now has a good idea of how the test turned out. He is probably most impressed with the average (mean) as an indicator, but it is important that the majority of the grades clustered about that average. It would be possible, for example, to have a set of grades containing mostly A's and F's and still have a C average.

EXAMPLE 2

Find the mean of the following grades:

$$72, 50, 48, 61, 94, 93, 49, 98, 89, 100, 42$$

The sum is 796 for the 11 scores, so the mean is 72, approximately; Table 4 is the frequency table.

Table 4   Frequency table for Example 2

	Interval	Frequency
F	Below 60	4
D	60–69	1
C	70–79	1
B	80–89	1
A	90–100	4

If you arrange these last grades in increasing order, you will find the median to be 72. The range is 58. It should be evident that there are two modes; one in the below-60 range, and the other in the 90–100 range. The students probably would assume that the results were distorted; even though the average was 72, only three students were close to that average. These grades do not cluster about the mean.

In Section 10.2 we investigate the clustering of data about the mean by investigating a measure of scatter or dispersion.

Exercise Set 10.1

1 Construct a histogram for the grades in Example 2.

2 (a) Find the number of pages in each chapter of this book.
   (b) Compute the mean, median, and mode (if any).
   (c) What is the range for this set of data?
   (d) Would you say the chapters are of uniform length?

3 For the data given for 30 can prices in this section, change the price intervals to 10–39, 40–69, and 70–99. (See Table 1.)
   (a) With these new intervals, make a frequency table and cumulative frequency table.
   (b) Construct a histogram.
   (c) What is the new interval mode?
   (d) Does the median change? Does the mean change?
   (e) Would the store owner now say that "More can prices fall into the 20 to 29-cent category than in any other"?

4 A histogram like the one shown in Figure 1 suggests that the can prices are *random*. That is, there is no clustering about any particular price interval between 10 and 100 cents. Investigate the following set of numbers to see whether they might have been chosen at random.

35	44	13	18	80
37	54	87	30	43
94	62	46	11	71
00	38	75	95	79
77	93	89	19	36

Hint: To do this, make a frequency table of the numbers in intervals of 0–9, 10–19, 20–29, and so on, and then make a frequency table of the numbers in intervals of 0–19, 20–39, 40–59, and so on.

5 In a math class, the following was the grade distribution: two students got 56, three got 61, five got 67, seven got 75, five got 77, four got 82, three got 84, and one got 93. What is the mean grade for the class? What is the median grade?

6 Average temperatures for Miami, Florida, for a 30-year period by months run

$$67, 68, 71, 74, 78, 81, 82, 82, 81, 78, 72, 68$$

(Figures are in order from January to December. Source: Weather Bureau, United States Department of Commerce.)
(a) What is the mean temperature in Miami?
(b) Make a histogram, using 10-degree intervals starting with 0–9 and ending with 90–99.
(c) Make a histogram using 2-degree intervals starting with 67–68 and ending with 81–82.
(d) Which histogram makes the most sense for these data?
(e) Besides knowing that the mean temperature is 75, what else would you like to know about the temperature statistics before you move to Miami?

7 While the mean temperature for New York City is 54, the range is 121. The maximum temperature recorded in New York is 106. What is the minimum?

8 Most college students are familiar with calculating "quality-point averages" (Q.P.A.) or "grade-point averages" (G.P.A.) based upon the following idea: Let A = 4, B = 3, C = 2, D = 1, and F = 0. Then the semester-hour credits for each course are multiplied by the numerical values assigned to each grade received; these products are added and their sum divided by the number of semester hours (s.h.) taken. For example; John takes Physics (4 semester hours), Chemistry (4 s.h.), English (3 s.h.), Math (3 s.h.), and Phys Ed (2 s.h.) and gets these grades:

Physics	B	B = 3	4(3)
Chemistry	A	A = 4	4(4)
English	C	C = 2	3(2)
Math	A	A = 4	3(4)
Phys Ed	B	B = 3	2(3)

Thus we have

$$\text{G.P.A.} = \frac{4(3) + 4(4) + 3(2) + 3(4) + 2(3)}{16} = 3.25$$

We call this a *weighted mean* (a B in a 4-semester-hour course is worth more than a B in a 3-semester-hour course.)
(a) Suppose the "grades" were not multiplied by the semester hours. What would be the divisor in calculating the average grade?
(b) What would John's unweighted average be?
(c) Suppose John got all A's. What would his G.P.A. be?
(d) Suppose John got C in Physics, B in Chemistry, F in English, A in Math, and A in Phys Ed. What would his weighted mean be?

10.2 STANDARD DEVIATION

In some of the previous examples the data seemed to be clustered about the mean; in others the data were more or less evenly distributed throughout the range. The data could also be concentrated at points far from the mean. Certainly if the data are clustered near the mean, the mean is a valuable piece of information. But if the data are evenly distributed or seem to be grouped far from the mean, then the mean does not tell us much. In this section we develop a measure of the dispersion or the variability of data from the mean.

Suppose we have two samples, A and B, that produce the data given in Table 5. Simply by scanning the data we can see that more data in sample B are clustered about the mean than in sample A. But the means are nearly the same, and the ranges are exactly the same. Suppose, now, we calculate the deviation, or distance from the mean, for each data value. That is, we calculate $\bar{x} - x$ (or $x - \bar{x}$, whichever is positive) for each data point in A and B. The results are shown in Table 6.

For example, the first data point in sample A is 1, so we calculate $4.9 - 1 = 3.9$. The last data point in sample A is 9, so we calculate $9 - 4.9 = 4.1$. Mathematicians frequently use $|\bar{x} - x|$, (read "the *absolute value* of the difference of $\bar{x}$ and x") to stand for a nonnegative difference. Thus $1 - 4.9 = -3.9$, but $|1 - 4.9| = 3.9$.

In Table 6 we have also averaged the ten deviations from the mean for each sample. The average deviations of 3.1 and 1.3 show that the data points in sample A are farther from the mean, on the average, than those of sample B.

The average deviation is called the *mean deviation*. It gives us a measure of how the data are scattered about the mean. Statisticians could

Table 5 Data and means for two samples

Sample	Data	$\bar{x}$
A	1, 1, 2, 2, 3, 5, 8, 9, 9, 9	4.9
B	1, 3, 4, 5, 5, 5, 5, 6, 7, 9	5

Table 6 Deviations from the mean for two samples

Sample	$\bar{x}$	Deviations from mean	Average deviation
A	4.9	3.9, 3.9, 2.9, 2.9, 1.9, .1, 3.1, 4.1, 4.1, 4.1	3.1
B	5	4, 2, 1, 0, 0, 0, 0, 1, 2, 4	1.3

work with the mean deviation, but they do not. Instead they work with a measurement that is *like* the mean deviation, but that simplifies other statistical calculations.

The measure of scatter or dispersion most commonly used is called the *standard deviation*. In words, the standard deviation is the square root of the average of the squares of the deviations from the mean. This seemingly difficult sentence can be clarified if we take a small sample and form the deviations from the mean, their squares, the average, and then the square root.

EXAMPLE 1

Let sample C be $\{\, 8, 9, 10, 11, 12 \,\}$ and find the standard deviation.

Sample:	8, 9, 10, 11, 12
Mean $(\bar{x})$:	10
Deviations from $\bar{x}$:	2, 1, 0, 1, 2
Squares of deviations:	4, 1, 0, 1, 4
Sum of squares:	10
Average of squares:	$10/5 = 2$
Square root (standard deviation):	$\sqrt{2} = 1.4$ (approximately)

In Example 1, note that the deviations are the unsigned distances to the mean, that is, the absolute value of the differences of the data points and the mean ($|8 - 10| = 2$, $|9 - 10| = 1$, $|10 - 10| = 0$, $|11 - 10| = 1$, $|12 - 10| = 2$). In calculating the standard deviation, we can use deviations from the mean that are signed *or* unsigned, since in the next step these deviations are squared. Squaring $\bar{x} - x$ always results in a nonnegative number.

If we let sample $D = \{\, a, b, c, d, e \,\}$, where a, b, c, d, and e are any numbers, the standard deviation for five data points can be written as a formula. We go through the same steps as we did in Example 1.

Sample:	a, b, c, d, e
Mean:	$\bar{x}$ (Calculated first)
Deviations from $\bar{x}$:	$a - \bar{x}, b - \bar{x}, c - \bar{x}, d - \bar{x}, e - \bar{x}$
Squares:	$(a - \bar{x})^2, (b - \bar{x})^2, (c - \bar{x})^2, (d - \bar{x})^2, (e - \bar{x})^2$
Sum:	$(a - \bar{x})^2 + (b - \bar{x})^2 + (c - \bar{x})^2 + (d - \bar{x})^2 + (e - \bar{x})^2$
Average:	$\dfrac{(a - \bar{x})^2 + (b - \bar{x})^2 + (c - \bar{x})^2 + (d - \bar{x})^2 + (e - \bar{x})^2}{5}$
Standard deviation:	$\sqrt{\dfrac{(a - \bar{x})^2 + (b - \bar{x})^2 + (c - \bar{x})^2 + (d - \bar{x})^2 + (e - \bar{x})^2}{5}}$

It should be clear that if there were six pieces of data we would have six squared terms in the numerator under the square-root sign and would get the average by dividing by 6. For any number n in the sample there would be n squared terms in the numerator under the square-root sign and a denominator of n.

While comprehension of the next formula is not mandatory, we will state the general formula for calculating standard deviation, since it is so commonly seen in the literature.

Let s be standard deviation, let $x_1, x_2, x_3, x_4, \ldots, x_n$ be n pieces of data and let $\bar{x}$ be the mean of the x_n pieces of data. Then

$$s = \sqrt{\frac{(x_1 - \bar{x})^2 + (x_2 - \bar{x})^2 + (x_3 - \bar{x})^2 + \ldots + (x_n - \bar{x})^2}{n}}$$

EXAMPLE 2

Compute standard deviation for sample A of this section.

Sample: 1, 1, 2, 2, 3, 5, 8, 9, 9, 9
$\bar{x}$:   4.9
Deviations: −3.9, −3.9, −2.9, −2.9, −1.9, 0.1, 3.1, 4.1, 4.1, 4.1
Squares: 15.21, 15.21, 8.41, 8.41, 3.61, .01, 9.61, 16.81,
                     16.81, 16.81
Sum: 110.9
Average: 11.09
Standard deviation: $\sqrt{11.09} = 3.3$, approximately

EXAMPLE 3

Compute the standard deviation for sample B of this section.

Sample: 1, 3, 4, 5, 5, 5, 5, 6, 7, 9
$\bar{x}$:   5
Deviations: −4, −2, −1, 0, 0, 0, 0, 1, 2, 4
Squares: 16, 4, 1, 0, 0, 0, 0, 1, 4, 16
Sum: 42
Average: 4.2
Standard deviation: $\sqrt{4.2} = 2.05$, approximately

We can now compare the standard deviation with the mean deviation for samples A and B (see Table 7).

Table 7   Comparing standard deviation to mean deviation for two samples

Sample	Mean deviation	Standard deviation
A	3.1	3.3
B	1.3	2.05

Standard deviation is obviously not the same as mean deviation, but it does measure dispersion the same way. A higher standard deviation implies greater dispersion about the mean.

The square of the standard deviation is called the *variance*. Thus, if s is the standard deviation, s^2 is the variance. In Example 3 the standard deviation is $\sqrt{4.2}$, so the variance is 4.2. In words, the variance is the average of the squares of the deviations from the mean. We will not use the term variance very often, but we could define the standard deviation as the square root of the variance.

Exercise Set 10.2

In performing statistical calculations, it is common today to use desk calculators or computers (perhaps a computer terminal). You may find these machines easy to operate; they can save a great deal of time. In addition, we note that many statistics books contain tables of squares and square roots useful in calculating standard deviation.

1 Find the mean (average) deviation for the data

$$3.5, 3.7, 3.4, 3.5, 3.3, 3.5, 3.7, 3.9, 3.4, 3.3$$

2 Find the standard deviation for the same data.

3 Find the mean deviation for the set of data 8, 9, 10, 11, 12, and compare it to the standard deviation calculated in the text for sample C, Example 1.

4 What is the standard deviation for the data 1, 2, 3, 4, 5? For 13, 14, 15, 16, 17?

5 (a) Find the standard deviation for 2, 4, 6, 8, 10 and relate it to the answers to question 4.
(b) What do you think the standard deviation for 20, 40, 60, 80, 100 is?

6 By inspection, which set of grades below has the greatest standard deviation? Why?

Set *A*:   40, 60, 62, 63, 77, 78, 79, 90, 91, 99
Set *B*:   57, 62, 73, 74, 75, 75, 78, 78, 83, 85

7 Find the standard deviation for the grades 62, 72, 75, 67, 84, 89, 83, 83, 80, 91, 96, 59, 67, 73, 79, 67, 93, 98, 52, 73, 79.

8 Find the standard deviation of the scores 72, 50, 48, 61, 94, 93, 49, 98, 89, 100, 42.

9 An instructor scores a certain quiz from 0 to 10. The following are the results:

Score	0	1	2	3	4	5	6	7	8	9	10
Frequency	1	1	2	2	3	7	20	6	3	3	2

(a) Find the mean score.
(b) What is the median score?
(c) What is the mode?
(d) What is the range?

(e) Make a histogram dividing the range into 11 parts.
(f) Find the mean (average) deviation.
(g) Find the standard deviation.

10 (a) The average monthly temperatures for Miami are

$$67, 68, 71, 74, 78, 81, 82, 82, 81, 78, 72, 68$$

Calculate the variance and the standard deviation for these data.
(b) The average monthly temperatures for San Francisco are

$$49, 51, 53, 56, 58, 61, 63, 63, 64, 61, 55, 50$$

Calculate the variance and the standard deviation for these data.
(c) Which temperatures, those for Miami or those for San Francisco, are more greatly dispersed about the mean?

11 The average monthly temperatures for New York City are

$$33, 33, 41, 51, 62, 71, 77, 75, 69, 58, 47, 36$$

(a) Do you expect the standard deviation to be greater or smaller than the standard deviation for the San Francisco data of 10(b)? Explain.
(b) Calculate the standard deviation.

12 If two samples, A and B, have the same number of data points and the same mean, and if sample A has a greater range than sample B, which sample usually has the greater standard deviation? Why?

13 For a set of data that are all the same (say 5, 5, 5, 5, 5, 5, 5, 5), what is the standard deviation?

14 (a) What is the minimum number a standard deviation can be?
(b) Is there a maximum?

15 The following are the days of the month on which 30 people, chosen at random, were born:

22	17	25	30	14
1	5	23	3	3
29	26	4	19	24
9	18	7	7	5
14	6	3	8	8
25	30	30	16	8

(a) Make a frequency table using intervals 1–5, 6–10, and so on.
(b) Construct a corresponding histogram.
(c) What is the range?
(d) What is the interval mode (or modes)?
(e) What is the median day?
(f) What is the mean of these data (to the nearest day)?
(g) Find the variance.
(h) Find the standard deviation.
(i) How many birthdays fall between the 5th and the 24th, inclusive? (The significance of this question will become clear later.)

10.3   INTERPRETING STANDARD DEVIATION – NORMAL DISTRIBUTIONS

Since the standard deviation is a measure of closeness of the data to the mean, an obvious question arises. *How* close? That is, how many of the data points lie within any given interval of the mean? How wide would an interval (whose center is at the mean) have to be to include, say, 3/4, or 75% of the data points?

While it may seem that an exact answer to such questions could not be given, just the opposite is true. The Russian mathematician, Pafnuti Tchebycheff (1821–1894), proved that *at least* $1 - (1/n^2)$ of the data points lie within n standard deviations of the mean.

EXAMPLE 1

For the data 1, 3, 4, 5, 5, 5, 5, 6, 7, 9 the mean is 5 and the standard deviation is 2.05 (See Example 3, Section 10.2). We now apply Tchebycheff's theorem for two different values of n.

(1) Let the number of standard deviations be 1, that is, $n = 1$. Evaluating $1 - 1/n^2$ for $n = 1$, we get $1 - 1/1^2 = 0$. The theorem says there are at least 0 data points between $5 - 2.05$ and $5 + 2.05$ or between 2.95 and 7.05. Since there are 8 data points (3, 4, 5, 5, 5, 5, 6, and 7) between 2.95 and 7.05, then, of course, there are at least 0 data points in that interval.

(2) Let $n = 2$; then $1 - 1/n^2 = 1 - 1/2^2 = 1 - 1/4 = 3/4$, The theorem says that there are at least 3/4 or 75% of the data points within two standard deviations of the mean. "Within two standard deviations of the mean" is the interval from $5 - 2(2.05)$ to $5 + 2(2.05)$, or from $5 - 4.10$ to $5 + 4.10$, or from 0.9 to 9.1. Certainly *at least* 75% of the data points lie between 0.9 and 9.1, since *all* the data are in that interval.

We can plot the data of Example 1 as points or dots above the real number line (Figure 3). Repeated data points are plotted one above the other. From this graph (sometimes called a *dot frequency diagram*), we can easily see how Tchebycheff's theorem applies. In Figure 3 the mean is $5 (\bar{x} = 5)$. Given the standard deviation $s = 2.05$, the interval 2.95 to 7.05 (interval for $n = 1$) contains those data points within one standard deviation of the mean. The interval 0.9 to 9.1 (interval for $n = 2$) contains those data points within two standard deviations of the mean. Certainly, at least 0% of the data points lie in the interval for $n = 1$, and 75% of the data points lie in the interval for $n = 2$.

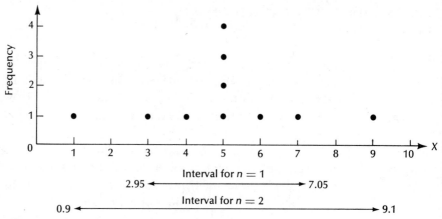

Figure 3   The dot frequency diagram for the data of Example 1.

EXAMPLE 2

Apply Tchebycheff's theorem to the data of exercise 7, Exercise Set 10.2. These 21 grades have a mean of 77.2 and a standard deviation of 12.1. Thus, within 2 standard deviations of the mean [that is, between $77.2 - 2(12.1)$ and $77.2 + 2(12.1)$ or between 53.0 and 101.4] the theorem asserts that there are at least 75% of the grades. Since 20 of the 21 grades are in that interval, the assertion is true.

Tchebycheff's theorem is a *guarantee* about the number of data points falling within n standard deviations of the mean. For $n = 1$ we always get 0% of the data within one standard deviation of the mean. This is not very surprising. Furthermore, our two examples show substantially more than 75% of the data within two standard deviations of the mean.

It has been shown that for a large set of data approximately 68% of the data points fall within one standard deviation of the mean. Also, approximately 95% of the data points will fall within two standard deviations of the mean, and almost all the data (99.7%) fall within three standard deviations of the mean. Let us compare these approximations with the guarantees of Tchebycheff's theorem.

Table 8   Percentages of data in intervals about the mean

Interval	Guaranteed Tchebycheff percent of data in interval	Approximate percent of data in interval
$\bar{x} - s$ to $\bar{x} + s$	0%	68%
$\bar{x} - 2s$ to $\bar{x} + 2s$	75%	95%
$\bar{x} - 3s$ to $\bar{x} + 3s$	89%	99.7%

The approximate percentages given in Table 8 are more accurate if there are a large number of data points. The next example applies these percentages to the data in exercise 15, Exercise Set 10.2.

EXAMPLE 3

For the data in Exercise Set 10.2, question 15, the mean is calculated as 14.6 and the standard deviation as 9.6. The interval $\bar{x} - s$ to $\bar{x} + s$ is 5 to 24.2. There are 18 data points in this interval. We expect about 68% of 30 or about 20 data points to be in the interval. The interval $\bar{x} - 2s$ to $\bar{x} + 2s$ is −4.6 to 33.8, and there are 30 data points in this interval. We expect about 95% of 30 or about 28 of the data points to be in the interval.

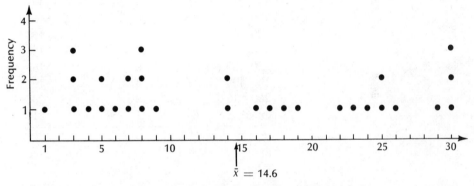

Figure 4   The dot frequency diagram for the data of Example 3.

The 68% and 95% figures give us an approximate number of points to expect within one and two standard deviations of the mean. In Example 3, we notice that the percentages do not work exactly; this is why we use the word "approximate" in connection with the percentages. If we were to take a larger sample, say of 100 or 1000 data points, the percentages of distribution about the mean would become more accurate.

The data for Example 3 are not concentrated about the mean. If we construct a dot frequency diagram, Figure 4, this becomes apparent. We say that the data for Example 3 are not *normally distributed*. In a normal distribution there is a concentration of data points close to the mean.

Any distribution of data in which the percentages 68–95–99.7 work *exactly* is called a *normal distribution*. There are many examples of normal distributions of data: "heights of 8-year-olds," "miles per gallon of 8-cylinder cars," "weights of leaves on maple trees," "time to assemble a specific object on an assembly line," "intelligence quotients of Americans," and "SAT (scholastic aptitude test) scores" are all examples of normally distributed data. Let us take a set of data that ought to be normally distributed and examine its histogram.

EXAMPLE 4

The following are the reported weights of 100 boys in a certain senior class, categorized in intervals of 10 pounds.

Table 9 Data for Example 4

Weight interval in pounds	Frequency	Assume weight at this point
Below 110	1	105
110–119	2	115
120–129	5	125
130–139	8	135
140–149	14	145
150–159	20	155
160–169	19	165
170–179	15	175
180–189	9	185
190–199	4	195
200–210	2	205
Above 210	1	215

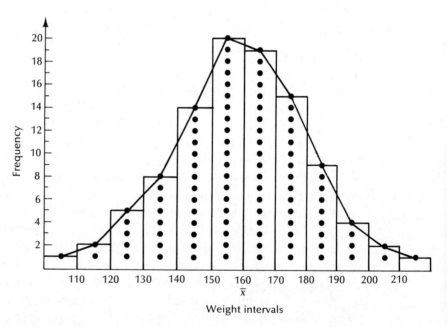

Figure 5 Histogram, dot frequency diagram, and frequency polygon for Example 4.

We can display the data in Example 4 in a histogram, using the weight intervals as widths of the rectangles, and using frequencies as heights of the rectangles. Assuming the weights of the hundred boys to be 105, 115, 125, and so on, we can superimpose a dot frequency diagram on the histogram (see Figure 5).

Figure 5 also shows a series of lines connecting the dots at the highest frequency in each interval. Such a broken line is called a *frequency polygon*. Later we will see the reason for drawing the frequency polygon.

With the weights assumed at the values indicated in each interval, we calculate the mean by dividing the following sum by 100: $1(105) + 2(115) + 5(125) + 8(135) + 14(145) + 20(155) + 19(165) + 15(175) + 9(185) + 4(195) + 2(205) + 1(215) = 16,000$. The mean is 160 pounds. Calculating the standard deviation, we get 20.8. Assuming the weights at points in the interval (instead of as exact weights) probably does not affect the accuracy of our calculations very much. But there may be some error introduced when we placed the weights of the boys below 110 and above 210 at 105 and 215 pounds, respectively. If we disregard the weights above 210 and below 110, the mean is still 160, but the standard deviation is 19.5. Thus, either way we calculate it, the standard deviation is close to 20. We assume it to be 20.

The $\bar{x} - s$ to $\bar{x} + s$ interval is $160 - 20$ to $160 + 20$, or 140 to 180 pounds. In that interval there are $14 + 20 + 19 + 15 = 68$ boys, which is exactly 68% of our hundred-boy sample.

The $\bar{x} - 2s$ to $\bar{x} + 2s$ interval is $160 - 40$ to $160 + 40$, or 120 to 200 pounds. In that interval there are $5 + 8 + 14 + 20 + 19 + 15 + 9 + 4 = 94$ boys, which is very close to 95% of our sample.

We now admit that we contrived the data of Example 4 to work the way they do, yielding the proper percentages for the intervals about the mean. Yet we do expect weights of senior boys to be normally distributed. In a real experiment involving weight distributions a much larger sample could be taken, and the 68–95–99.7 percentage distributions in intervals about the mean should be observed. The larger the sample, the greater the chances of observing a normal distribution.

You can make an interesting device illustrating a normal distribution by interpreting the schematic shown in Figure 6. Hundreds of BB's are started at A and allowed to fall through a slot C; they hit nails driven into a board in the pattern at B. There will be a normal distribution of BB's in the channels at the bottom of the board. If the channels are wide, the accumulation at the bottom will look like a histogram. You could actually calculate the areas and approximate 34% of the total histogram area on each side of the mean (for a total of 68% within one standard deviation of the mean). Then if the distribution is normal, you would expect 95% of the histogram (95% of the BB's) to be within two standard deviations of the mean.

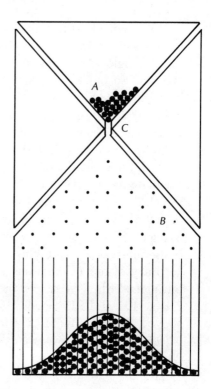

Figure 6
A schematic for illustrating
a normal distribution.

If the channels are narrow, perhaps just a bit wider than the BB's themselves, the histogram rectangles will look like lines of BB's. The upper boundary will closely resemble a smooth curve.

Such curves are called *normal curves*. If the histogram of a normal distribution is drawn for a sample with a range of 100 with intervals of 10 units, a certain histogram results. If the same data are distributed over 20 intervals (width 5 units), the histogram changes. Continuing the process until the interval width approaches 0, the upper boundary of the histogram will approach a smooth normal curve. Of course, to get an actual curve we would have to have an infinite amount of data and we would have to divide the range into an infinite number of "histogram widths" approaching 0. At this point we can assert that 68% of the histogram area means 68% of the area under the normal curve.

Normal curves look like the curve of Figure 7.

A normal curve has these properties:

1. The mode (highest frequency) occurs at the mean ($\bar{x}$), and so does the median.
2. The curve is bell-shaped and is symmetrical about a vertical line through the highest point. (The left branch will land on the right branch if the paper is folded on the dashed line through $\bar{x}$.)
3. Thirty-four percent of the area under the curve lies between $\bar{x}$ and $\bar{x}$ minus one standard deviation, and 34% lies between $\bar{x}$ and $\bar{x}$ plus one standard deviation (68% within one standard deviation of the mean).

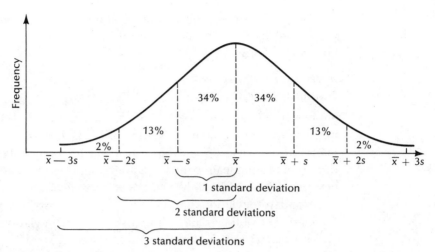

Figure 7   A normal curve.

4. Of the area under the curve 13% + 34% = 47% lies between $\bar{x}$ and $\bar{x}$ minus two standard deviations. The same percentage of area lies between $\bar{x}$ and $\bar{x}$ plus two standard deviations.

In Figure 5 we drew a frequency polygon for a set of data having 100 data points. A frequency polygon is similar to the normal curve provided that the data are normally distributed. Since the data for Example 4 are normally distributed, we can approximate the normal curve by drawing a smooth curve (Figure 8) through all the data points on the frequency polygon.

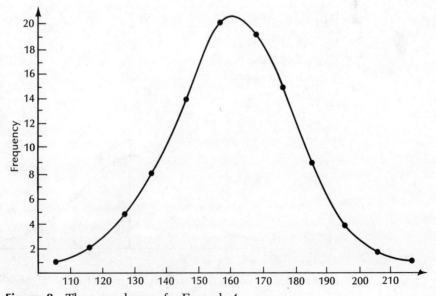

Figure 8   The normal curve for Example 4.

We can say that the frequency polygon becomes the normal curve as the number of data points increases indefinitely and as the interval width approaches zero.

EXAMPLE 5

Suppose the grades at a certain college are normally distributed, with a mean of 75 and a standard deviation of 10. If there are 1000 grades, how many are between 65 and 85? Since the interval within one standard deviation of the mean is the 65–85 interval, there are 68% or 680 grades between 65 and 85. How many are between 55 and 95? This is the two-standard-deviation interval. The normal distribution implies that there are 95% of the grades within two standard deviations, so 950 grades are between 55 and 95. How many are between 55 and 65? We expect 13% (47–34) to be in this interval, so the answer is 130 grades.

For normal curves the percentage areas are obtained in advanced work involving calculus. We have approximated the percentages, but tabulate more exact values in Table 10.

Nearly every statistics textbook contains extensive tables of the normal distribution. They give the areas under the curve from the line of symmetry (at $\bar{x}$) out to Z, where Z is a fractional part of one standard deviation. We show some of the Z values and corresponding percentages of areas under the normal curve in Table 11.

Table 10   More exact percentages of areas under normal curves

Within $\bar{x}$ +	Lies this % of area	Or more exactly
1 standard deviation	34%	34.13447%
2 standard deviations	47%	47.72499%
3 standard deviations	49%	49.86501%

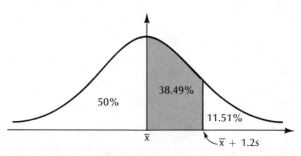

Figure 9   The shaded portion is 38.49% of the total area under the normal curve. (Z = 1.2)

If Z = 1.2, then 38.49% of the total area under the curve lies between the mean $\bar{x}$ and the mean plus 1.2 standard deviations, $\bar{x} + 1.2s$. Since the curve (Figure 9) is symmetrical about the line through $\bar{x}$, 50% of the total area lies on each side of the line of symmetry. Thus 11.51% of the area is to the right of the shaded portion.

Table 11   A partial table of Z values and % of area

Decimal part of one standard deviation (Z value)	Percentage of area under the normal curve
.0	0.00
.1	3.98
.2	7.93
.3	11.79
.4	15.54
.5	19.15
.6	22.57
.7	25.80
.8	28.81
.9	31.59
1.0	34.13
1.1	36.43
1.2	38.49
1.3	40.32
1.4	41.92
1.5	43.32
1.6	44.52
1.7	45.54
1.8	46.41
1.9	47.13
2.0	47.72
2.1	48.21
2.2	48.61
2.3	48.93
2.4	49.18
2.5	49.38
2.6	49.53
2.7	49.65
2.8	49.74
2.9	49.81
3.0	49.87
3.2	49.93
3.4	49.97
3.6	49.98
3.7	49.99
4.0	50.00

EXAMPLE 6

If a normal grade distribution has a mean of 75 and a standard deviation of 10, then 38.49% of the grades lie between 75 and 75 + 1.2(10) = 75 + 12 = 87. Also, 11.51% of the grades are higher than 87. If there are 1000 grades, 385 of them (approximately) are between 75 and 87, and 115 of them are over 85. And, of course, 500 of the grades are less than 75.

Suppose John gets a grade of 90 in the same grade distribution as in Example 6. How much area is contained between the mean (75) and John's grade of 90? To find out, we can write $\bar{x} + Z \cdot s = 90$, substitute for $\bar{x}$ and s, and then solve for Z.

$$75 + Z \cdot 10 = 90$$
$$Z \cdot 10 = 15$$
$$Z = 1.5$$

From Table 11, a Z value of 1.5 corresponds to 43.32% of the area under the curve. There is 43.32% of the area between the mean and John's grade of 90. Since 50% of the area is below the mean, 50 + 43.32 = 93.32% of the grades were lower than John's. Only 6.68% of the grades were higher than John's grade.

EXAMPLE 7

On a standard IQ test the mean, $\bar{x}$, is 101.8 and the standard deviation, s, is 16.4. If Mary scores 125, what percents of those tested got lower and higher scores? We let $\bar{x} + Z \cdot s = d$ (where d is a particular data point) to determine the percentage of area between the mean and the data point $d = 125$. Since $\bar{x} = 101.8$ and $s = 16.4$, we can solve for Z:

$$101.8 + Z(16.4) = 125$$
$$Z(16.4) = 23.2$$
$$Z = 1.4$$

This corresponds to an area of 41.92 (see Table 11). Thus 41.92 + 50 = 91.92% of those tested got lower scores than Mary, and 100 − 91.92 = 8.08% got higher scores.

You can see now why statisticians are so concerned with standard deviation and its relationship to the mean. The mean itself is a measure of central tendency, but application of standard deviation techniques enables us to determine how close the data are to the mean. For normal distributions, a particular data point can be compared to all others by way of area considerations and the normal curve.

Exercise Set 10.3

1 Consider the following set of data:

100	74	68	89	87	83
84	91	97	70	67	100
89	86	78	58	70	66
73	100	65	63	63	85
72	94	100	68	93	67

(a) Find the mean of the data.

(b) Find the standard deviation, correct to the nearest whole number.

(c) Tchebycheff's theorem requires at least 75% of the data to be within two standard deviations of the mean. What interval is this referring to for this data?

(d) What percentage of the data falls in the two-standard-deviation interval?

(e) Approximately 68% of the data should be found in the interval $\bar{x} - s$ to $\bar{x} + s$. What percentage *does* fall in that interval?

(f) Are the data normally distributed?

2 (a) For the data in exercise 1, make a frequency table, using intervals 55–59, 60–64, 65–69, ..., 100–104.

(b) Make a dot frequency diagram, plotting the points at the midpoint of each interval.

(c) Superimpose the frequency polygon on the dot frequency diagram.

(d) Does the frequency polygon appear to be close to a normal curve?

3 In Example 4 data on the weights of 100 boys were given. The mean is 160 pounds and the standard deviation is 20.

(a) How many data points are there between 160 and $160 + 1.7(s)$?

(b) Assuming a normal curve for this data, what percentage of the area under the curve lies between the mean and the mean plus $1.7(s)$?

(c) How many boys weigh less than 194 pounds?

4 If a set of data are normally distributed and have a mean equal to 45, what is the mode? The median?

5 A set of 1500 data points is normally distributed and has a mean of 60 and a standard deviation of 12. Use Table 11 to answer these questions.

(a) How many of the points lie between 48 and 60? Between 36 and 84? Between 36 and 72?

(b) How many of the points lie below 36? Above 84?

(c) How many of the points lie below 86?

(d) How many of the points lie between 65 and 86?

6 A set of grades is 62, 72, 75, 67, 84, 89, 83, 83, 80, 91, 96, 59, 67, 73, 79, 67, 93, 98, 52, 73, 79. Investigate the grades to see if it is reasonable to say they are normally distributed. To do this, calculate $\bar{x}$ and s, find out how many grades lie between $\bar{x} + s$ and $\bar{x} - s$, and compare the result with the percentages given in Table 11.

7 Consider the statistics given in Example 7. If a person scored 115 on the IQ test, what percentage of those tested got lower and higher scores?

8 (a) From an almanac or other source, find the age at death of each of the first 36 Presidents of the United States.
  (b) Compute the mean and standard deviation of these data.
  (c) If the data are normally distributed, how many Presidents should have died within one standard deviation of the mean?
  (d) Count the number of Presidents who did die within one standard deviation of the mean.
  (e) Is it reasonable to assume that these data are normally distributed?

9 For the same data as in exercise 8,
  (a) Construct a frequency table, using intervals 45–49, 50–54, 55–59, and so on.
  (b) Construct a dot frequency diagram, plotting the data at the midpoint of the intervals.
  (c) Superimpose a frequency polygon on the dot frequency diagram.
  (d) Does this polygon approach a normal curve?

10 (a) Find a set of data containing more than 25 data points that you think should be normally distributed.
  (b) Calculate the mean and standard deviation.
  (c) How many of the data fall within one standard deviation of the mean?
  (d) If the data are normally distributed, how many of your data points *should* fall within one standard deviation of the mean?
  (e) Are your data normally distributed?

SUMMARY

The goal of this chapter is to show how data can be organized and how certain statistics can be used to interpret the data.

The frequency table was used to construct a histogram or bar graph. Such graphs enable us to scan the data to see groupings or central tendency of the data. This histogram converts numerical data into areas of rectangles.

Two statistics indicating central tendency are the median and the mean. Informally, the median is the middle value of the data; half the data fall below the median and half fall above. The mean is simply the average of the data values. Of the two, the mean is much more important in statistical work.

Two other statistics that we mentioned are the range and the mode. The range indicates how widely the data vary. The mode is that data value occurring most frequently.

To describe how the data are scattered or dispersed about the mean, we use standard deviation. Although the calculation of standard deviation is time-consuming, we hope our examples and exercises make it seem worthwhile. Tchebycheff's theorem gives us a guarantee that at least 75% of the data fall within a two-standard-deviation interval about the mean. But empirically, statisticians have found stronger results, namely, that approximately 68% of the data will fall within one standard deviation of the mean and 95% within two standard deviations of the mean.

It is important to realize that the larger the set of data, the more accurate these approximations become.

For normally distributed data the percentages work exactly. No finite set of data has a distribution about the mean that is exactly normal. But for all practical purposes, we can assume a set of data are normally distributed when the percentages hold approximately.

We concluded this chapter with a study of the normal curve and percentages of area under the normal curve about the mean. Table 11 gives percentages of area from the mean out to a given decimal part of a standard deviation.

Statistics has a tremendous impact on life today. As a practical matter, we now locate shopping centers, or determine school bus routes, or anticipate election outcomes, or analyze a baseball player's performance by statistical means. Almost every field of research has some aspect that is statistical in nature: nuclear physics asserts a statistical location for molecules; economic theorists collect data to make inferences for the future; government funds are allocated to the states on the basis of statistics gathered by the Bureau of the Census; life insurance premiums, medical plans, and retirement benefits are all cost-analyzed by statistical means.

To the mathematician, statistics provides a sound way of analyzing data that can be used productively in making inferences about uncertain outcomes. Today, he is involved in statistical problems having to do with the design of experiments, with decision-making and with hypothesis-testing. Statistics is probably the branch of mathematics that most applies to other fields. As evidence we list some relatively new fields of study involving statistics: biometrics, sociometrics, cybernetics, econometrics, psychometrics, and operations research. You might like to investigate just what statistics has to do with these fields.

Before attempting the Review Test, be sure you are familiar with the following:

1. Making frequency tables, histograms, dot frequency diagrams, and frequency polygons.

2. Calculating median, mean, and range.

3. Finding the mode (or modes).

4. Finding standard deviation.

5. Using Tchebycheff's theorem.

6. Using the percentages, 68–95–99.7, of approximate distribution about the mean.

7. Using a table of Z values to get percent of area under a normal curve.

8. Using $\bar{x} + Z \cdot s = d$ in connection with a given data point d.

REVIEW TEST

1 To the nearest degree, the following are average January temperatures of 21 U.S. cities:

$$23, 35, 45, 35, 10, 29, 30, 25, 16, 37, 43,$$
$$25, 26, 34, 28, 47, 30, 21, 46, 28, 20$$

(a) What is the mean temperature?
(b) What is the median temperature?
(c) Setting up intervals 10–14, 15–19, and so on, construct a frequency table.
(d) Construct either the corresponding histogram or a dot frequency diagram.
(e) What is the interval mode?

2 For the same data given in question 1, find each of the following:
(a) The set of deviations from the mean.
(b) The squares of the deviations.
(c) The sum of the squares.
(d) The average of the squares of the deviations from the mean.
(e) The standard deviation.

3 (a) For the same data, what percentage of the temperatures are expected to be within one standard deviation of the mean?
(b) How many pieces of the data *do* fall within one standard deviation of the mean?

4 Given the weights A = 4, B = 3, C = 2, D = 1, F = 0 and semester hours English (3), History (3), Math (3), Biology (4), Phys Ed (2) find Ann's grade point average if she receives an A in Math, a B in English, a C in Biology, a B in Phys Ed, and an A in History.

5   A light-bulb manufacturer determines that the average bulb burns 980 hours and that the standard deviation is 40.
    (a) In a shipment of 1000 bulbs how many are expected to burn between 940 and 1020 hours?
    (b) How many should burn between 940 and 980 hours?
    (c) How many should burn between 900 and 940 hours?

6   A set of 1000 data points is normally distributed with mean $\bar{x} = 200$ and standard deviation 25.
    (a) How many data points are between the mean and the data point $d = 240$? (A table of Z values is needed.)
    (b) How many data points are greater than 240?
    (c) How many data points are smaller than 300?

Statistics provide a way to analyze data, to observe
central tendencies, and to predict knowledge
of a population from observations of only a
representative sample. However the famous line
"Figures don't lie—liars figure" hints that statistics are
often misused. (Who has not passed judgment on a
restaurant on the basis of only *one* meal?) The *Time*
magazine essay (September 8, 1967) below urges us
to be aware of the *misuse* of statistics.

THE SCIENCE & SNARES OF STATISTICS
TIME MAGAZINE

Americans believe in numbers. As a democracy, the U. S. chooses its
leaders statistically, so to speak, by the simple process of counting votes.
Numbers measure the economy, record social progress, identify people
on credit-card rolls and bank accounts. "In a numerically conscious
society," says Rand Corp. Researcher Amrom H. Katz, "progress is
measured by numbers, not by quality."

In fact, the American appetite for statistics seems insatiable, and the
statisticians obligingly crank out an unending supply, ranging from the
annual per-capita consumption of paper (540 lbs.) to the number of
dishes (nine) that the typical family breaks in the course of a year. Sports
fans are longtime lovers of the well-tempered statistic. To know that
Roger Maris replaced Babe Ruth as the home-run king through a fluke in
total games played, is to be an *aficionado* instead of an amateur. For the
average American, to be told that a lofting astronaut has threaded a
celestial needle of time and place and reached orbit is to be faced with
the incomprehensible. But to know that he is traveling at 17,500 m.p.h.
is a measure that means something to an earthling who must watch the
"60 m.p.h." speed-limit signs.

But this dedication to numbers has created its own pit-falls for the
innocent—and opportunities for the purveyors. There is an an air of
certainty about the decimal point or the fractionalized percentage—even
in areas where the measurement is statistically absurd or the data
basically unknowable. A classic example is a survey made some years

ago, which solemnly reported that 33⅓% of all the coeds at John Hopkins University had married faculty members. True enough. Johns Hopkins had only three women students at the time, and one of them married a faculty member. The American Medical Association announces not that very few people dream in color, but that "only 5% of Americans" dream in color. New York City has 8,000,000 rats. How does anybody know? Statisticians have a phrase for this, borrowed from the computer industry on which they now rely. The phrase is "garbage in, garbage out"—meaning that the result that comes out is only as good as the material that is fed in.

SALES & STRIKES

For the sake of drama or publicity, numbers are slapped on nearly everything—and the bigger the number the better. During July's two-day rail strike, the Chicago Association of Commerce and Industry issued an instant statistic that the city was losing $40 million to $60 million a day, into which total were cranked lost railroad fares and freight revenues, reduced restaurant and hotel receipts, smaller store sales, and presumably the money that visiting butter-and-egg conventioneers or traveling salesmen might spend on tours and girls. Overlooked was the probability that most of the businessmen made their visit anyway the minute the strike had ended. "What can you say about a strike," says DeVer Sholes, the association's director of research and statistics, "except that they're striking? But the news media are anxious to build up the story, so you have to fluff it up some way."

Fluffing for or by news media is the root cause for many an abused statistic. Newsmen during the Detroit race riots pressed a harried fire chief for damage estimates. His guess: $500 million. So far, in the cooling aftermath of riot, insurance companies are processing only $84 million worth of damage claims, and the overall loss is now put at $144 million. For newsmen, the National Safety Council issues forecasts of expected highway deaths over holiday weekends, usually with a prediction tacked on of "record fatalities." What the forecast never says is that the record is due to population increases and wider use of automobiles, and that the fatality rate is usually just as high proportionately on other weekends—holiday weekends are just a bit longer.

"The knowledge industry," as education is now called, is touted as a $200 billion industry—one of the nation's biggest. Presumably every expenditure, down to janitors' salaries and the cost of the new gymnasium (also computed as part of the construction industry), is figured in. But then there is the $126.7 billion of Government spending, the $189 billion service industry, the $21 billion annual economic loss through crime, and the $25 billion that Vance Packard says is spent on disposable packages each year. The grand total has soon soared past the gross national product.

HUNGRY TO BED

In an irreverent study of the numbers game called *How to Lie with Statistics*, Author Darrell Huff coined the word "statisticulation"—the art of lying with statistics while seeming objective. One trade that statisticulates regularly is advertising. A nostrum for piles or pyorrhea is endorsed by eight out of ten physicians because the ad agency has tirelessly spent time and money assembling panel after panel until it finds one that shows an eight-out-of-ten result.

Politicians are equally susceptible to the selected statistic. John F. Kennedy's campaign claim that "17 million Americans go to bed hungry every night" was based on nothing more than a 1955 Department of Agriculture study on eating habits, which reported that, along with predictable diet deficiencies in lower-income groups, 13% to 17% of U. S. households with incomes of $10,000 or more also suffered nutritional shortages. In fact, it can easily be proved that most people in a given group make less than the average income. Sample: In a group of ten, nine make $10,000 a year and one has an income of $1,000,000. Their average yearly income is $109,000. This reminds many statisticians, who have some humor about their shortcomings, of the fellow with his head in a refrigerator and his feet in an oven, who declares, "On the average, I feel fine."

Since it is obviously impractical to poll the nation on anything less important than the selection of a President, one cherished statistical tool is the sample. Not even statisticians can agree on how big or good a sample can be relied upon as representing the whole. Dr. Alfred C. Kinsey's celebrated reports were criticized by statisticians not so much for their moral implications but because they made sweeping presumptions on the basis of too small a sample (in the male study, only 5,300 men provided data). The Nielsen ratings, by which television programs live or die, have been justly attacked because Nielsen recorders are necessarily hooked to the sets of those viewers willing to have a recorder—a special class by definition, whose tastes may or may not correspond with those of the unpolled millions of the total TV audience.

Still the state of the art of statistics has come a long way since 1661, when its founding father, London Haberdasher John Graunt, began a careful count and found that more boy babies died in infancy than girls, and concluded that therefore there must be more women than men in Britain. Today's scientists who no longer believe that anything is absolutely certain, also believe that many things are predictably probable. And it is the computer, fed with vast amounts of past data, that can project or at least outline the alternatives of several possible futures. "The computer has enshrined statistics," says M. I. T.'s Professor Harold A. Freeman. "Without it, statistics would still be a grubby business." Where once all they had to do was count, and perhaps draw graphs, statisticians are now "programmers," with a mystique all their own. Unquestionably, for the moment, numbers are king. But perhaps the time has come for society to be less numerically conscious and therefore less willing to be ruled by statistics.

11 INFINITY

We have made several references to infinity. In Chapter 1 we said that the set of natural numbers is infinite; we also mentioned the cardinality of infinite sets. In Chapter 7 we discussed the infinitude of the set of primes.

Indeed, our basic understanding of counting implies that we can count indefinitely far (where we mean, by indefinite, "as far as we wish"). It is obvious that it is impossible to identify the largest natural number that exists; merely adding 1 gives a larger number.

Perhaps our use of the term "infinite" has been too loose. There is a popular tendency to think of infinity in terms of a quantity, or to place infinity at some point in the universe. Minds delight in such questions as "How big is infinity?" or "How many points *are* there on a straight line?" or "Do parallel lines meet at infinity?" or "Is our universe infinitely long, wide, and high?"

Religion, philosophy, science, and mathematics have been concerned with the infinite for thousands of years. Euclid proved that there are infinitely many primes in about 300 B.C. Nearly every treatment of infinity mentions a famous paradox by Zeno of Elea (about 500 B.C.) concerning a race between Achilles and a tortoise. Basically the problem is this:

If Achilles and a tortoise are to race one another, and if the tortoise is given a lead, then how can Achilles ever catch up? When Achilles gets to the spot where the tortoise started, the tortoise is no longer there. Then when Achilles gets to the tortoise's later position, the tortoise has moved ahead again. Achilles is always coming up to a previous position held by the tortoise. As slow as the tortoise may be, and as swift as Achilles may be, it would appear that the tortoise would always be ahead, and in the end must win the race. But certainly the swift-footed Achilles can outrun a mere tortoise.

A rather loose way of explaining the Zeno paradox is to say that Achilles closes the gap between himself and the tortoise at a faster and faster rate. Then, just as he passes the tortoise, the "gap-closing-rate" is infinite. This last remark is worth some careful thought.

It has taken thousands of years for mathematics to develop an adequate theory of the infinite. The invention of the calculus by Newton and Leibniz in the late seventeenth century provided the basis for understanding problems of motion involving velocities or rates. Students of calculus work a great deal with problems involving infinity. The work of Georg Cantor (1845–1918) dealt with infinite sets; his theory of the transfinites enables us to speak of infinities of different sizes or of different cardinalities.

In this chapter we will explore infinity as a phenomenon; we will investigate a bit of Cantor's theory of transfinites; and we will discuss infinity and the universe. While we cannot deal with all the mathematical complexities that will come up, we should be able to clarify some of the intuitive notions about infinity.

11.1   INFINITE SETS

In Chapter 1 we defined "one-to-one correspondence" and showed such a correspondence between the counting numbers, N, and the even counting numbers, E.

$$N = \{\ 1,\ 2,\ 3,\ 4,\ \dots,\quad n,\dots\}$$
$$\uparrow\ \uparrow\ \uparrow\ \uparrow\qquad\quad \uparrow$$
$$E = \{\ 2,\ 4,\ 6,\ 8,\ \dots,\ 2n,\dots\}$$

Though we cannot list all the counting numbers, nor all the evens, the pairing of the elements of N and E is complete. Any counting number, n, is paired with the even, $2n$, and vice versa. So, the sets N and E are equivalent and there are just as many evens as counting numbers. Yet the evens are contained in the set of counting numbers (E is a *proper* subset of N).

It would seem that there ought to be more counting numbers than evens, but the fact that the sets are equivalent assures us that this is not so. This strange property is used as a criterion for defining an infinite set.

DEFINITION 1   Infinite set

If there exists a one-to-one correspondence between a set S and a proper subset of S, then S is an *infinite set*. (If a set is not infinite, it is finite.)

This definition can be used either to prove that a set is infinite or to prove that it is finite.

EXAMPLE 1

Show that the set of counting numbers, N, is infinite. We arbitrarily choose a proper subset A to be the odd numbers, and establish the following one-to-one correspondence:

$$N:\quad 1\ 2\ 3\ 4\ \dots\quad n\quad\dots$$
$$\uparrow\ \uparrow\ \uparrow\ \uparrow\qquad\quad \uparrow$$
$$A:\quad 1\ 3\ 5\ 7\ \dots\ (2n-1)\dots$$

Is it clear that every element in N corresponds to an element in A, and vice versa? For example, the counting number 15 corresponds to $2(15) - 1$ or 29, and 29 in set A corresponds to 15 in set N. The correspondence is complete and yet A is a *proper* subset of N. (*Why* is A a proper subset of N?) Thus the set N is infinite, according to Definition 1.

The set of counting numbers is taken as a reference set in dealing with all other infinite sets. Any set equivalent to N is itself infinite and is called *countably infinite*, or *countable*. We have shown that the set of evens and the set of odds are countably infinite sets.

Thus a set is infinite if it is equivalent to a proper subset of itself, or if it is equivalent to the set of counting numbers. However, there are infinite sets that are not equivalent to the counting numbers; we will see some examples later.

Let us now apply Definition 1 to some other situations.

EXAMPLE 2

Show that the set of squares of counting numbers is infinite. Shown below is the required correspondence between the squares and a proper subset of the squares.

Squares:	1	4	9	16	25	...	n^2 ...
	↕	↕	↕	↕	↕		↕
Proper subset:	4	9	16	25	36	...	$(n+1)^2$ ...

The trick in Example 2 is to delete the first element from the set of squares in order to form a proper subset. Then each square is associated in a one-to-one way with the next highest square.

EXAMPLE 3

Show that the set $\{1, 2, 3\}$ is not infinite. The proper subsets are $\{1, 2\}$, $\{1, 3\}, \{2, 3\}, \{1\}, \{2\}, \{3\}, \{\ \}$. It is impossible to set up a one-to-one correspondence between $\{1, 2, 3\}$ and a proper subset of $\{1, 2, 3\}$. Thus $\{1, 2, 3\}$ is not infinite.

It is possible to list all the proper subsets of a given finite set. Then each proper subset in turn may be tested to see whether a one-to-one correspondence can be established with the given set. Of course, the test will always fail, because every proper subset of a finite set will have at least one less element than the set itself. But some proper subsets of a given infinite set do not have fewer elements than the given set.

Since it is impossible to list all the proper subsets of an infinite set, it may be difficult to find a proper subset suitable for establishing the required one-to-one correspondence. For example, the set of points on this page is infinite. This may be difficult for you to prove if you do not think of an appropriate subset and a one-to-one correspondence. The next example should provide a hint.

EXAMPLE 4

Prove that the set of points on any line segment, such as AB in Figure 1, is infinite. It is clear from Figure 2 that CD is a proper subset of AB. We can meet the requirements of our definition of infinite set if we show how the points of AB can be placed in one-to-one correspondence with those of CD. First, we construct the rectangle $CDD'C'$. Then we draw the triangle ABP so that D' is on BP and C' is on AP, as shown in Figure 2.

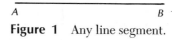

A   B

Figure 1   Any line segment.

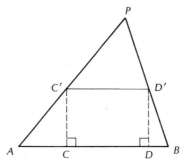

Figure 2   Showing that AB contains an infinity of points.

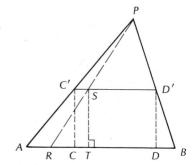

Figure 3   Point R is in one-to-one correspondence with point T.

Let R be any point of AB, and draw RP; this line intersects $C'D'$ in S, and we decide that R is in one-to-one correspondence with S. Then if we drop a perpendicular to AB from S, a point T is determined and we place S in one-to-one correspondence with T. Thus we can take R to be in one-to-one correspondence with T (by way of S). A similar construction can be performed for any other location of point R on AB. This establishes a one-to-one correspondence between all the points of AB and all the points of CD.

Showing that the set of points on a line segment is infinite, then, is not really difficult. The surprising thing is that short line segments contain just as many points as longer ones.

Figures 4 and 5 show how a line segment contains the same number of points as an "infinitely long" line.

Take the line segment as the diameter of a semicircle (Figure 4). We can set up a one-to-one correspondence between the points of the segment and the points of the semicircle simply by dropping perpendiculars from the segment so that they intersect the semicircle. In this way every point

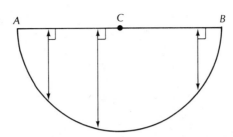

Figure 4   Associating the points of *AB* with those of a semicircle.

of the segment corresponds to some point on the semicircle, and every point on the semicircle corresponds to some point on the segment. Thus there are just as many points on the segment as there are on the semicircle.

   Next we associate every point on the semicircle with a point on a line in a one-to-one way (Figure 5). The arrows indicate that the line continues without end in both directions, which is what we mean by "infinitely long."

   Let the semicircle be tangent to the line at *T*. Radial lines extended beyond the semicircle eventually intersect the line. Point *A* corresponds to *A'*, point *B* to *B'*, and so on. Thus there are just as many points on the semicircle as there are on an infinitely long line.

   The combination of these two established one-to-one correspondences shows that there are just as many points on a segment as there are on an entire line. Infinite sets behave very strangely!

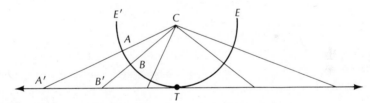

Figure 5   Associating points of a semicircle with those of an infinitely long line.

Exercise Set 11.1

(The first three questions have to do with very large numbers.)

°1 (a) If we assume that a grain of sand takes up the space of a cube measuring 1/32 of an inch on a side, how many grains of sand are there per cubic inch?

 (b) If a sandy beach is 1 mile long (1 mile = 5280 ft), 500 ft wide, and 6 ft in average depth, how many cubic inches of sand does it contain?

(c) Under the conditions of both (a) and (b), approximately how many grains of sand are on the beach? The answer is nearly 900 trillion.

(d) Are there more grains of sand on that beach than there are molecules in a cubic inch of gas?

(e) One inch equals 2.54 centimeters. Therefore there are about 16 cubic centimeters per cubic inch. Look up the meaning of the term *Avogadro's number* (or ask your Chemistry teacher). Reconsider part (d).

°2 It has been estimated that the number of grains of sand on the beach at Coney Island is 10^{20} or 100000000000000000000, that is, a 1 with 20 zeros after it. The number of words uttered by everyone who ever existed is estimated to be 10000000000000000 or 10^{16}. The mathematician Edward Kasner first used the term "*googol*" (named by his nine-year old nephew) to stand for a 1 with 100 zeros after it, or 10^{100}. A "*googolplex*" is 10 to the googol power, or a 1 with a googol of zeros after it.

(a) How many zeros are there in the number 10^2? 10^6? 10^{50}?

(b) How many zeros are there in the number $10^{(10^2)}$? What is another name for this number?

(c) How many zeros are there in the number $10^{googol} = 10^{(10^{100})}$? Could you write this number showing all the zeros in 2 hours? In 24 hours?

°3 (a) Are any of the numbers involved in questions 1 and 2 infinite?

(b) Write down a number larger than a googolplex.

4 Consider the set N with general element n. Suppose every natural number is corresponded with $2n + 1$. Then 1 corresponds to $2(1) + 1$ or 3, for example.

(a) To what number does 2 correspond? 3? 4? 5?

(b) Describe, in words, the numbers of the form $2n + 1$.

(c) Are the numbers of the form $2n + 1$ a subset of N?

(d) Are they a proper subset of N?

(e) If we also correspond $2n + 1$ with n, have we established the criterion for proving N infinite?

5 Prove that the set of odd numbers is infinite

(a) by using Definition 1.

(b) by showing that they are equivalent to the set N.

6 (a) Prove that the set of multiples of 5, $\{5, 10, 15, 20, 25, \ldots\}$, is infinite.

(b) Is this set countably infinite?

(c) Are there just as many elements in $\{5, 10, 15, \ldots\}$ as there are in $\{1, 2, 3, 4, 5, \ldots\}$?

7 Prove that the set $\{1, 8, 27, 64, \ldots, n^3, \ldots\}$ is infinite.

8 Use Definition 1 to prove that the set $\{a, b, c, d, e\}$ is not infinite (see Example 3).

9 Prove that the set of points in any square is infinite. It will suffice to describe how to obtain a one-to-one correspondence with a proper subset.

10 Tell how to show that the set of points on this page is infinite.

11 Show that the set of integers $\{\ldots, -3, -2, -1, 0, +1, +2, \ldots\}$ is countably infinite. Hint: Can you think of another way to write the set so that there is a first element, a second element, and so on?

°12 Table 5 in Chapter 4 displays the rational numbers in such a way that every rational can be found in some row or column. Show that the rationals can be counted, thus establishing a one-to-one correspondence with N and proving that the rationals are infinite.

13 Draw two concentric circles, that is, two circles with the same center but different diameters. Show a construction that proves that there are just as many points on the larger circle as on the smaller.

14 Are there fewer points on a perfect map of your state than there are locations in the state itself?

15 Are there more points in a plane than there are ordered pairs of real numbers? (See Section 8.4.)

16 Which of the following are properties shared by an orange and the planet earth?
   (a) They have the same volume.
   (b) Their circumferences are the same.
   (c) Their diameters are equal in length.
   (d) Their diameters are equivalent.
   (e) They contain the same number of interior points.
   (f) Their surface areas are equal.

17 The estimated total number of atoms in the universe is given in the book *One, Two, Three, Infinity* by George Gamow. Look up this number and report it to class.

18 The German mathematician David Hilbert (1862–1943) is credited with the following story called "Hilbert's Hotel."

A traveler stops at the Hilbert and asks for a room. He is told that every room is taken, but "No matter. We will simply put you in Room 1 after moving the occupant to Room 2, and then move the guest in Room 2 to Room 3, and so on. The person is room n will shift to room $n + 1$, and everyone will have a room."

How many rooms must there be in the Hilbert Hotel?

11.2   TRANSFINITE CARDINAL NUMBERS

In Chapter 1 we introduced the concept of the cardinality of a set. If a set is finite, its cardinality is simply the number of elements in that set. For example, if $A = \{a, e, i, o, u\}$, then we say that the cardinality of A is 5. This idea is symbolized as $n(A) = 5$. The cardinality of a set is a measure of its size. If $n(A) = 5$ and $n(B) = 8$, set B has more elements than set A. To say "set B is greater than set A" means that the cardinality of B is greater than the cardinality of A.

The question of cardinality for infinite sets is slightly different. There is no number (natural, integer, rational, or irrational) that can represent the cardinality of an infinite set. Georg Cantor, working in the 1870s and 1880s, proved that not all infinities are of the same "size." Thus it was necessary to invent the so-called *transfinite cardinals* to speak of the different "magnitudes" of infinite sets.

We stated that the set N of counting numbers is the reference point in dealing with infinite sets. Cantor chose the symbol $\aleph_0$, read "aleph null," to stand for the cardinality of the set N. That is,

$$n(N) = \aleph_0$$

He proved that N is the smallest infinite set, so $\aleph_0$ is the smallest transfinite cardinal.

We have already seen that the sets of even and odd counting numbers (as well as others) are equivalent to N. All the sets below have cardinality $\aleph_0$:

$$\{1, 2, 3, 4, 5, \ldots\}$$
$$\{2, 4, 6, 8, \ldots\}$$
$$\{1, 3, 5, 7, \ldots\}$$
$$\{5, 10, 15, 20, \ldots\}$$
$$\{3, 6, 9, 12, 15, \ldots\}$$

You should be able to think of many others with the same cardinality.

But Cantor proved that the set of real numbers, $R^{\#}$, does not have cardinality $\aleph_0$. We can show N to be equivalent to a subset of the real numbers (see below), but the real numbers are not equivalent to a subset of N. This implies that the cardinality of the reals is greater than the cardinality of N.

$$N: \quad 1 \quad 2 \quad 3 \quad 4 \quad 5 \quad \ldots \quad n \quad \ldots$$
$$\updownarrow \quad \updownarrow \quad \updownarrow \quad \updownarrow \quad \updownarrow \qquad \updownarrow$$
$$\text{Subset of } R^{\#}: \quad 1/3 \quad 2/3 \quad 3/3 \quad 4/3 \quad 5/3 \quad \ldots \quad n/3 \quad \ldots$$

The same idea works for finite sets. That is, if finite set A is equivalent to a proper subset of finite set B, but set B is not equivalent to a subset of set A, then the cardinality of B is greater than the cardinality of A.

EXAMPLE 1

Let $A = \{a, b, c\}$ and $B = \{1, 2, 3, 4, 5\}$. To show that A is equivalent to a proper subset of B, we need only pick the subset $\{1, 2, 3\}$ from B and match A with that subset.

$$a \quad b \quad c$$
$$\updownarrow \quad \updownarrow \quad \updownarrow$$
$$1 \quad 2 \quad 3$$

But we *cannot* show that B is equivalent to a subset of A. So the cardinality of B is greater than the cardinality of A.

Of course, in Example 1, the cardinality of B is 5 and the cardinality of A is 3, so it is obvious that $n(B)$ is greater than $n(A)$. But we turned to finite sets to explain our remark that the cardinality of the reals is greater than the cardinality of the counting numbers (both infinite sets).

The symbol used for the cardinality of the real numbers is c; that is,

$$n(R^{\#}) = c$$

We now know that $c > \aleph_0$. In Chapter 4 we explained how the real numbers are placed in one-to-one correspondence with the points of a line. So the set of points on a line has cardinality c. In fact, Cantor developed a proof that the set of numbers between 0 and 1 has cardinality c. If a line segment AB is chosen as a unit length in constructing a real number line, A could correspond to 0 and B to 1. Thus the set of points on any line segment has cardinality c.

A remarkable result of this is that there are more points on a line segment measuring 1/16 inch (or even smaller) than there are numbers to count them with. (Remember, c is greater than $\aleph_0$.)

So far we have only two transfinite cardinals, $\aleph_0$ and c. An obvious question comes to mind. Are there others? Maybe $\aleph_0 + 1, \aleph_0 + 2$, and so on are different from $\aleph_0$. The next example shows that this is *not* so.

EXAMPLE 2

Show that $\aleph_0 + 1 = \aleph_0$. To do this we note that $\aleph_0$ is the cardinality of the set $\{1, 2, 3, 4, \ldots\}$. A set that has cardinality 1 is $\{0\}$. If we form the union of these two sets, we get $\{0, 1, 2, 3, 4, \ldots\}$, which still has cardinality $\aleph_0$, since we can establish the following correspondence:

$$\begin{array}{ccccccc}
1 & 2 & 3 & 4 & \ldots & n & \ldots \\
\updownarrow & \updownarrow & \updownarrow & \updownarrow & & \updownarrow & \\
0 & 1 & 2 & 3 & \ldots & (n-1) & \ldots
\end{array}$$

So $\{0\} \cup \{1, 2, 3, 4, \ldots\} = \{0, 1, 2, 3, 4, \ldots\}$ implies that $1 + \aleph_0 = \aleph_0$.

By reasoning similar to that used in Example 2 we can show that $\aleph_0 + 2 = \aleph_0$, $\aleph_0 + 3 = \aleph_0$, and so on. In fact, since the union of the even and the odd counting numbers is the set of all counting numbers, we can see that $\aleph_0 + \aleph_0 = \aleph_0$. In this very strange world of transfinite "algebra" we have facts like these:

$$\aleph_0 + 1 = \aleph_0$$
$$\aleph_0 + 2 = \aleph_0$$
$$\aleph_0 + n = \aleph_0, \text{ where } n \text{ is any counting number}$$
$$\aleph_0 + \aleph_0 = \aleph_0$$

which is the same as
$$2\aleph_0 = \aleph_0$$
$$3\aleph_0 = \aleph_0$$
$$n\aleph_0 = \aleph_0, \text{ where } n \text{ is a counting number}$$

In the search for other transfinite cardinals, Cantor noticed that there are more proper subsets of a finite set than there are elements in the set. Thus the set $\{\,1, 2, 3\,\}$ has cardinality 3, but has more than 3 proper subsets, namely $\{\quad\}, \{\,1\,\}, \{\,2\,\}, \{\,3\,\}, \{\,1, 2\,\}, \{\,1, 3\,\},$ and $\{\,2, 3\,\}$. In general, a set with n elements has 2^n subsets and $2^n - 1$ proper subsets. Cantor was able to prove that the same is true for infinite sets. Thus a set with cardinality $\aleph_0$ has more than $\aleph_0$ proper subsets. If we collect all these proper subsets into a new set, we get a set with cardinality $\aleph_1$ (as Cantor called it).

The process can be continued. Thus the set of subsets of a set with cardinality $\aleph_1$ has cardinality $\aleph_2$, and so on. While you may not appreciate this kind of argument, suffice it to say that there are infinitely many transfinite cardinals. In fact, there are more transfinite cardinals than there are counting numbers.

So we assert the existence of a series of transfinite cardinals $\aleph_0, \aleph_1, \aleph_2, \ldots$ and we have the transfinite cardinal c. We can think of the counting numbers along with the transfinite cardinals in a list:

$$1, 2, 3, 4, 5, \ldots, n, \ldots, \aleph_0, \aleph_1, \aleph_2, \ldots$$

But in the list, where does c belong?

Cantor asserted that $c = \aleph_1$, that is, that the cardinal number associated with $R^{\#}$ (the real numbers) is the next highest cardinal number after $\aleph_0$. This is referred to as the *continuum hypothesis*. In 1940 Kurt Gödel proved that Cantor's assertion cannot be proved or disproved. That is, we are free to accept or reject the idea that c is the next highest cardinal beyond $\aleph_0$. Either way the resulting mathematics will be consistent.

Cantor's revolutionary mathematical advances of the 1870s and 1880s created shock waves among mathematicians. Two who opposed Cantor's work were Poincaré (who referred to the theory as a disease among mathematicians) and Hermann Weyl (who called transfinite algebra a "fog upon a fog"). Yet David Hilbert said "No one shall expel us from the paradise which Cantor created for us." Indeed Cantor's legacy of transfinite theory has provided a rich source for mathematical research for the last 100 years.

Exercise Set 11.2

1 Let $A = \{a, b, c, d, e\}$, $B = \{e, f, g\}$, and $C = \{f, g, h\}$.
   (a) What is the cardinality of each set? That is, find $n(A)$, $n(B)$, and $n(C)$.
   (b) Find $n(A \cup B)$ and $n(A \cup C)$.
   (c) True or false: $n(A \cup B) = n(A) + n(B)$.
   (d) True or false: $n(A \cup C) = n(A) + n(C)$.

2 In general, when is $n(R \cup S) = n(R) + n(S)$?

3 Give three sets with cardinality $\aleph_0$, other than those mentioned in this section.

4 In this section we showed that N is equivalent to a subset of $R^\#$. Find another subset of $R^\#$ to which N is equivalent. How many such subsets of $R^\#$ do you think there are?

5 Show that $T = \{1, 2, 3, \ldots, 15\}$ has greater cardinality than $S = \{a, b, c, d, e\}$ by the method of Example 1 in this section.

6 What is the cardinality of the set of points on the
   (a) real number line?
   (b) real number line between 0 and 10?
   (c) real number line between any two different points, P and Q?

7 Show that $\aleph_0 + 2 = \aleph_0$.

8 Show that $\aleph_0 - 3 = \aleph_0$.

9 Let A be the set of even counting numbers, and B be the set of odd counting numbers.
   (a) What is the cardinality of A, $n(A)$?
   (b) What is $n(B)$?
   (c) Form $A \cup B$.
   (d) What is $n(A \cup B)$?

10 Is the cardinality of $A = \{a, e, i, o, u\}$ greater than or less than the cardinality of the set of proper subsets of A?

°11 Cantor proved that $2^{\aleph_0} = c$. In view of the definition of $\aleph_1$ (the next highest cardinal number after $\aleph_0$), why is Cantor's assertion that $c = \aleph_1$ reasonable? Hint: If a finite set has n elements, it has 2^n subsets. How many subsets of N ought there to be?

°12 Try to show that the set of all ordered pairs of counting numbers has cardinality $\aleph_0$. If you are unsuccessful, find an ingenious method for doing this in James R. Newman's *The World of Mathematics* (New York: Simon & Schuster, 1956).

°13 Find a proof that the decimals between 0 and 1 cannot be counted. How does this prove that the cardinality of $R^\#$ is not $\aleph_0$?

°14 Find out why c is used as a symbol to stand for the cardinality of the real numbers.

11.3 IS SPACE INFINITE?

The Cantor theory of the infinite has to do with infinite *sets*. Insofar as it has to do with sets of numbers and statements of equality, like $\aleph_0 + 1 = \aleph_0$, we can speak of the theory as a transfinite algebra. When we associate numbers with lines or line segments we begin to think geometrically about infinity. Certain infinite sets of points can be organized into lines, circles and curves, planes, and solid figures.

While it is common to think of geometric figures as occupying some part of the universe in which we live, they are abstractions. No one has ever seen a point or a line or a plane. We sketch figures to represent these abstract geometric objects; they can be thought of as models with which to study the objects. For example, the "lines" in Figure 6 assist our minds in conceiving of parallelism.

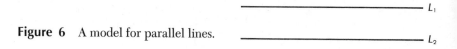

Figure 6 A model for parallel lines.

Yet neither L_1 nor L_2 is a line. First of all, lines extend indefinitely far in two directions. Secondly, the lines are assumed to be "straight," but we can only approximate straightness in the physical world. Furthermore, L_1 and L_2 are not lines at all, but streaks of graphite or ink on the paper. With a magnifying glass it is easy to see that a sketch of a line has thickness. If the streak is say, one-hundredth of an inch wide, then we have to accept the fact that there are infinitely many points in the width of the line.

Euclidean geometry (dating from about 300 B.C.) is a study of objects that have no real existence. Nonetheless, our sketches of geometric figures do assist us in thinking about them. And in the physical world properties of parallelism, perpendicularity and so on, are *used* in such trades as surveying, carpentry, and machine design. Thus Euclidean geometry can be interpreted, approximately, in the physical universe.

If lines are not physical objects, then, of course, parallel lines are not physical objects either. Conceptually, the parallel lines of Euclidean geometry never meet; they are always equidistant from one another. Does this mean that two parallel lines run off into space forever? Does this imply that space must be infinite in extent?

In the next few pages we will shed some light on these questions. First, let us consider some ancient and modern theories of the universe.

A thousand years ago it was common to conclude that the world is flat, that the sun travels around the earth, and that the heavens are like a huge bowl pierced with the stars. Observations by Ptolemy (A.D. 100–170) supported a view that the planets and the sun were spheres that revolved about the earth (a geocentric system). The universe was cast in an enormous sphere containing the stars and planets with the earth at the center. It was thought that the sphere containing the stars rotated about the earth once every 24 hours. This view persisted for 1400 years or so until about 1530 when Copernicus (1473–1543) proposed the heliocentric theory, wherein the sun became the center of our universe. To some, space was infinitely dotted with stars.

Our own galaxy, the Milky Way, was thought by others to be all that existed, floating about in an infinite ocean of space. Scientists pondered heavy questions about whether or not there were infinitely many stars or whether there were infinitely many galaxies. In 1924 other star systems beyond our own were discovered by the Mount Wilson 100-inch telescope.

One current theory of the universe is that it is expanding; that the totality of all galaxies is moving outward from some central point into the vastness of space. There is some scientific evidence for this in the so-called "galactic red shift," a phenomenon comparable to the change in pitch of the sound of a horn of a passing automobile or train. The rate of expansion of the universe has even been calculated to be in the neighborhood of 60,000 kilometers per second (about 37,280 miles per second).

We can still ask "What are the boundaries of space?" To our minds it is inconceivable that there is an endless or infinite set of stars because stars are physical objects. There can be only so much matter. It remained for Einstein in his Theory of Relativity, to construct a model of the universe in which space is *endless yet finite.*

The phrase "endless yet finite" does not fit into a model of space that is stated in terms of the lines and planes of Euclidean geometry. Euclidean parallelism, we have said, seems to require space to be endless and infinite. A concept of curved space is needed before we can see how a "line" can be both finite and yet without end. A rubber band is endless yet finite in extent; the equator is endless yet finite. But are they lines? Can lines (or straight lines, anyhow) exist in a curved space? After all, a straight line is not a curved line.

Mathematicians generalize the concept of "line" by using the term "geodesic." A geodesic on a surface is the path of shortest distance between two points. On a plane the geodesic is a straight line; in fact, people have tried to define a line as the shortest distance between two points. But clearly, on a sphere, the geodesic is a great circle (like the equator, or like the imaginary lines of longitude running through the poles). We speak of the "great circle route," for example, between New York and the British Isles.

If space, as vast as it may be, is curved, then a straight line will not do for a geodesic. If we could withdraw ourselves from the universe (impossible) we could look back at space and might observe that geodesics are curves that wander about in some fashion. If they were closed curves, then space could be endless yet finite.

It is true that this requires a concept of geometry other than our traditional Euclidean one. In Euclidean geometry lines are "straight lines" with no curvature; planes are determined by pairs of intersecting lines; space is determined by intersecting planes. In particular, parallel lines exist. Euclid's so-called fifth axiom implies that: Given a line L and a point not on it, exactly one line L' can be drawn through the point and parallel to the given line.

Figure 7   A model for the Euclidean fifth axiom.

This has been found to typify Euclidean geometry in such a way that any mathematical system in which the axiom is not true is termed *non-Euclidean geometry*. For thousands of years, it was thought that the Euclidean fifth axiom could be proved. But the ingenious work of Riemann, Gauss, Lobachevsky, and Bolyai all working in the early nineteenth century, resulted in consistent geometries in which the fifth axiom specifically did not hold. No logical contradictions resulted. In other words, the new non-Euclidean geometries that were invented were just as good as the Euclidean.

In the Lobachevsky non-Euclidean geometry *two* lines can be drawn through point P parallel to the given line L. In the Riemann non-Euclidean geometry no lines can be drawn through point P parallel to line L.

Certainly we cannot imagine either of these geometries if we restrict our thinking to a plane. But recall that on a sphere the geodesic is a great circle. On a sphere, then, all "lines"—that is, all great circles—intersect. (Lines of longitude, or meridians, intersect at the poles. Every meridian intersects the equator, another great circle.)

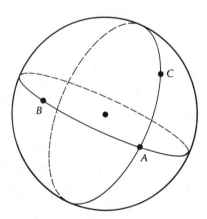

Figure 8
Lines on a sphere are great circles.

In Figure 8 the great circle through A and C can be taken as a given "line." Point B is not on that great circle, but any great circle through B intersects the first one somewhere. Thus there are no parallels to the great circle AC through a point B not on AC.

Another interesting phenomenon can be explained here. In Euclidean geometry we can prove that the sum of the angles of a triangle is 180°. The proof uses the Euclidean fifth axiom. A parallel to AB is drawn through point C. Then $\measuredangle\, a + \measuredangle\, b + \measuredangle\, c = \measuredangle\, d + \measuredangle\, e + \measuredangle\, c = 180°$. (You might like to reconstruct the details of this proof.) See Figure 9.

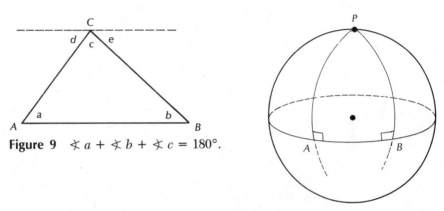

Figure 9 $\measuredangle\, a + \measuredangle\, b + \measuredangle\, c = 180°$.

Figure 10 $\measuredangle A + \measuredangle B + \measuredangle P > 180°$.

In Riemann non-Euclidean geometry the corresponding theorem states that the sum of the angles is *greater* than 180°. This is believable if we consider the sphere as a model and take two meridians along with the equator. These three great circles form a triangle, with angles A, B, and P. The angles at A and B are right angles, so their sum is already 180°. Thus, the excess over 180° is due to the angle at P. See Figure 10.

In the Lobachevsky non-Euclidean geometry the sum of the angles of a triangle is less than 180°. A model for this kind of geometry is the so-called pseudo-sphere. A pseudo-sphere is sketched in Figure 11; geodesics (or "lines" of shortest distance between two points) are spirals that twist around the figure. There are many "lines" passing through a given point that do not intersect a given line.

Each of the three geometries mentioned is useful in some context. Euclidean geometry is good enough for the small world of our experience. The non-Euclidean geometries are used in explaining problems having to do with time, force, or motion, where the distances involved are very large (billions and trillions of miles). Einstein's Theory of Relativity makes use of non-Euclidean geometry for the universe, in which space is conceived as endless yet finite. The universe, as enormous as it is, does not have to be infinite in extent.

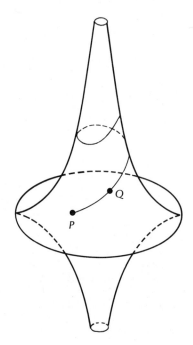

Figure 11
Sketch of a pseudo-sphere.

To date no one has proved that the universe is either finite or infinite. It may well be that we will never know. Perhaps different kinds of geometry will always be with us and we will be forced to accept two concepts of space; one holding that it is finite and one holding that it is infinite. With alternatives, we might be able to use either one to explain various scientific observations.

There is experimental evidence that space *is* curved. Many predictions have been made concerning physical happenings in the universe on the basis of the non-Euclidean aspects of the Theory of Relativity. Some of these predictions have later led to discovery of the phenomena. For example, Relativity predicts that light does not travel in a straight line. During a total eclipse of the sun, astronomers at different points on the earth actually measured the displacement or bending of light from distant stars as it passed by the sun. The predicted phenomenon was found to be true.

Even though we may accept that space is curved and finite in extent, we cannot escape infinity. The set of counting numbers would still be infinite; the set of real numbers between any two reals would still be infinite. If numbers are but abstractions, then we can mention the measurement of time; any time interval can be divided into infinitely many sub-intervals.

The question in the title of this section remains unanswered. But we hope we have sparked some interest and enthusiasm for your own thoughts about infinity. There are probably few other concepts so interesting or so profound.

Exercise Set 11.3

1 Explain what might be meant by "There is no such thing as a straight line."

2 (a) If the earth were flat, would there have to be boundaries or edges?
  (b) In 1520 Magellan circumnavigated the earth. How does this prove that the earth is not flat?
  (c) Is the surface of the earth boundless? Is it infinite in extent? Does it contain an infinite number of points?

3 If the universe is endless yet finite in extent, does this mean it does not contain an infinity of points?

4 (a) On a globe, are lines of latitude great circles?
  (b) True or false: Great circles always have a center that is also the center of the sphere.

5 In a direct flight from New York to Bangkok, Thailand, a jet plane heads north but eventually must be heading south. Explain.

6 Perform the following thought experiment. From a point on the equator travel exactly north on a great circle until you come to the north pole. Then turn 90° and travel south on a great circle until you come to the equator again. Then turn 90° toward your starting point and travel to that starting point. (A globe or a polar projection map will help you to visualize this trip.)
  (a) In the spherical triangle corresponding to your trip, what is the sum of the angles?
  (b) Is there such a thing as a spherical rectangle?

7 In Euclidean geometry two lines intersect in at most one point. In Riemannian geometry (with a sphere as a model), in how many points do two "lines" intersect?

8 In Euclidean geometry a straight line is conceived of as infinite in length. Are the "lines" in Riemannian geometry infinite? Are they all the same length?

9 Report to the class on one of the following:
  (a) Sir Arthur Eddington, *Space, Time and Gravitation*. New York: Harper & Row, 1959, prologue, "What is Geometry?"
  (b) *Ibid.*, Chapter X, "Towards Infinity."
  (c) *Ibid.*, Chapter XII, "On the Nature of Things."
  (d) Sir Arthur Eddington, *The Nature of the Physical World*. Ann Arbor, Mich.: Ann Arbor Paperbacks, 1963, Chapter IV, "The Running-Down of the Universe."
  (e) *Ibid.*, Chapter VIII, "Man's Place in the Universe."
  (f) Sir Arthur Eddington, *The Expanding Universe*. Ann Arbor, Mich.: Ann Arbor Paperbacks, 1962, Chapter I, "The Recession of the Galaxies."
  (g) Edwin A. Abbott, *Flatland: A Romance of Many Dimensions*, 5th rev. ed. New York: Barnes & Noble, 1963. (Abridged in James R. Newman, *The World of Mathematics*, Volume IV. New York: Simon & Schuster, 1956).
  (h) George Gamow, *One, Two, Three, Infinity*. New York: Viking Press, 1961, Chapter IV, "The World of Four Dimensions."
  (i) Archimedes, "The Sand Reckoner," in James R. Newman, *The World of Mathematics*. New York: Simon & Schuster, 1956.

Suggested further readings

1 Moses Richardson, *Fundamentals of Mathematics*, Third Edition. New York: Macmillan, 1966, Chapter 17.

2 *Mathematics, Life* Science Library. New York: Time, Inc., 1963, Chapter 7, "A Logical Leap Into the Wild Blue Yonder."

3 *Mathematics in the Modern World, Readings from Scientific American.* San Francisco: W. H. Freeman, 1968, Chapters 16, 17, 18, 30, 33.

4 Hans Hahn, "Infinity," in James R. Newman, *The World of Mathematics*, Vol. 3. New York: Simon & Schuster, 1956.

5 Lillian R. Lieber, *Infinity*. New York: Rinehart & Co., 1953.

SUMMARY

A set is infinite if it can be placed in one-to-one correspondence with a proper subset of itself. This definition becomes the criterion for an infinite set when we realize that the correspondence is impossible for a finite set. Many surprising results follow. For example, there are as many odd and as many even counting numbers as there are counting numbers. There are as many points on a one-inch line as there are on a line five million miles long. There are as many decimals between 0 and 1 as there are points in a plane.

Cantor devised a system for dealing with the cardinality of infinite sets. Just as we say we have five fingers, we can say there are aleph null counting numbers. Furthermore, there are different "sizes" of infinity. We mentioned that the transfinite c is larger than $\aleph_0$ and that there are many others; $\aleph_1$, $\aleph_2$, $\aleph_3$, and so on. Though Cantor's work was the result of genius, he created many problems that are still unsolved.

Infinite sets of points lead us naturally to a geometric consideration of infinity, which in turn, makes us wonder if space itself is infinite. Once the parallel axiom of Euclid was questioned, new geometries were invented. These non-Euclidean geometries are as consistent as Euclidean geometry. Though we mentioned only the Riemannian and Lobachevskian geometries, others exist. In fact there is a hierarchy of geometries wherein we find that Euclidean and non-Euclidean geometries are special cases of the so-called "projective" geometry.

We know we cannot prove Euclidean geometry any more correct than any other kind. Mathematicians have created models in which each geometry is workable. This leaves open the question about the shape and extent of the universe. Perhaps the universe itself is a model wherein an all-inclusive kind of geometry prevails. When (and if) we discover that, we will have answered the questions about infinite space.

It may very well be that the universe is finite, that reality is finite. Infinity would then exist only in the abstract. There are those who would then conclude that it does not exist at all!

The Review Test assumes an understanding of the following topics:

1. Showing that sets of numbers are infinite.

2. Showing that a set is finite.

3. Showing geometric figures to be infinite sets of points, or that two different figures have the same number of points.

4. $\aleph_0$ and c as infinite cardinal numbers.

5. The concept "endless yet finite."

6. The difference between Euclidean and non-Euclidean geometry in regard to the parallel axiom.

REVIEW TEST

1   (a) Show a one-to-one correspondence between the following two sets:
$$A = \{1, 2, 3, 4, 5, \ldots\}$$
$$B = \{5, 6, 7, 8, 9, \ldots\}$$

(b) To what element in B does the element 12 in A correspond?
(c) To what element in A does the element 20 in B correspond?
(d) Is B a proper subset of A?
(e) Does the correspondence show that B is infinite or that A is infinite?

2   Prove that the set $\{3, 7, 11, 15, 19, \ldots\}$ is infinite.

3   Explain why $\{r, s, t, u, v\}$ is not an infinite set.

4   (a) Show that the square and the circle below have the same number of points.

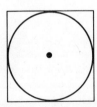

(b) What is the appropriate cardinality of the points comprising either the square or the circle?

5   What is the cardinal number of the infinite set of numbers $\{10, 20, 30, 40, \ldots\}$?

6 (a) On the number line below how many points are there between 0 and 1?

$$\begin{array}{ccccc} \vdash & \vdash & \vdash & \vdash & \vdash \\ -1 & 0 & 1 & 2 & 3 \end{array}$$

(b) True or false: There are c points between 0 and 2.

(c) True or false: There are c points between 1 and 2.

7 True or false:

(a) Nothing can be endless yet finite in extent.

(b) In Euclidean geometry there is exactly one line parallel to a given line through a point not on the given line.

(c) In Riemann or Lobachevsky non-Euclidean geometries the sum of the angles of a triangle is 180°.

(d) Non-Euclidean geometry is more correct than Euclidean geometry.

(e) The sum of the angles of a spherical triangle must be greater than 180°.

8 Land surveyors assume that the earth is flat. Why is this a reasonable assumption?

Maurits Cornelis Escher (1898–1972) was an artist intrigued by optical illusions and the limitations inherent in the laws of visual perspective and two-dimensionality. Though he holds no high rank with the great masters, he is considered by some to be the precursor of "op art." His works, such as *Catalogue 119: Study of the Regular Division of the Plane with Reptiles,* have a special appeal to mathematicians. Escher provides a visual portrait of the problems of space and ordered planes. His essay "Approaches to Infinity" is nonmathematical but conveys an artist's concept of infinity.

APPROACHES TO INFINITY
M. C. ESCHER

Man is incapable of imagining that time could ever stop. For us, even if the earth should cease turning on its axis and revolving around the sun, even if there were no longer days and nights, summers and winters, time would continue to flow on eternally.

It is no easier for us to imagine that somewhere, past the farthest stars in the nocturnal heavens, there is an end to space, a borderline beyond which "nothing" exists. The concept "empty" does have some meaning for us, because we can at least visualize a space that is empty, but "nothing," in the sense of "spaceless," is beyond our capacity to imagine. This is why, since the time when man came to lie, sit and stand on this earth of ours, to creep and walk on it, to sail, ride, and fly over it (and now fly away from it), we have clung to illusions—to a hereafter, a purgatory, a heaven, and a hell, a rebirth or a nirvana, all existing eternally in time and endlessly in space.

Has a composer, an artist for whom time is the basis on which he elaborates, ever felt the wish to approach eternity by means of sounds? I do not know, but if he has, I imagine that he found the means at his disposal inadequate to satisfy that wish. How could a composer succeed in evoking the suggestion of something that does not come to an end? Music is not there before it begins or after it ends. It is present only while our ears receive the sound vibrations of which it consists. A stream

"Approaches to Infinity" by M. C. Escher, from *The World of M. C. Escher,* edited by J. L. Locher, New York: Harry N. Abrams, 1971, pp. 37–39. Reprinted by permission of the Escher Foundation—Haags Gemeentemuseum—The Hague.

of pleasant sounds that continues uninterrupted through an entire day does not produce a suggestion of eternity but rather fatigue and irritation. Not even the most obsessive radio listener would ever receive any notion of eternity by leaving his set on from early morning to late in the night, even if he selected only lofty classical programs.

No, this problem of eternity is even more difficult to solve with dynamics than with statics, where the aim is to penetrate, by means of static, visually observable images on the surface of a simple piece of drawing paper, to the deepest endlessness.

It seems doubtful that there are many contemporary draftsmen, graphic artists, painters, and sculptors in whom such a wish arises. In our time they are driven more by impulses that they cannot and do not wish to define, by an urge which cannot be described intellectually in words but can only be felt unconsciously or subconsciously.

Nevertheless, it can apparently happen that someone, without much exact learning and with little of the information collected by earlier generations in his head, that such an individual, passing his days like other artists in the creation of more or less fantastic pictures, can one day feel ripen in himself a conscious wish to use his imaginary images to approach infinity as purely and as closely as possible.

Deep, deep infinity! Quietness. To dream away from the tensions of daily living; to sail over a calm sea at the prow of a ship, toward a horizon that always recedes; to stare at the passing waves and listen to their monotonous soft murmur; to dream away into unconsciousness. . .

Anyone who plunges into infinity, in both time and space, further and further without stopping, needs fixed points, mileposts, for otherwise his movement is indistinguishable from standing still. There must be stars past which he shoots, beacons from which he can measure the distance he has traversed.

He must divide his universe into distances of a given length, into compartments recurring in an endless series. Each time he passes a borderline between one compartment and the next, his clock ticks. Anyone who wishes to create a universe on a two-dimensional surface (he deludes himself, because our three-dimensional world does not permit a reality of two nor of four dimensions) notices that time passes while he is working on his creation. But when he has finished and looks at what he has done, he sees something that is static and timeless; in his picture no clock ticks and there is only a flat, unmoving surface.

No one can draw a line that is not a boundary line; every line splits a singularity into a plurality. Every closed contour, no matter what its shape, whether a perfect circle or an irregular random form, evokes in addition the notions of "inside" and "outside" and the suggestion of "near" and "far away," of "object" and "background."

The dynamic, regular ticking of the clock each time we pass a boundary line on our journey through space is no longer heard, but we can replace it, in our static medium, by the periodic repetition of similarly shaped figures on our paper surface, closed forms which border on each other, determine each other's shape, and fill the surface in every direction as far as we wish.

What kind of figures? Irregular, shapeless spots incapable of evoking associative ideas in us? Or abstract, geometrical, linear figures, rectangles or hexagons at most suggesting a chess board or honeycomb? No, we are not blind, deaf, and dumb; we consciously regard the forms surrounding us and, in their great variety, speaking to us in a distinct and exciting language. Consequently, the forms with which we compose the divisions of our surface must be recognizable as signs, as distinct symbols of the living or dead matter around us. If we create a universe, let it not be abstract or vague but rather let it concretely represent recognizable things. Let us construct a two-dimensional universe out of an infinitely large number of identical but distinctly recognizable components. It could be a universe of stones, stars, plants, animals, or people.

What has been achieved with the orderly division of the surface in *Study of Regular Division of the Plane with Reptiles* (Cat. 119)? Not yet true infinity but nevertheless a fragment of it, a piece of the universe of the reptiles. If the surface on which they fit together were infinitely large, an infinitely large number of them could have been represented.

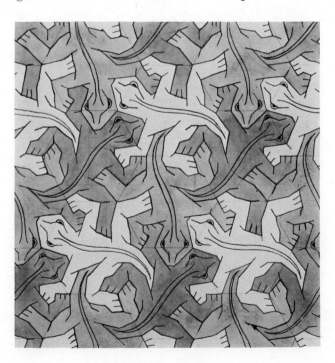

Study of Regular Division of the Plane with Reptiles by M. C. Escher.

But this is not a matter of an intellectual game; we are aware that we live in a material, three-dimensional reality, and we are unable in any way to fabricate a flat surface extending infinitely on all sides. What we *can* do is to bend the piece of paper on which this reptilian world is represented fragmentarily and make a paper cylinder of it so that the animal figures on that cylindrical surface continue without interruption to interlock while the tube revolves around its longitudinal axis. In this way, endlessness is achieved in one direction but not yet in all directions, because we are no more able to make an infinitely long cylinder than an infinitely extending flat surface.

12 THE COMPUTER AGE

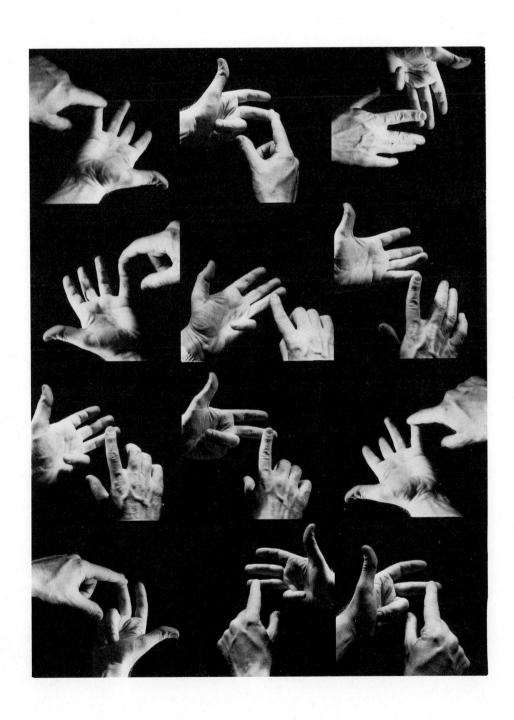

We are in the midst of an information-handling revolution comparable in its impact on society to the Industrial Revolution of the early nineteenth century.

Data processing by computers, as commonplace as it is today, is a recent innovation, but other automatic machines were used to handle information as early as the nineteenth century. Punched cards for the processing of data were invented in the 1890s by Herman Hollerith, a statistician who first used machinery to handle the United States Census. The cards have been with us ever since.

There were computers in the late 1930s, but it took World War II to give momentum to the development of computing machinery. These wartime computers were constructed to handle artillery firing problems and to produce mathematical tables used by gunners in the battlefield. Probably few people knew of their existence.

Between 1946 and 1951 only 20 computers were built. IBM entered the electronic computer field in the early 1950s with the expectation that there might be 50 customers for their Model 650 computer. They sold over 1000. Meanwhile other companies got on the computer bandwagon during the 1950s and 1960s. In 1969 it was estimated that there were about 60,000 computers in use. There are many thousands more in operation at the present time.

In the last 15 or 20 years there has been a boom in the computer field. Before 1955 few people were affected by computer printouts or by punched cards. Today we know little else; we are truly in the computer age.

In this chapter we will take a brief look at the history of calculating and computing machinery. We will examine, in broad terms, the operation of computers, and we will see how to write instructions so that a computer will perform calculations for us.

12.1 FROM CALCULATORS TO COMPUTERS

Although the evolution of computers has been rapid and recent, calculating devices have had a long history. The abacus, consisting of several columns of movable beads in a frame, and used for arithmetic computation, had its beginnings 5000 years ago in ancient Babylonia. The first such devices were probably no more than dust-covered boards on which marks were made to keep track of computations. In Asia one still finds the abacus in common use by clerks and merchants.

An early contraption called Napier's rods or Napier's bones, invented by the Scottish mathematician John Napier (1550–1617), was capable of performing multiplication. This simple machine consisted of lengths of

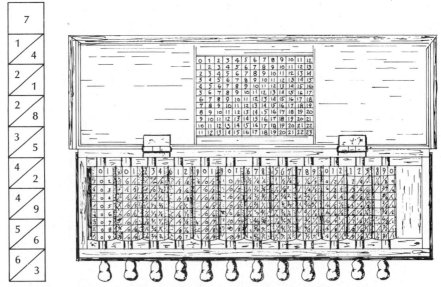

Figure 1   (a) The scheme for a Napier rod. (b) Napier's rods.

bone or ivory, each engraved with one of the numbers from one to nine. Below each number were the corresponding multiples of that number, arranged in columns. A typical rod, showing the multiples of seven, is shown in Figure 1. These rods were manipulated in relation to one another in ways that gave correct products. Napier is also credited with developing much of the first mathematics concerning logarithms (exponents). He invented a primitive multiplication machine that led to the later development of the slide rule. We can think of the slide rule as a computer that calculates products, quotients, and powers, operating according to the principles of logarithms.

The French mathematician Blaise Pascal (mentioned elsewhere in connection with probability) invented an adding machine in 1640. This mechanical wonder was housed in a box about 14 by 5 by 3 inches and operated by a system of gears not unlike the odometers (mileage indicators) in our automobiles. Today some housewives add up prices in a supermarket on small flat adding machines operated by a stylus. These little devices are as capable as Pascal's machine and are available for as little as $1.98.

Leibniz, creator of the calculus along with Newton, improved on Pascal's adding machine in 1671. Leibniz's machine was capable of multiplication, division, and extraction of roots. The principle of operation was about the same as in Pascal's machine, where multiplication was accomplished as repeated addition ($3 \cdot 4 = 4 + 4 + 4$) and division as repeated subtraction.

Nearly 150 years later the Englishman Charles Babbage proposed a so-called "difference or analytical engine" to calculate and print out logarithmic and astronomic tables. Only part of the machine was built, because of the lack of funds and technology. We can almost imagine his feeling when the funds, which had been supplied by the British government, were cut off and his machine consigned to a government museum. Babbage's invention dates from about 1812 but in theory used the same principles as a modern computer. It used steam as a source of power, had its operating instructions stored on punched cards, had a kind of memory and arithmetic unit, and would have printed results.

Joseph Marie Jacquard successfully used punched cards to control the operation of a machine in 1804. Jacquard built a loom that could be operated by one man and was capable of producing the most complex designs. Instructions to the machine were punched into cards, each line of woven fabric being controlled by one card. The presence or absence of holes controlled mechanical fingers that raised certain strands of the background just prior to the fly of the shuttle. For the next weave another card would determine the particular strands to be lifted. The cards were hooked together and arranged in a loop to allow repeating patterns.

A commentator at the time wrote "We may say most aptly that the Analytical Engine (of Babbage) *weaves algebraical patterns* just as the Jacquard loom weaves flowers and leaves."°

The use of punched cards to handle information was the brainchild of Dr. Herman Hollerith. The 1880 census had taken 7 years to analyze by hand, and it was projected that the next census would take 12 years. Hollerith devised machinery to mechanically sort and tabulate punched cards, thereby enabling the U. S. Bureau of Census to complete the 1890 census in only 2 years. In fact, Hollerith standardized the size of the present-day punched card. We often call them IBM cards, but a more accurate name is the Hollerith card. The code used to punch cards is named the Hollerith code in his honor.

From 1900 to 1940 many changes were made in equipment built to handle punched cards. Entire accounting systems were built upon this *unit record equipment.* "Unit record" is virtually synonymous with "punched card," but the term implies that each bit of information is stored on a separate card. An analogy is the card catalog in a library, in which each book or set of books has its title entered on a card, and then the cards are filed alphabetically. If a library contains 50,000 books (excluding sets and duplicates) there are 50,000 title cards.

The impetus for faster card-handling equipment came from an ever more complex society whose population was increasing dramatically.

°From Philip and Emily Morrison, eds., *Charles Babbage and His Calculating Engines.* New York: Dover, 1961.

Imagine the need for record keeping and data processing spawned by the Social Security Act of 1935. At the same time, our impending involvement in World War II put great emphasis on the need for data handling. That war became, more than any previous war, a battle between the technologies of opposing nations. New weapons and aircraft came off the drawing boards in more and more rapid succession. Quality control, inventory analysis, logistics, and an endless invention of complex apparatus demanded faster information processing.

The first large-scale electronic computers were developed by teams from universities and government military agencies. The ENIAC (an acronym for Electronic Numerical Integrator and Calculator) was built between 1943 and 1946 by Professors J. P. Eckert and John Mauchly of the University of Pennsylvania for the Ballistic Research Laboratories of the U.S. Army Ordnance Corps. It was used in calculating trajectories and firing tables. This computer was quite a success in spite of vacuum-tube failures; it contained about 18,000 tubes, which would frequently malfunction. The "down time" for maintenance and repairs was enormous, and the periods of "up time" (time of satisfactory operation) rarely lasted more than 30 minutes.

Even before the ENIAC, Howard H. Aiken began working on the MARK I, a mechanical computer that contained myriads of telephone components. A mechanical computer depends upon moving wheels and gears to perform calculations, while an electronic computer uses electric circuits to do the same thing. The MARK I project, given financial support by IBM, was finished in 1943 and produced data for military use for 5 years.

The MARK I and the ENIAC were decimal machines; that is, they dealt with numbers in base ten. The ENIAC was improved in design in a new model called the EDVAC (for Electronic Discrete Variable Automatic Computer). Eckert and Mauchly were joined in this project by the noted Princeton mathematician John von Neumann. The EDVAC was different from the ENIAC in that it stored its own operating instructions and operated in binary, or base two. It was in operation by the late 1940s. With EDVAC, binary became the preferred number base in the operation of electronic computers.

In today's jargon, a machine that is capable of storing its own instructions, instead of needing some external control system, is called a computer. Calculators do not have a stored program of instructions, but operate from punched cards or from a keyboard. Thus a desk calculator or a slide rule or an abacus is a calculator, since each receives its instructions from an operator. There is no way to store instructions in a calculator so that the operator may walk away and return to find the problem completed. This self-operating feature distinguishes the computer from the calculator.

Business and industrial use of computers began in the early 1950s. A business-oriented model of the ENIAC was created by Remington Rand and called the UNIVAC (Universal Automatic Computer). IBM's first business computer, the IBM 650, was introduced in 1954, and there was a rash of related but ever more powerful models; the 704, 705, 707, and 709. All of these early, or *first-generation* electronic computers used vacuum tubes.

In 1948 the transistor was invented. While computer manufacturers were quick to see its advantages over vacuum tubes, several years were required to plan and produce transistorized computers. Vacuum tubes generated heat and consumed a large amount of power, but transistors did all the work of tubes, yet were faster, generated little heat, required much less energy, and were much smaller, making possible smaller computers. Computers built with transistors were called *second-generation* computers. In 1954 the first transistorized computer, the TX-O, was started at the Lincoln Laboratories. Four years later, no computers were being built with vacuum tubes.

Between 1959 and 1963, IBM alone sold 10,000 computers. But there were other companies competing for a share of the market; RCA, Remington Rand, Burroughs, Honeywell, General Electric, NCR (National Cash Register), Philco, and CDC (Control Data Corporation).

The second-generation computer performed a basic operation in a few microseconds (millionths of a second), while the earlier vacuum-tube models required a few milliseconds (thousandths of a second). In the *third generation* of computers (1965 and later), clusters of transistors were attached to small ceramic chips or were deposited photochemically on the chips. The circuits formed are called *monolithic integrated circuits*. The chips are so small that they are wired to each other under a microscope—50,000 chips will fit in a thimble. The result of this new microelectronic technology is a computer that is smaller and faster. Basic arithmetic operations are now performed in a few nanoseconds (billionths of a second).

An interesting way to survey computer evolution is to compare the cost of performing basic operations. Let us say that solving a certain problem requires 100,000 additions. Table 1 shows how the cost of solving such a problem is going down.

Table 1   Comparing the cost of computing since 1940.

Year	1940–45	1945–50	1952–57	1959–63	1965–67	1972
Cost per 100,000 computations	$450	$25	$2.50	25¢	2.5¢	0.7¢

These costs are only approximate and would depend upon the particular computer involved. But the change from $450 to 0.7¢ is dramatic. At first only supercorporations or governmental agencies could afford the great expense of operating a computer. Today even small companies can afford to use computers; in fact, business competition is such that they cannot afford *not* to use computers.

Almost every college has the use of a computer. We mean by "use of a computer" that either the college has its own computer center, or else it has terminals that are hooked up to a computer by telephone lines. These terminals have the appearance of a typewriter. The operator types his instructions on the keyboard; the computer, which may be hundreds of miles away, does the computation and sends back the results over the telephone line. The input typewriter also performs the output function; results are printed out by the same machine.

Many different terminals can be operated simultaneously with the remote computer. This is accomplished by a time-sharing system. The computer works a bit on person A's problem, and then abandons it while it works on person B's problem, and so on. Each person seems to have the sole use of the computer because the switching time from problem to problem is so short. Time-sharing systems exist in which more than 500 terminals are connected to the same computer.

The last 20 years have brought fantastic changes in computer design and use. The next 20 years will probably make present systems seem prehistoric. Probably no other area of technology has developed as rapidly as the computer field. It is estimated that 3 million time-sharing terminals will be in use by 1980. The telephone companies predict that over 50% of the use of their lines will be for communications between terminals and computers, or between computers and other computers.

Exercise Set 12.1

1 Write a brief report on one of the following:
   (a) I. Adler, *Thinking Machines*. New York: John Day, 1961.
   (b) Jeremy Bernstein, *The Analytical Engine*. New York: Random House, 1964.
   (c) W. R. Corliss, *Computers*. Oak Ridge, Tenn.: U. S. Atomic Energy Commission, 1967.
   (d) P. Morrison and E. Morrison, eds., *Charles Babbage and His Calculating Engines*. London: Dover Publications, 1961.
   (e) John McCarthy, "Information," *Scientific American*, Sept. 1966. (Also reprinted in *Computers and Computation*, San Francisco: W. H. Freeman, 1971.)
   (f) Saul Rosen, "Electronic Computers: A Historical Survey," *Computing Surveys*, Vol. 1, No. 1 (March 1969).

(g) D. S. Halacy Jr., *Computers, the Machines We Think With.* New York: Dell Publishing Company, Inc., 1964.

(h) John G. Kemeny, *Man and the Computer.* New York: Charles Scribner's Sons, 1972.

2 Report on one or two issues of one of the following periodicals:

(a) *Computerworld*   (b) *Business Automation*

(c) *Computer Decisions*   (d) *Computers and Automation*

(e) *Datamation*   (f) *Data Processing Magazine*

(g) *Computing Surveys*   (h) *EDP Weekly*

3 Research one or two of the following:

(a) Charles Babbage   (b) John von Neumann

(c) Automata or computing automata (d) Unit-record equipment

(e) Time-sharing   (f) Cybernetics

(g) Joseph M. Jacquard

4 Write to one of the companies below to obtain information of a general nature about the history and present characteristics of the computers they manufacture.

(a) International Business Machines   (b) Sperry Rand Corporation

(c) Control Data Corporation   (d) National Cash Register Company

(e) Burroughs Corporation   (f) General Electric

12.2 MAKING A COMPUTER WORK

In broad terms, a system for computing consists of two parts, frequently referred to as the "hardware" and the "software." The hardware is the physical apparatus that takes up the space in a computer center. The software is the set of human-written instructions that direct the computer, along with any other operating documents like flowcharts and manuals. In this section we describe briefly the hardware aspects of a computer in terms of function, and we will see what flowcharts are.

There are three major components in computer machinery: an input device, a processing device, and an output device. But many other machines could be described the same way. An ordinary typewriter could be described as having these three parts. (The input device would be the keys; the processing device would be the internal parts that select the corresponding typebar; and the output device would be the typebar itself.) The components of a home music system can be similarly described. (The input device might be a turntable, a tuner, or a tape deck; the processing device would be the amplifier, and the output device would be the speakers.)

Of course, unlike typewriters or music systems, a computer calculates. Calculations are made in an *arithmetic unit.* So that these calculations are not made at random there must be a *control unit* to handle data and to monitor the operation of the arithmetic unit. The control unit keeps track of proper sequencing. Also in the central processing machinery are *storage devices* or *memories*. A computer's memory apparatus consists of either magnetic cores in planes, or the so-called monolithic storage array modules, or magnetic tapes, or magnetic disks. Most computers have a combination of different kinds of memory equipment.

Cores are tiny doughnut shaped pieces of ferromagnetic material (about the size of a pinhead), which are threaded on wires crossing each other at right angles. A core memory contains thousands of cores that are capable of being magnetized in one of two directions. A core magnetized in a counterclockwise direction can be interpreted as a 1; one magnetized in the clockwise direction would be interpreted as a 0. Thus, by magnetizing a string of cores in one direction or the other, it is possible to store a number in binary.

Third-generation computers use small chips about one-eighth inch square having a storage capacity equivalent to 256 cores. Two of these chips with appropriate wiring are mounted on a module one-half inch square, giving the module the same capacity as 512 cores. These monolithic storage array modules comprise an important part of the memory space of many large-scale computers today.

Magnetic tape units are also used to record data. Typically the tape reel contains 2400 feet of half-inch tape and is capable of storing the equivalent of 250,000 punched cards. Access to this storage is slow, but tapes can store enormous quantities of data. Magnetic disks, stacked like phonograph records, are also used in recording and reading data.

Input to the arithmetic, control, and storage sections of a computer is accomplished by a variety of equipment. Card readers, paper tape readers, optical mark readers, console typewriter keyboards, and tape and disk units are all used as input devices. An operator can sit at a terminal and type instructions into a computer many miles away.

The computer receives instructions in the form of a programming language along with the decimal numbers and perhaps certain English words. These are processed through a *compiler*, which converts the instructions into the binary numbers used by the computer internally. Thus it is not necessary for the operator to convert a simple problem like $7 + 13 = ?$ into the binary $111 + 1101 = ?$ We communicate with computers in decimal numbers, but the computer operates in binary numbers.

The *output* is also in decimal numbers. Before reaching the output stage, the internal binary information is changed to decimal form by a decoder section. Thus we would see 20 for the answer to $7 + 13$. (Internally, the computer has found that $111 + 1101 = 10\ 100$.) Output machinery is as varied as input machinery. The computer may respond to us by operating a typewriter, or a printer, or a card punch, or a graph plotter, or a TV tube display unit.

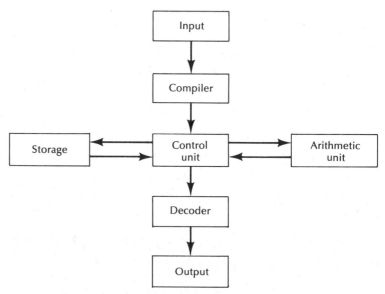

Figure 2   The operation of a computer in block form.

In block form, Figure 2 shows the sequence of equipment as it is interrelated.

Although it is helpful to know the various functions performed by a computer, it is not necessary to be aware of the components while using a computer. The same is true when you use the telephone or operate a record player. The user's prime concern is with the input. How do you start the machine? How do you make it work for you?

Before we begin to instruct a computer, it is helpful to have a plan for the execution of the steps involved in solving a problem. The plan can be diagrammed in a *flow chart* illustrating each sequential step. Figure 3 is a flow chart of the steps involved in making a telephone call.

In flow charts we ordinarily use circles or ovals to indicate the "start" and "stop" instructions, rectangles to indicate actions or operations to be performed, and diamond-shaped figures to indicate decisions to be made.

The flow chart in Figure 4 solves the equation $2x - 7 = 3$ with reference set $R = \{\,1,2,3,4,5,6\,\}$. If you study this flow chart, you will see that the first x is 1. The answer to the first question is "no," so the new x becomes 2. This is tried, then 3, then 4; $x = 5$ yields an answer of "yes." Then the 5 is printed. But we don't want to stop until all numbers in R are exhausted. So after "print 5" we try $x = 6$. When $2x - 7$ is not 3, we add 1, getting 7. Since 7 is not in R, the procedure stops.

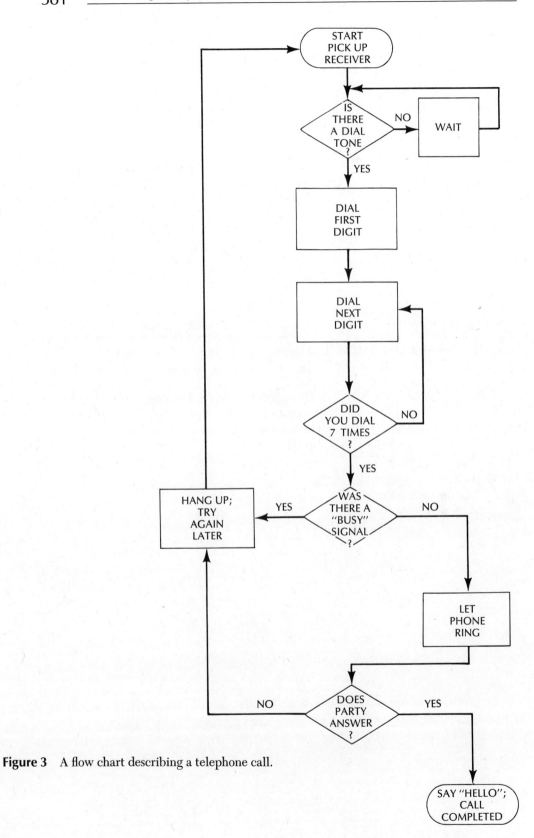

Figure 3   A flow chart describing a telephone call.

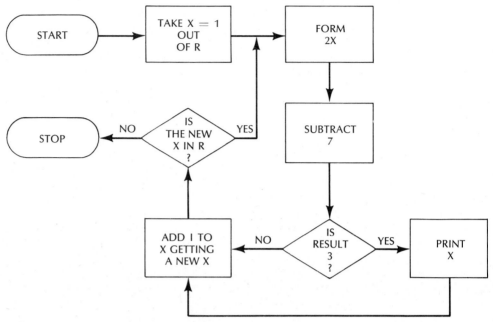

Figure 4   A flowchart to solve an equation.

A flow chart is usually part of the necessary planning done before actually writing instruction steps for the operation of the computer. Writing instructions is called *programming*; we will look briefly at programming in Section 12.3.

Exercise Set 12.2

1 Examine the following flow charts and tell what they do.

(a)

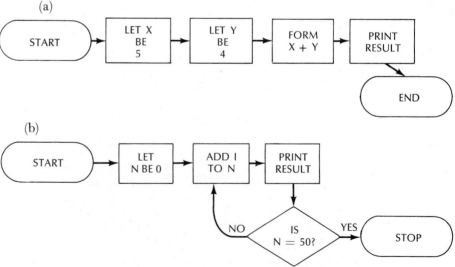

(b)

(c)

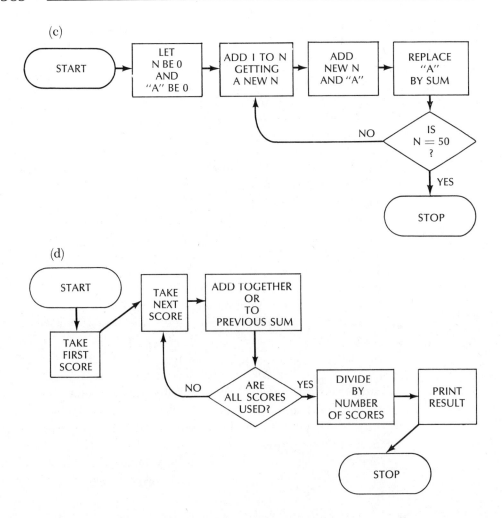

(d)

2 Construct your own flow chart for each of the following situations:
   (a) Sequentially drawing cards from a bridge deck until the queen of spades is drawn.
   (b) Sorting 25 names of students into 2 groups according to sex.
   (c) Checking a customer out at a supermarket, given that there is a 5% tax on certain items.
   (d) Deciding which of two numbers is larger.
   (e) Finding the smallest number in a list of numbers, all of which are different. Hint: The smallest number in a list is the only one that can be subtracted from each of the others to yield a positive result.

3 Set up a flow chart to plan how to solve each equation below with reference set { 1, 2, 3, 4, 5, 6, 7 }.
   (a) $2x - 1 = 9$ (b) $3x^2 = 12$
   (c) $(x - 1)(x - 4) = 0$ (d) $x^3 - 8 = 0$
   (e) $2x + 3x = 15$ (f) $(n - 3)n = 10$

12.3   WHAT IS PROGRAMMING?

We have indicated that the flow chart is a guide to the sequence of steps used in solving a problem. The next thing to be done is to write the program (or set of instructions) that will tell the computer what to do. There are many kinds of languages used to instruct a computer, each having been designed for a specific purpose. Some computers will accept several languages, but no computer accepts them all. A programming language is the code used to write instructions; once a program is written, it is fed into the computer, where it is converted to the 1's and 0's with which the machine operates.

Among the more common programming languages are FORTRAN, RPG, ALGOL, COBOL, PL/1, APL, and BASIC. One language may be more suited to a particular application than another. For example, FORTRAN (FORmula TRANslation) is commonly used in mathematical and scientific applications, while COBOL (COmmon Business-Oriented Language) is useful in solving business problems.

We will introduce BASIC (Beginners All-purpose Symbolic Instruction Code) because of its simplicity. It takes only a few hours to learn BASIC, whereas some other languages take a lot longer. BASIC was developed by John Kemeny and Thomas Kurtz at Dartmouth College especially for undergraduate students using time-sharing computer systems. The language has a small vocabulary and only a few rules of grammar. By "vocabulary" we mean the actual words or symbols used, and by "grammar" we mean the rules for combining the words and symbols. One other advantage of starting with BASIC is that it provides a good introduction to the more advanced and powerful FORTRAN, should you desire to study further.

The goal here is not to become expert programmers using BASIC, but to survey what it is and to see what a program in BASIC looks like. Nonetheless, the language is so simple that we will be able to use it to solve some familiar problems. We start with the following program:

EXAMPLE 1

```
10      REM  EXAMPLE 1
20      LET N = 3
30      READ  X, Y, Z
40      LET  A = (X + Y + Z) / N
50      PRINT  A
60      DATA  81, 84, 72
70      END
RUN
```

Probably without knowing any BASIC you can tell that this program will average three numbers. It will also print the average. We make the following points about the program:

1. Only capital letters are used. Every statement in a program must have a line number, arbitrarily chosen from the numbers 1 to 99999. The program in our example has 7 statements, and we have used 10, 20, 30, 40, and so on, as line numbers. We could have used any others, provided that they were in numerical order. However, it is common to employ multiples of 10 for a program containing only a few steps, so that additional steps can be inserted if needed.

2. REM (line 10) stands for "Remark." This provides information to someone looking at the program. The computer ignores anything in an REM statement as far as computation is concerned. Thus, REM provides a way to name or identify a program in English.

3. The LET command (line 20) is a substitution sentence. LET $N = 3$ means "replace N by 3." Thus, LET $N = N + 1$ means "replace N by $N + 1$." This is a different use of the equals sign from the one we find in algebra; the algebraic statement $N = N + 1$ is not true for any real number. In line 40, LET $A = (X + Y + Z) / N$, means "divide the sum $X + Y + Z$ by N, and replace A by the result."

4. READ and DATA (lines 30 and 60) go together. That is, for every READ statement, there must be a DATA statement. When a computer senses the READ instruction, it scans down the rest of the program until it finds the first DATA instruction. The first data value is assigned to the first variable following READ, the second data value to the second variable following READ, and so on. In the program of Example 1, with line 30 as it is, we could also have the following DATA statements:

$$
\begin{array}{ll}
60 & \text{DATA } 81 \\
61 & \text{DATA } 84 \\
62 & \text{DATA } 72
\end{array}
$$

The computer treats these three lines exactly as it treats line 60 in the given program.

5. PRINT A (line 50) means simply to print the value of A. If you want to print words, you can do so; we might have written line 50 as

50   PRINT "THE ANSWER IS", A

The computer recognizes quotation marks as enclosing words to be printed; the comma tells what to print next.

6. END (line 70). Every BASIC program must have an END statement to signal the computer that you are finished entering a program. After an END statement the computer enters a mode wherein it is ready to run the program, if you so desire.

7. RUN typed into the input keyboard (after all the instructions of the program, with their line numbers) is a system command and belongs after an END statement. RUN is used without a line number and tells the computer to execute the program as listed.

Table 2 lists the BASIC symbols and expressions and compares them to operations you are familiar with.

Table 2   Comparing BASIC symbols to ordinary algebra.

BASIC symbol	Example of a BASIC expression	Ordinary algebraic equivalent
+	X + Y	$x + y$
−	X − Y	$x - y$
$\circ$	X $\circ$ Y	xy
/	X / Y	$\dfrac{x}{y}$
↑	X ↑ 2	x^2

Where possible, the BASIC symbols are the same as those in ordinary algebra. $\circ$ is used for multiplication to differentiate between X as a number and × as an operation. The symbol / is used to enable division to be typed on one line; the same is true for ↑, which means "raising to a power."

In Example 2 we give several BASIC expressions and their ordinary algebraic counterparts. This example is worth studying carefully in order to appreciate the use of parentheses.

EXAMPLE 2

(a) 2 + 8/2	means	$2 + \dfrac{8}{2}$,	which is 6
(b) (2 + 8)/2	means	$\dfrac{2 + 8}{2}$,	which is 5
(c) 2 + 8 ↑ 2	means	$2 + 8^2$,	which is 66
(d) (2 + 8) ↑ 2	means	$(2 + 8)^2$,	which is 100
(e) 3 $\circ$ 4 ↑ 2	means	$3(4^2)$,	which is 48
(f) (3 $\circ$ 4) ↑ 2	means	$(3 \cdot 4)^2$,	which is 144

(g) $3 \uparrow (2 \cdot 2)$ means $3^{2 \cdot 2}$,   which is 81

(h) $(2 + 5)/(6 + 1)$ means $\dfrac{2 + 5}{6 + 1}$,   which is 1

(i) $(2 + 5)/6 + 1$ means $\dfrac{2 + 5}{6} + 1$,   which is $1\dfrac{7}{6}$

In the next example we show how to write common algebraic statements.

EXAMPLE 3

(a) $y = x^2 + 3$ might be written in BASIC as

           020   LET   Y = X ↑ 2 + 3

(b) $y = (x + 3)^2$ can be written as

           030   LET   Y = (X + 3) ↑ 2

(c) $y = \sqrt{a + 3}$ can be written in three ways:

           040   LET   Y = (A + 3) ↑ (1/2)
           040   LET   Y = (A + 3) ↑ .5
           040   LET   Y = SQR(A + 3)

In Example 3(c) our knowledge of algebra tells us that raising $a + 3$ to the 1/2 power does yield the square root of $a + 3$. The 1/2 requires parentheses when written in BASIC, since

$$(A + 3) \uparrow 1/2$$

raises $A + 3$ to the first power and divides the result by 2, which is not $\sqrt{A + 3}$. We can avoid parentheses by writing $(A + 3) \uparrow .5$, since .5 is the decimal equivalent of 1/2. Another convenient way to get the square root of $A + 3$ is to write SQR $(A + 3)$. If used in a program, SQR calls on one of the BASIC *library functions*, which are stored in the computer memory. Thus, for

           050   LET   B = SQR(6.25)

the computer calculates the square root of 6.25 (which is 2.5) by a method stored in the computer. It is important to note how parentheses are used when the SQR command is employed.

We conclude this brief introduction to BASIC programming with some sample programs.

EXAMPLE 4

A formula that yields the amount of money S, after interest is compounded on an original principal P for N periods, at an annual interest rate R, is

$$S = P\left(1 + \frac{R}{M}\right)^N$$

Write a program to calculate the amount S, for a principal of $1000 at an annual rate of 5% compounded quarterly ($M = 4$), for 1 year ($N = 4$). The program will evaluate

$$S = 1000\left(1 + \frac{0.05}{4}\right)^4$$

```
010    REM COMPOUND INTEREST
020    READ P, R, M, N
030    LET S = P * (1 + R/M) ↑ N
040    PRINT S
050    DATA 1000, .05, 4, 4
060    END
RUN
```

In Example 4 the computer will print 1050.95. That is, $1000 invested at 5% compounded quarterly for four quarters will amount to $1050.95 at the end of one year.

EXAMPLE 5

Write a program to find the area of a circle, given the formula $A = \pi r^2$ (where A is area, $\pi = 3.14159$, and r is the radius). Find the area of a circle with $r = 6$.

```
010    REM AREA OF CIRCLE
020    READ B, R
030    LET A = B * R ↑ 2
040    PRINT A
050    DATA 3.14159, 6
060    END
RUN
```

The computer will print 113.09724.

EXAMPLE 6

Write a program to calculate the length of the hypotenuse of a right triangle, where the legs are 5 and 12. The formula is $h = \sqrt{a^2 + b^2}$, where h is the hypotenuse and a and b are legs.

```
010    LET  A = 5
020    LET  B = 12
030    LET  H = SQR(A ↑ 2 + B ↑ 2)
040    PRINT  H
050    END
RUN
```

The computer output will be 13.

Exercise Set 12.3

1 The following are BASIC expressions. Translate them into ordinary algebraic symbols.
  (a) 3 ∘ 4   (b) 2 ∘ (4 + 5)
  (c) 3 ↑ 2   (d) SQR(3)
  (e) 3 ↑ (1/2)   (f) 3 ↑ .5
  (g) 7 − 2 ∘ 3   (h) (7 − 2) ∘ 3
  (i) 3 + 9/12   (j) (3 + 9)/12
  (k) Z + 5 ∘ (3 + 6)   (l) (Z + 5) ∘ (3 + 6)

2 Express each of the following in BASIC:

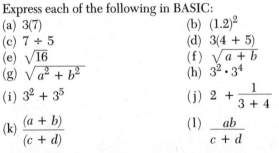

  (a) 3(7)   (b) $(1.2)^2$
  (c) 7 ÷ 5   (d) 3(4 + 5)
  (e) $\sqrt{16}$ (f) $\sqrt{a + b}$
  (g) $\sqrt{a^2 + b^2}$ (h) $3^2 \cdot 3^4$
  (i) $3^2 + 3^5$ (j) $2 + \dfrac{1}{3 + 4}$
  (k) $\dfrac{(a + b)}{(c + d)}$ (l) $\dfrac{ab}{c + d}$

3 Evaluate each of the following for $A = 2$, $B = 3$, and $C = 4$.
  (a) A ∘ B ↑ 2   (Raising to a power, or exponentiation, is always done before
                      ∘, +, −, or /.)
  (b) (A ∘ B) ↑ 2   (Expressions in parentheses are calculated first.)
  (c) A + B ∘ C
  (d) (A + B) ∘ C
  (e) A ↑ (A + B)
  (f) A ↑ A + B
  (g) A + B ∘ B + C
  (h) (A + B) ∘ B + C
  (i) A + B ∘ (B + C)
  (j) (A + B) ∘ (B + C)

4 Using the LET command, write each of the following:
  (a) $y = 4 + 3$ (b) $y = x \cdot z$
  (c) $a = \sqrt{3b}$ (d) Replace N by N + 1.
  (e) Replace X by Y + Z.   (f) Store N + 1 in place of N.

5 What is the result of the following program?

```
010      READ R
020      READ H
030      LET G = R ∘ H
040      PRINT R
050      PRINT H
060      PRINT G
070      DATA 1.25
080      DATA 40
090      END
```

6 What is the output of the following program (which computes ordinary simple interest on a principal P, at interest rate R, for T days)?

```
010      READ P, R, T
020      LET I = (P ∘ R ∘ T)/360
030      DATA 1000, .03, 60
040      PRINT I
050      PRINT I + P
060      END
```

7 Write a program that will
   (a) Calculate the area of a rectangle, given $L = 30.5$ and $W = 5.4$.
   (b) Calculate the area of the same rectangle and print A, W, and L.
   (c) Average any four grades received on tests.
   (d) Calculate $y = x^2 + 3x - 18$ for $x = -8$, $x = -7$, $x = -6$, $x = -5$, $x = -4$, and so on through $x = 5$.
   (e) Calculate $(a + b)^2$ for $a = 1$ and $b = 0.002$.
   (f) Calculate $S = \dfrac{a}{1 - r}$ for $a = 1$ and $r = 1/5$.
   (g) Calculate $L = a + (n - 1)d$ for $a = 3$, $n = 7$, and $d = 4$.

8 A first-degree equation in x is of the form $ax + b = c$, and the formula solution is $x = \dfrac{c - b}{a}$.

Write a program that solves
   (a) $2x - 3 = 7$
   (b) $2x - 3 = 8$
   (c) $(1/3)x - 2/3 = 5/7$
   (d) $0.003x - 1.47 = 7.98$

9 A formula solution for the system of equations

$$\begin{cases} Ax + By = C \\ Dx + Ey = F \end{cases}$$

is $x = \dfrac{CE - BF}{AE - BD}$ and $y = \dfrac{AF - CD}{AE - BD}$

Write a program that will calculate and print either x or y, using these formulas, where the given system is

$$\begin{cases} x + 2y = 1 \\ -x + 3y = 4 \end{cases}$$

CONCLUDING REMARKS

Before the advent of time-sharing in the 1960s, computer use was restricted to those experts who had learned relatively difficult programming languages. In addition, the cost of computing was so high that only "worthy" problems could reasonably be programmed and run on the computer. Time-sharing places terminals in the hands of millions of persons, and features a language that is simple enough to learn in a short time.

A time-sharing terminal is a very private machine. You can write a program and try to run it. The machine will print out error messages when some part of your program has gone wrong. You can try a program many, many times until you get it to run correctly. Since you are using only a few seconds of computer time, the cost is low. We emphasize that the machine is private. No one is counting your mistakes or criticizing your technique. Your communication is only with the computer.

In fact, for many time-sharing installations you can refer to a stored program in the computer library that teaches you a computer language. In a sense the computer can teach you to creatively operate itself.

The impact of time-sharing is tremendous. For example, a college could now require all students to take a certain computer course. All that is needed to provide "hands-on" computer use is a set of terminals available to the student. He can learn to write programs easily, and he can try out his own programs. Before time-sharing this would have been impossible. Each student would have had to wait for a turn at the computer or would have had to send his program off to a center for processing. Time-sharing has created a positive psychological setting for everyone to learn something about computers.

One aspect of an advance in computer technology is Computer-Assisted Instruction (CAI). Entire courses, or parts of courses, are programmed into the computer's memory. The student is directed step by step through the intricacies of a problem. He is asked questions by the computer. He gives his answer, and the computer tells him whether or not he is right. One great advantage of CAI is that students work at their own rates. If your college has computer terminals, chances are that you can call from the library programs, for instance, on "Algebra drill" or "Practice with Spanish verbs." Many state universities are establishing catalogs of CAI programs available at various centers throughout the state.

In 1972 Professor John Kemeny (co-inventor of BASIC at Dartmouth) predicted that every home would become a "mini-university" by the year 1990. By that he meant that university lectures would be available on TV cassettes, books would provide the background material, and the student's own computer terminal would provide access to library research. The terminal would also test the student and calculate his grades.

The technology for computer terminals in the home exists now. The monthly rental for some terminals is as low as $100, although it may be as high as $1000 for some models. It is probable that the cost will come down considerably. If we consider the inflationary cost of a college education, CAI in the home may become a necessity.

Any discussion of the future use of computers sounds like science fiction, but even their present use is almost unbelievable. We are all aware of how well computers predict results of elections, often with only 10 or 15 per cent of the returns in. Computers monitor the life-support systems on spacecraft and make minute navigational corrections. Anyone who has ever visited a large airport knows how complicated the various take-offs and landings are; computers keep track of arrivals and departures and also handle ticket sales and reservations. The banking industry may seem to some like a huge computer system. Our checks are read by computers from the special characters imprinted at the bottom, and our balances are credited and debited by the same machine.

Motor vehicle departments are now computerized. It takes seconds to verify your registration numbers when you apply for new forms. Lost or stolen licenses can be duplicated, for a small fee, from the computer's central files. Often local offices transmit and receive such information on a terminal connected by telephone to the central motor vehicle office.

Crime detection on a nationwide scale is assisted by computers. The National Crime and Information Center of the FBI maintains computerized files of criminal histories, fingerprints, and arrest records. Other governmental agencies also keep computer lists. We probably do not want to be reminded about the Internal Revenue Service and their computers. But there are also computers providing information to the Department of Defense, to the Social Security Administration, and to the Bureau of the Budget. Virtually every Federal agency has its computer. There are more computers used in all phases of American government than in any other enterprise—more than in business, more than in private research, more than in education.

Hospitals use computers to calculate their inventory and payroll, to account for patient bills, to control the paperwork necessitated by insurance claims. Computers also monitor patients by reporting irregularities in heart beat, temperature, and respiration rate to central nurses' stations. In some areas individual medical records are computerized so that doctors can have immediate and accurate access to data that are helpful in making a diagnosis.

The speed, accuracy, storage capacity, and nontiring capabilities of computers have made them indispensable. In the later years of the twentieth century they will be working on problems not even imagined now.

You should be familiar with the following before attempting the Review Test:

1. Following through a flow chart and indicating what it does.

2. Examining a program written in BASIC, and telling what the outcome will be.

3. Translating BASIC expressions into common algebraic notation.

4. Expressing algebraic expressions in BASIC notation.

5. Using the commands LET, READ, DATA, PRINT, END, and RUN in writing simple programs.

REVIEW TEST

1 (a) Examine the following flow chart and tell what it does.

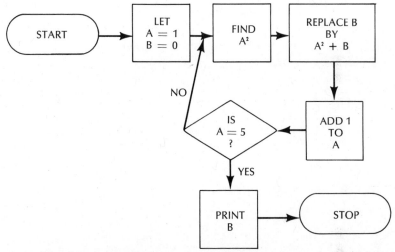

(b) What number would be printed if the flow chart in 1(a) were to be followed to its completion?

2 Evaluate each of the following BASIC expressions:
   (a) $3 * 4 - 2$
   (b) $3 * (4 + 2)$
   (c) $7 \uparrow 2$
   (d) $SQR(3 \uparrow 2 + 4 \uparrow 2)$
   (e) $3 \uparrow (1 + 2)$

3 Write each of the following in BASIC notation:
   (a) $a \div (b + c)$
   (b) $ab + c$
   (c) $\dfrac{a + b + c}{3}$
   (d) $5^2 + 12^2$
   (e) $\sqrt{a + b^2}$

4 What is the result of the following program?

```
010   READ A
020   READ B
030   READ C
040   LET X = A ↑ 2 + B ↑ 2 + C
050   LET Y = SQR(X)
060   PRINT Y
070   DATA 2, 2, 8
080   END
```

5 Write a program that will average the scores 87, 72, 91, 79, and 95.

6 Write a program that will print the solution to the equation $37x - 41 = 59$, given that $x = (c - b)/a$ is the formula solution to $ax + b = c$.

7 (a) Write a program that prints the volume V of a sphere with radius R, given that $V = (4/3)\pi R^3$. (Use $\pi = 3.14$.)
   (b) How would you alter the program to find V when $R = 7$?

Computers have invaded nearly every branch of human endeavor. In some areas the handling of sheer volumes of numerical data has demanded the use of computers (for example, in Census Bureau statistical work); in others, people have devised ways to expand their own jobs or level of performance by making use of them. This article from the September 1, 1974 *New York Times* shows how the field of journalism is beginning to employ the computer.

COMPUTER JOINS THE TYPEWRITER AMONG NEWS REPORTER'S TOOLS
ROBERT REINHOLD

Until fairly recently, all a bright young man or woman needed to know to get ahead in the news business was how to write clear sentences, find the way to City Hall and handle a fearsome city editor.

Today, the young reporter is more likely to make it if he knows something about chi square, Kendall's tau, variance and standard deviations. If he does know about such things, it is he who will probably scare the city editor.

All these strange terms are the lingo of statistics, and statistical method is quickly becoming almost as much a part of journalism as the note pad, typewriter and television camera.

Twenty years after political science, sociology and the other behavioral sciences began to move heavily into "quantification" as a means of testing armchair intuition against reality, journalism is cautiously adapting some of these "scientific" techniques of polling, surveys and data analysis for its own purposes.

SEARCH FOR CREDIBILITY

"One of the main reasons for what is happening is that editors want to improve their credibility," said Phillip Meyer, who is probably the leading exponent of the new trend by virtue of his book, "Precision

Journalism," published by the Indiana University Press. Mr. Meyer is a national correspondent of the Knight newspaper chain, which has pioneered in survey journalism.

At large and small newspapers all over the country, political reporters are doing sophisticated samplings of voter opinions to help them interpret election trends and results. Crime reporters are sifting criminal justice records with computers to detect patterns in law enforcement. Others have rummaged through census data and analyzed traffic accident patterns, the background of rioters, political campaign contribution lists and countless other kinds of data to tell readers more about their cities and what makes people act the way they do.

At the root of all this is a growing dissatisfaction among many journalists with the traditional tools of their trade. Reporting is sometimes a haphazard enterprise, the writer drawing inferences about reality from a few interviews, official statements and plain intuition. Sometimes the conventional wisdom has proved to be wrong. The vision of even the most penetrating journalistic eye can be distorted by its own experiences.

Through exact measurement and computer speed, the new techniques enable reporters to be more precise and accurate. These techniques also hold out the possibility of testing the truthfulness of official pronouncements or the claims of political candidates.

A PHILADELPHIA CASE

Last year, for example, The Phildelphia Inquirer, part of the Knight chain, explored allegations by the local District Attorney, Arlen Specter, that too-soft judges were letting criminals loose. Through a massive computer analysis of some 1,000 criminal cases, the newspaper was able to demonstrate that inefficiencies and failures in the prosecutor's office, not lenient judges, were the main reasons so many accused criminals went free. Mr. Specter lost his bid for re-election last fall.

Such successes have not been lost on other newspapers. Last week 22 reporters and editors completed a three-week crash course in statistics and experimental method at Northwestern University, where they formed the first organization of "precision journalists." Some other reporters, like Jay Harris of the The Wilmington News-Journal and Howard Covington of The Charlotte Observer, have taken leaves to study survey methods at major universities.

Mr. Harris discovered the power of computers two years ago while trying to do an article on heroin traffic in Delaware. Court records yielded about 2,000 names. So much information could be digested only by computer.

VIEWS OF CITY RESIDENTS

He is now completing a major statewide public-attitude survey in collaboration with the University of Delaware.

"We realized how very little we knew about the electorate in general, and we thought it would be nice to have people tell us what their concerns were rather than have the candidates tell us," Mr. Harris said.

The use of statistics in reporting the news, particularly for pre-election polling, is not entirely new. What is new is the aggressive use of such methods for analytical, not just descriptive purposes.

The New York Times and The Washington Post both earlier this year carried series based on surveys of New Yorkers and Washingtonians, respectively. The articles explored what pleased and displeased city residents, and how different ethnic and social groups differed in their concerns and perceptions.

Last week, Time magazine began a quarterly feature, called "Soundings," designed to monitor statistically the changing national mood on such factors as economic distress, social resentment and political conservatism.

Computer journalism is not confined to big-city national organs. Even such a small daily as The Dubuque Telegraph-Herald has mounted a sophisticated effort using its business computers. The paper's managing editor, James Geledas, has been handing out copies of Mr. Meyer's book to some of his reporters. One of them, John McCormick, is now analyzing the records of thousands of traffic accidents in Dubuque to find out why and where accidents occur.

All of this new fascination with numbers and "precision" is not without problems. Social measurement is at best a crude science.

Once the numbers—really just gross approximations of reality—are cranked into computers and emerge in neat tables, they begin to take on a reality of their own and tend to give a false sense of scientific accuracy. But what is statistically significant is not necessarily socially significant or newsworthy.

JUMPING TO CONCLUSIONS

Already some papers have fallen into mechanistic and simplistic uses. One small Midwestern daily recently polled the seven members of the City Council and created a numerical "irritation" rating to gauge what bothered the city fathers most. The figures were then averaged and carried out to two decimal places, with "public apathy" coming out at the top of the list. The article said nothing about how irritated the public was with the Council members.

Still another problem is when to draw the inference that one social factor directly causes another. While it is possible to find all kinds of statistical correlations in human behavior, it is hazardous to jump to causal inferences without first examining an enormous number of "variables." Scholars spend months or even years digesting data; reporters may have only hours or days.

Harvey Kabaker of The Washington Star-News concedes that "I may be accused of going beyond the data" in his analyses, but he contends that it is a lot better than the sheer guesswork that marks a good deal of political analysis.

Another difficulty is that numbers are dull. Few editors would want to let numbers substitute for well-written impressionistic reporting filled with color and human interest. For these and other reasons, most newspapers are moving with caution.

"We have been making use of these techniques in a variety of ways," said Seymour Topping, assistant managing editor of The New York Times. "We are now addressing ourselves to a more systematic application because new opportunities have been opened up in terms of technology."

Like many other editors, John G. Craig Jr., executive editor of The Wilmington News-Journal, feels that statistical methods are valuable but should not be oversold. He believes they should be used to validate customary reporting and suggest new ideas.

"Statistics are just a check on things," he said. "We just start from there."

ANSWERS TO SELECTED EXERCISES

CHAPTER 1 **FUNDAMENTAL CONCEPTS OF NUMBER AND COUNTING**

Exercise Set 1.1

1 The boy's mark list is longer.
3 The comparison can be made visually; there is no need to count marks in each list.
5 (a) ☆O/ (b) O/ (c) /☆ (d) OOO☆☆☆☆/☆ (e)☆☆ (f) O☆☆/☆

Exercise Set 1.2

1 (a) ◄▼   (b) ◄ ▼▼▼ ▼▼   (c) ◄◄

  (d) ◄◄◄◄   (e) ▼►◄
3 (a) 8542   (b) 15   (c) 594

Exercise Set 1.3

1 (a) numeral, number, numeral   (b) number
  (c) number, number (or numeral)   (d) number
  (e) All are numerals.
3 (a) 1 2 3 4 5 6 7 8 9 10 11 12 13 14 15
  (b) ten   (c) I II III IV V VI VII VIII IX X XI XII XIII XIV XV; only 3 different symbols are used.   (d) Possibly as MMMM, but the system suggests a new symbol should be introduced for 5000. If 5000 was denoted by P then 4000 could be MP.
5 1975

Exercise Set 1.4

1 (a) seventy-five, 70 + 5
   (b) thirteen, 10 + 3
   (c) four hundred ninety-nine, 400 + 90 + 9
   (d) five thousand six hundred thirteen, 5000 + 600 + 10 + 3
   (e) ten thousand three, 10,000 + 3
   (f) seven hundred five, 700 + 5
   (g) three hundred forty-two and five tenths, 300 + 40 + 2 + .5
   (h) thirteen and forty-two hundredths, 10 + 3 + .4 + .02
   (i) twenty-nine and three hundred fifty-seven thousandths, 20 + 9 + .3 + .05 + .007

3 (a) 7524   (b) 563   (c) 4032   (d) 8.29
   (e) 765.43   (f) 3460.9

5 (a) 144   (b) one hundred forty-four   (c) $1(10^2) + 4(10^1) + 4(10^0)$
   (d) $1(100) + 4(10) + 4(1) = 100 + 40 + 4 = 144$ (e) twelve dozen
   (f) one gross   (g) one hundred

Exercise Set 1.5

1 The numeral at the start (city A) is one of the sixteen.
3 8   5 334 (decimal)
7 Question 6(e) is impossible since there is no "5" in base 5.
9 (a) 8   (b) a decimal odometer   (c) 4444
11 (a) 00001, 00010, 00011, 00100, 00101, 00110, 00111, 01000, 01001, 01010,
        01011, 01100, 01101, 01110, 01111, 10000, 10001, 10010, 10011, 10100,
        10101, 10110, 10111, 11000, 11001, 11010, 11011, 11100, 11101, 11110,
        11111, 100000, 100001, 100010, 100011, 100100
    (b) 1. 250   2. 1594   3. 35   4. 229
        5. 2337   6. 6   7. 4.25
    (c) 1. 1101   2. 11101   3. 110010
        4. 10000000   5. 10000001   6. 10011000
        7. 11001000   8. 10110111   9. 1111101000
°13 (a) *b, c, ba, bb, bc, ca, cb, cc, baa, bab, bac, bba, bbb, bbc, bca, bcb, bcc, caa,*
        cab, cac, cba, cbb, cbc, cca
    (b) the numeral *a*   (c) 102   (d) *baaa*
°15 No. In a proposed base 1 the only numeral would be 0, so the only number we
     can represent is 0.

Exercise Set 1.6

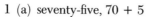

1 1 2 3   1 2 3   1 2 3   1 2 3   1 2 3   1 2 3

   a b c   *a b c*   *a b c*   *a b c*   *a b c*   *a b c*
°3 (a) 24   (b) 120   (c) none
5 $n(S) = 50$
7 (a) the letters in the set { e,d,r }   (b) three   (c) as { e,d,r }

CHAPTER 2   SETS—THE ABC'S OF MATHEMATICS

Exercise Set 2.1

1 (a), (e), (f), (h) are not well-defined.

2 (a) true   (b) false   (c) true   (d) false
   (e) true   (f) false   (g) true

3 (a) $\{\, x \mid x$ is a counting number less than 4 $\}$
   (b) $\{\, z \mid z$ is a counting number and a multiple of 3 $\}$
   (c) $\{\, y \mid y$ is a counting number, a multiple of 5, and less than 76 $\}$
   (d) $\{\, x \mid x$ is a primary or secondary color $\}$
   (e) $\{\, x \mid x$ is the symbol for a counting number $\}$
   (f) $\{\, x \mid x$ is a kind of tree $\}$
   (g) $\{\, x \mid x$ is the square of a counting number $\}$

Exercise Set 2.2

1 (a) 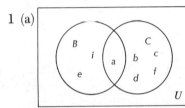   (b) 4   (c) 3   (d) 4

3 It does not allow for an element to be in A, in B, and in C.

5 (a) 10   (b) 30   (c) It is impossible to answer question (a) because there are more data than students, that is, the data are inconsistent with the total number of cards.

7 $10(10)(10)(10)(10)(10)(10) = 10^7 = 10,000,000$
   $8(10)(10)(10)(10)(10)(10) = 8(10^6) = 8,000,000$

Exercise Set 2.3

1 $w \in M$ but $w \notin L$. Therefore $M \neq L$.

2 yes, yes   3 true   4 yes

5 (a) true   (b) true   (c) false   (d) false
   (e) true   (f) true   (g) true

7 (a) $\{\ \}, \{1\}, \{2\}, \{3\}, \{4\}, \{1,2\}, \{1,3\}, \{1,4\}, \{2,3\}, \{2,4\}, \{3,4\}, \{1,2,3\},$
     $\{1,2,4\}, \{1,3,4\}, \{2,3,4\}, \{1,2,3,4\}$
   (b) $\{1,2,3,4\}$

9 (a) 4   (b) $16, 32, 64, 2^{100}$

Exercise Set 2.4

1 (a) $\{1,2,3,4,5,6,7\}$ (b) $\{3,4,5\}$ (c) $\{1,2,3,4,5\} = R$
   (d) $\{1,2,3,4,5,6,7,8,9,10\} = U$
   (e) $\{6,7,8,9,10\}$ (f) $\{1,2,8,9,10\}$ (g) $\{1,2,6,7,8,9,10\}$
   (h) $\{8,9,10\}$ (i) $\{8,9,10\}$ (j) $\{1,2,6,7,8,9,10\}$
   (k) $\{1,3,4,5,7,9\}$ (l) $\{1,3,4,5,7,9\}$ (m) $\{2,4,6,7,8,9,10\}$
   (n) $\{2,4,6,7,8,9,10\}$ (o) $\{6,8,10\}$ (p) $\{6,8,10\}$

3 (a)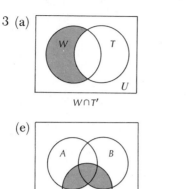

(c)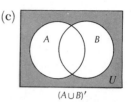

(e) 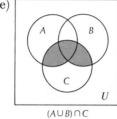   (g)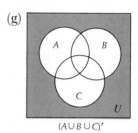

5 (a) yes   (b) AB (c) l
7 (a) $A \cap B' \cap C'$ (region 6), $A' \cap B \cap C$ (region 5)
       $A' \cap B \cap C'$ (region 8), $A' \cap B' \cap C$ (region 2)
       $A' \cap B' \cap C'$ (region 1).
  (b) 1. Regions labeled 3, 4, 5   2. Regions labeled 2, 3
      3. Regions labeled 1, 2, 3, 5, 6, 7, 8   4. Region labeled 1
      5. Region labeled 4   6. Regions labeled 3, 4, 5, 7
      7. Regions labeled 1, 2, 3, 6, 7, 8   8. Regions labeled 1, 6

Exercise Set 2.5

1 Theorem 1: $A \cup (B \cap C) = (A \cup B) \cap (A \cup C)$
  Step 1   A corresponds to 1, 5, 6, 7
             $B \cap C$ corresponds to 1, 3
             So, $A \cup (B \cap C)$ corresponds to 1, 3, 5, 6, 7
  Step 2 $A \cup B$ corresponds to 1, 2, 3, 5, 6, 7
             $A \cup C$ corresponds to 1, 3, 4, 5, 6, 7
             So, $(A \cup B) \cap (A \cup C)$ corresponds to 1, 3, 5, 6, 7
  Step 3   Since both $A \cup (B \cap C)$ and $(A \cup B) \cap (A \cup C)$ correspond to
             regions 1, 3, 5, 6, 7 we conclude that the theorem is true.
3 Theorem 3: $(A \cup B)' = A' \cap B'$
  Step 1 $A \cup B$ corresponds to 1, 2, 4
             So, $(A \cup B)'$ corresponds to region 3
  Step 2 A' corresponds to regions 3, 4
             B' corresponds to regions 2, 3
             So, $A' \cap B'$ corresponds to region 3
  Step 3   The conclusions of steps 1 and 2 prove the theorem.
5 $(A \cap B)'$ is not equal to $A' \cap B'$
  Assume a Venn diagram labeled as in exercise 3.
  Step 1 $A \cap B$ corresponds to region 1
             So, $(A \cap B)'$ corresponds to regions 2, 3, 4
  Step 2 A' corresponds to regions 3, 4
             B' corresponds to regions 2, 3
             So, $A' \cap B'$ corresponds to region 3

Step 3   Since regions 2, 3, 4 are not the same as region 3, we conclude that the sets are not equal.

Exercise Set 2.6

1

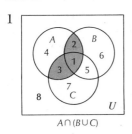

A ∩ (B ∪ C)

Exercise Set 2.7

1 $A \cup B = \{a,c,d,e,f,g\}$ so $(A \cup B)' = \{b\}$
   $A' = \{b,d,f\}$ and $B' = \{a,b,c\}$, so $A' \cup B' = \{a,b,c,d,f\}$.
   Thus $(A \cup B)' \neq A' \cup B'$.

3

A	B	(A ∩ B)'		A' ∩ B'			A' ∪ B'			(A ∪ B)'	
∈	∈	∈	∉	∉	∉	∈	∉	∉	∈	∈	∉
∈	∉	∉	∈	∉	∉	∈	∉	∈	∈	∈	∉
∉	∈	∉	∈	∈	∉	∉	∈	∈	∉	∈	∉
∉	∉	∉	∈	∈	∈	∈	∈	∈	∈	∉	∈
		1	2	1	3	2	1	3	2	1	2

answers: (a) (b) (c) (d)

5

A	B	A ∩ B'			A ∩ (B ∪ B')				
∈	∈	∈	∉	∉	∈	∈	∈	∈	∉
∈	∉	∈	∈	∈	∈	∈	∉	∈	∈
∉	∈	∉	∉	∉	∉	∉	∈	∈	∉
∉	∉	∉	∉	∈	∉	∉	∉	∈	∈
		1	3	2	1	5	2	4	3

answers: (a) (b)

(c)

A	A ∪ A'		
∈	∈	∈	∉
∉	∉	∈	∈
	1	3	2

A	B	C	A ∪ (B ∩ C)			(A ∪ B) ∩ (A ∪ C)		
∈	∈	∈	∈	∈	∈	∈	∈	∈
∈	∈	∉	∈	∈	∉	∈	∈	∈
∈	∉	∈	∈	∈	∉	∈	∈	∈
∈	∉	∉	∈	∈	∉	∈	∈	∈
∉	∈	∈	∉	∈	∈	∈	∈	∈
∉	∈	∉	∉	∉	∉	∈	∉	∉
∉	∉	∈	∉	∉	∉	∉	∉	∈
∉	∉	∉	∉	∉	∉	∉	∉	∉
			1	3	2	1	3	2

answers: (d) (e)

(f) $A \cup (B \cup C)$ results in ∈ ∈ ∈ ∈ ∈ ∈ ∈ ∉
(g) $(A \cup B) \cup C$ results in ∈ ∈ ∈ ∈ ∈ ∈ ∈ ∉
(h) $A \cap (B \cap C)$ results in ∈ ∉ ∉ ∉ ∉ ∉ ∉ ∉
(i) $(A \cap B) \cap C$ results in ∈ ∉ ∉ ∉ ∉ ∉ ∉ ∉

7 (a) 2   (b) 4   (c) 8   (d) 16

11 (a)

A	B	C	A	∩	(B	∪ C)'
∈	∈	∈	∈	∉	∈	∉
∈	∈	∉	∈	∉	∈	∉
∈	∉	∈	∈	∉	∈	∉
∈	∉	∉	∈	∈	∉	∈
∉	∈	∈	∉	∉	∈	∉
∉	∈	∉	∉	∉	∈	∉
∉	∉	∈	∉	∉	∈	∉
∉	∉	∉	∉	∉	∉	∈
			1	4	2	3

(b)

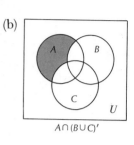

A∩(B∪C)'

CHAPTER 3   SETS OF ORDERED PAIRS

Exercise Set 3.1

1 (a) { (7,8), (5,3), (6,4) }   (b) { (8,7), (3,5), (4,6) }

3 (a) no   (b) no   (c) no   (d) yes

5 { (a,c), (a,d), (a,e), (b,c), (b,d), (b,e) }

Exercise Set 3.2

1 (a) { (−1,−2), (−1,2), (1,−2), (1,2) }

  (b) { (−2,−1), (−2,1), (2,−1), (2,1) }

  (c) { (−2,−2), (−2,2), (2,−2), (2,2) }

  (d) { (−1,−1), (−1,1), (1,−1), (1,1) }   (e) 4   (f) 4

3 (a) { (1,1), (1,2), (1,3), (1,4), (1,5), (1,6),   (b) 6   (c) 11

      (2,1), (2,2), (2,3), (2,4), (2,5), (2,6),   (d) 3   (e) 6/36 or 1/6

      (3,1), (3,2), (3,3), (3,4), (3,5), (3,6),

      (4,1), (4,2), (4,3), (4,4), (4,5), (4,6),

      (5,1), (5,2), (5,3), (5,4), (5,5), (5,6),

      (6,1), (6,2), (6,3), (6,4), (6,5), (6,6) }

5 none

7 (a) 100   (b) { (1,2), (1,4), (1,6), (1,8), (1,10),   (c) 25

               (3,2), (3,4), (3,6), (3,8), (3,10),

               (5,2), (5,4), (5,6), (5,8), (5,10),

               (7,2), (7,4), (7,6), (7,8), (7,10),

               (9,2), (9,4), (9,6), (9,8), (9,10) }

Exercise Set 3.3

1 (a) { (a,b), (a,c), (a,d), (a,e),   (b) { (a,b), (a,c), (a,d), (a,e),

     (b,b), (b,c), (b,d), (b,e),   (b,c), (b,d), (b,e),

     (c,b), (c,c), (c,d), (c,e) }   (c,d), (c,e) }

  (c) No, b does not precede b.   (d) { (b,b), (c,c) }

3 (a) the set of all possible names, first name followed by last name

  (b) the set of all possible names, last name first

Exercise Set 3.4

1 (a) (a,a) and (b,b) (b) (b,a)
3 (a) ref., sym., trans.   (b) none
   (c) ref., sym., trans.   (d) ref., sym., trans.
   (e) ref., sym., trans.   (f) ref., sym., trans.
   (g) trans.   (h) trans.
   (i) ref., trans.   (j) ref., trans.
5 (a) (c,a) (b) yes
7 (a) 64
   (b) $\{ (S_1,S_1), (S_2,S_2), (S_3,S_3), \ldots, (S_8,S_8) \}$
   (c) It is all three.
   (d) $\{ (S_1,S_1), (S_1,S_2), (S_1,S_3), (S_1,S_4), (S_1,S_5), (S_1,S_6), (S_1,S_7), (S_1,S_8), (S_2,S_2),$
       $(S_2,S_5), (S_2,S_6), (S_2,S_8), (S_3,S_3), (S_3,S_5), (S_3,S_7), (S_3,S_8), (S_4,S_4), (S_4,S_6),$
       $(S_4,S_7), (S_4,S_8), (S_5,S_5), (S_5,S_8), (S_6,S_6), (S_6,S_8), (S_7,S_7), (S_7,S_8), (S_8,S_8) \}$
   (e) It is reflexive. Eight ordered pairs so indicate.
   (f) No. If A is a subset of B, then B does not have to be a subset of A. The
       presence of (S_1,S_2) and the absence of (S_2,S_1) *proves* that the relation is not
       symmetric.
   (g) (S_1,S_4) and (S_4,S_7) are in the relation. Then transitivity requires (S_1,S_7) to
       be in the set, which is true.
9 (a) $\{\ \}, \{ 1 \}, \{ 2 \}, \{ 3 \}, \{ 4 \}$
       $\{ 1,2 \}, \{ 1,3 \}, \{ 1,4 \}, \{ 2,3 \}, \{ 2,4 \}, \{ 3,4 \}$
       $\{ 1,2,3 \}, \{ 1,2,4 \}, \{ 1,3,4 \}, \{ 2,3,4 \}$
       $\{ 1,2,3,4 \}$
   (b)

   (c) It is all three.
11 (a) four   (b) yes

Exercise Set 3.5

1 (a) "In the same area of study as" would be a relation that partitions the
       students into groups.
   (b) One group might be the business students, another might be the liberal
       arts students, etc.
3 (a), (c), (e) are equivalence relations.
5 (a) "means the same as" (b) "runs the same as" (c) "has the same
       cardinality as" (d) "is the same fraction as" (e) "has the same wealth
       as" (f) "has the same angles as" (g) "has the same cardinality as"
       (h) "was born on the same date as" (i) "has the same feeling as"
       (j) "has the same intelligence as"

7 (a) true   (b) true   (c) false
9 (a) (10,10), (15,15), (5,5), (20,20), (35,35), (105,105)
   (b) Remainder 1: (6,11), (16,11), (16,6), and so on
       Remainder 2: (7,12), (22,32), (12,7), and so on
       Remainder 3: (3,8), (8,3), (13,8), (23,53), and so on
       Remainder 4: (14,9), (4,14), (4,9), and so on
   (c)

5,10, 15,20, 25,30, 35,...	1,6,11, 16,21, 26,...	2,7,12, 17,22, 27,...	3,8,13, 18,23, ...	4,9,14, 24,...
Rem. 0	Rem. 1	Rem. 2	Rem. 3	Rem. 4

   (d) Five. Appropriate names are indicated under the rectangles. Mathematicians refer to them as the integers modulo five.
11 Yes. The partitioning is possible only because it is an equivalence relation.

Exercise Set 3.6

1 (a) false   (b) true   (c) true   (d) true
  (e) true   (f) false   (g) false   (h) false
3 (a) $\{ (1,8), (2,9), (3,10), \ldots \}$
  (b) $\{ (1,1), (2,4), (3,9), (4,16) \}$
5 (a) $f(1) = 3(1) + 7 = 10$; (1,10) is determined
  (b) $f(0) = 7; f(3) = 16; f(10) = 37$
  (c) $f(c) = 3c + 7; f(b) = 3b + 7; f(c) = 3c + 7$
  (d) $f(2) + f(3) = 13 + 16 = 29$ Note $f(2) + f(3) \neq f(5)$.
7 (a) $f(212) = 5/9(212-32) = 5/9(180) = 100$
  (b) same as (a)   (c) $f(32) = 5/9(32-32) = 0$
  (d) $f(50) = 5/9(50-32) = 5/9(18) = 10$
  (e) probably Celsius (Calculate $f(68)$.)

CHAPTER 4   CREATING NEW NUMBERS

Exercise Set 4.1

1 (a) $5 + (3 + 1) = 5 + 4 = 9$ (b) $(5 + 3) + 1 = 8 + 1 = 9$
  (c) $(6 \times 2) \times 2 = 12 \times 2 = 24$ (d) $(7 - 4) - 2 = 3 - 2 = 1$
  (e) $7 - (4 - 2) = 7 - 2 = 5$ (f) $(12 \div 3) \times 2 = 4 \times 2 = 8$
  (g) $12 \div (3 \times 2) = 12 \div 6 = 2$ (h) $A \cup (B \cap C)$ (i) $(A \cup B) \cap C$
3 yes   5 yes

Exercise Set 4.2

1 (a) yes   (b) yes   (c) yes
  (d) $+5 \div 4$ is not an integer
3 (a) yes   (b) no   (c) no
5 (a) $+6$ (b) $+13$ (c) -5 (d) $+3$ (e) 0   (f) -10

7 (a) -5 (b) -11 (c) -17 (d) 0 (e) -15
   (f) 7 (g) $+7$ (h) $+49$ (i) -49 (j) $+25$
   (k) -28 (l) $+9$ or $+18$ (m) $+14$ (n) -2
   (o) -4 (p) $+8$ (q) $+21$ (r) $+9$ (s) -7
   (t) $+144$

9 (a) $+6$ (b) 0 (c) $+18$ (d) -6 (e) $+6$

11 (a) no   (b) I^- (c) I^+ (d) $I^{+,0}$; $I^{-,0}$
   (e) I (f) $\{0\}$ (g) no

Exercise Set 4.3

1 (a) true   (b) false   (c) false
   (d) true   (e) false   (f) false

3 (a) an integer   (b) $3/5 + 2/9 = [3(9) + 5(2)]/(5)(9) = 37/45$

5 (a) $a/a = 1/1$ $(a \neq 0)$ because $a(1) = a(1)$
   (b) no

7 (a) $-4/-1$ $+8/+2,$ $-8/-2$
   (b) $+6/+4,$ $-6/-4,$ $+9/+6$
   (c) $+4/+6,$ $+2/+3,$ $-6/-9$
   (d) $+6/+5,$ $+12/+10,$ $-18/-15$
   (e) $-7/8,$ $14/-16,$ $70/-80$
   (f) $6/20,$ $30/100,$ $-3/-10$
   (g) $-6/20,$ $-30/100,$ $3/-10$
   (h) $0/2,$ $0/3,$ $0/7$
   (i) $(-1,-10),$ $(2,20)$ $(-3,-30)$
   (j) $(50,-100),$ $(-5,10),$ $(10,-20)$

9 (a) for positive integers a, b, c, and d (b) $9/15$
   $a/b > c/d$ if $ad > bc$

Exercise Set 4.4

1 (a) The squares of the multiples of 3 are themselves multiples of 3.

   (b) Assume $\sqrt{3} = a/b$ where a/b is a completely reduced rational. Then $3 = a^2/b^2$ (squaring both sides.) Then, $a^2 = 3b^2$, so a^2 is a multiple of 3 and a is a multiple of 3. Write $a = 3k$. Then $a^2 = 3b^2$ becomes $(3k)^2 = 3b^2$ or $3b^2 = 9k^2$. So $b^2 = 3k^2$ and b^2 is a multiple of 3 and we can write $b = 3m$. Since $a = 3k$ and $b = 3m$, the fraction a/b is $3k/3m$ and is *not* reduced completely. Our assumption is wrong.

3 irrational

5 $+12\sqrt{5}$, $+12 + \sqrt{5}$, $-6\sqrt{5}$, $-6 + \sqrt{5}$, $+9\sqrt{5}$, $+9 + \sqrt{5}$, $-15\sqrt{5}$, $-15 + \sqrt{5}$

7 (a) false   (b) true   (c) true   (d) true   (e) false

9 (a) no   (b) no   (c) no

Exercise Set 4.5

1 (a) 7.09   (b) 35.26   (c) 24.2   (d) 1.237
   (e) 0.98   (f) 0.44

3 (a) 0.636363 ...   (b) 0.7142857142 ...
  (c) 0.4444 ...   (d) 0.846153846 ...
  (e) 0.692307692 ...   (f) 0.882352941176470588 ...
  (g) 0.94117647058823529411 ...   (h) 0.541666 ...

5 (a)   6/17 = 0.35294117647058823529 ...
         7/17 = 0.41176470588235294117 ...
         8/17 = 0.4705882352941176470 ...
         9/17 = 0.5294117647058823529 ...
        10/17 = 0.5882352941176470588 ...
        11/17 = 0.6470588235294117647 ...
        12/17 = 0.7058823529411764705 ...
        13/17 = 0.7647058823529411764 ...
        14/17 = 0.8235294117647058823 ...
        15/17 = 0.8823529411764705882 ...
        16/17 = 0.9411764705882352941 ...

  (b) sixteen   (c) Yes. The pattern uses some sequential part of the list:
  0588235294117647058823529411764705

7 $s - 1$

9 Simply divide the particular cycle of repetition by the same number of nines.

11 (a) 1.01001000100001 ...
   (b) 0.2345678910111213141516 17 ...
   (c) 0.1141114111141111114111111 ...

13 $xyz/999$

Exercise Set 4.6

1 Use $\sqrt{2}$ as a unit, double it for $2\sqrt{2}$, triple it for $3\sqrt{2}$, and lay it off left of 0 for $-2\sqrt{2}$. All are irrational.

3 $\sqrt{5}$

5 0.281281128111 ...

CHAPTER 5   ORGANIZING MATHEMATICAL SYSTEMS

Exercise Set 5.1

1 (a) Yes. All the elements in the table belong to S.
  (b) #, □, and ☆   (c) $x \& y = y \& x$
  (d) 9
  (e) # & ☆ $\overset{?}{=}$ ☆ & #   □ & # $\overset{?}{=}$ # & □
        ☆ ≡ ☆   □ ≡ □
  (f) three pairs

3 (a) yes   (b) yes   (c) no; no   (d) no
  (e) yes   (f) no

5 (a) the hours on the clock
       { 12,1,2,3,4,5,6,7,8,9,10,11 }

(b)

$\oplus$	12	1	2	3	4	5	6	7	8	9	10	11
12	12	1	2	3	4	5	6	7	8	9	10	11
1	1	2	3	4	5	6	7	8	9	10	11	12
2	2	3	4	5	6	7	8	9	10	11	12	1
3	3	4	5	6	7	8	9	10	11	12	1	2
4	4	5	6	7	8	9	10	11	12	1	2	3
5	5	6	7	8	9	10	11	12	1	2	3	4
6	6	7	8	9	10	11	12	1	2	3	4	5
7	7	8	9	10	11	12	1	2	3	4	5	6
8	8	9	10	11	12	1	2	3	4	5	6	7
9	9	10	11	12	1	2	3	4	5	6	7	8
10	10	11	12	1	2	3	4	5	6	7	8	9
11	11	12	1	2	3	4	5	6	7	8	9	10

7 no

Exercise Set 5.2

1 (a) yes   (b) yes
 (c) $(x \# y) \# z = x \# (y \# z)$ (d) $(a \# o) \# e \overset{?}{=} a \# (o \# e)$
                                                                          $i \# e \overset{?}{=} a \# o$
                                                                          $i \equiv i$
                                                                 (Other checks are similar.)

 (e) the element e (f) yes; yes; yes
 (g) $(i \# u) \# a \overset{?}{=} i \# (u \# a)$
        $a \# a \overset{?}{=} i \# o$
        $o \neq u$

3 (a) $(c \circ b) \circ c \overset{?}{=} c \circ (b \circ c)$ (b) $(c \circ b) \circ c \overset{?}{=} c \circ (b \circ c)$
        $d \circ c \overset{?}{=} c \circ c$ $d \circ c \overset{?}{=} c \circ d$
        $b \neq a$ $b \equiv b$

5 (a) $(7 \oplus 6) \oplus 10 \overset{?}{=} 7 \oplus (6 \oplus 10)$ (b) $(9 \oplus 7) \oplus 6 \overset{?}{=} 9 \oplus (7 \oplus 6)$
        $1 \oplus 10 \overset{?}{=} 7 \oplus 4$ $4 \oplus 6 \overset{?}{=} 9 \oplus 1$
        $11 \equiv 11$ $10 \equiv 10$

7 (a) yes   (b) no

Exercise Set 5.3

1 (a) q (b) u, because $r \odot u = q$ (c) t (d) q (e) true (f) true
3 (a) $-i(+i) = -i^2 = -(-1) = +1$ (b) $+1$
 (c) -1 is the inverse of -1 because $(-1)(-1) = +1$.
      i is the inverse of $-i$ because $-i \times (i) = +1$

5 (a)

☆	1	−1
1	−1	1
−1	1	−1

(b) 1

7 No. There is no additive identity.

Exercise Set 5.4

1 (a) $4(5 + 6) = 4(11) = 44$
  $4(5 + 6) = 4(5) + 4(6) = 20 + 24 = 44$
  (b) $8(7 − 9) = 8(−2) = −16$
  $8(7 − 9) = 8(7) − 8(9) = 56 − 72 = −16$
  (c) $5(1/2 + 1/4) = 5(3/4) = 15/4$
  $5(1/2 + 1/4) = 5(1/2) + 5(1/4) = 5/2 + 5/4 = 15/4$
  (d) $1/2(3+4+5) = 1/2(12) = 6$
  $1/2(3+4+5) = 1/2(3) + 1/2(4) + 1/2(5) = 3/2 + 4/2 + 5/2 = 6$
3 (a) $1/2(3/4 + 1/4) = 1/2(1) = 1/2$
  $1/2(3/4 + 1/4) = 1/2(3/4) + 1/2(1/4) = 3/8 + 1/8 = 1/2$
  (b) $2/3 ÷ (1/3 + 2/3) = 2/3 ÷ (1) = 2/3$
  $2/3 ÷ 1/3 + 2/3 ÷ 2/3 = 2+ 1 = 3$
5 (a) $a \# (b \circ c) = (a \# b) \circ (a \# c)$ (b) $a \circ (b \# c) = (a \circ b) \# (a \circ c)$
  (c) yes
  (d) $7 \# (8 \circ 2) \overset{?}{=} (7 \# 8) \circ (7 \# 2)$
  $7 \# 8 \overset{?}{=} 8 \circ 7$
  $8 = 8$

(Others done similarly)

7 (a) $(x \# y) \# z = x \# (y \# z)$ yes
  (b) $(x \circ y) \circ z = x \circ (y \circ z)$ yes
  (c) $x \circ y$ selects x, so $(x \circ y) \circ z$ selects x. $x \circ (y \circ z)$ selects x regardless of $y \circ z$.
  (d) $(x \# y) \# z = x \# (y \# z)$
  Case 1: $x \# y = x$. Then $(x \# y) \# z = x \# z$. But $x \# (y \# z)$ selects either x or z, so it is also $x \# z$.
  Case 2: $x \# y = y$. Then $(x \# y) \# z = y \# z$. But $x \# (y \# z)$ selects either y or z, so it is also $y \# z$.

Exercise Set 5.5

1 (a) yes   (b) yes   (c) yes   (d) yes   (e) −3/4,
          7/5, and −15/13 respectively   (f) yes   (g) yes
3 yes
5

⊕	0	1	2	3	4	5
0	0	1	2	3	4	5
1	1	2	3	4	5	0
2	2	3	4	5	0	1
3	3	4	5	0	1	2
4	4	5	0	1	2	3
5	5	0	1	2	3	4

0 is the ⊕ identity; the inverse of 0 is 0, of 1 is 5, of 2 is 4, of 3 is 3, of 4 is 2, and of 5 is 1. Closure and associativity check out.

7

$\odot$	R_0	R_{60}	R_{120}	R_{180}	R_{240}	R_{300}
R_0	R_0	R_{60}	R_{120}	R_{180}	R_{240}	R_{300}
R_{60}	R_{60}	R_{120}	R_{180}	R_{240}	R_{300}	R_0
R_{120}	R_{120}	R_{180}	R_{240}	R_{300}	R_0	R_{60}
R_{180}	R_{180}	R_{240}	R_{300}	R_0	R_{60}	R_{120}
R_{240}	R_{240}	R_{300}	R_0	R_{60}	R_{120}	R_{180}
R_{300}	R_{300}	R_0	R_{60}	R_{120}	R_{180}	R_{240}

R_0 is the identity. The other group properties check out.

9 (a)

$\odot$	r	s	t	a	b	c
r	r	s	t	a	b	c
s	s	t	r	b	c	a
t	t	r	s	c	a	b
a	a	c	b	r	t	s
b	b	a	c	s	r	t
c	c	b	a	t	s	r

(b) It is identical.   (c) yes

CHAPTER 6   LOGIC AND THE STRUCTURE OF THOUGHT

Exercise Set 6.1

1 (a) $b \wedge t$ (b) $r \wedge b$ (c) $r \vee b$
  (d) $c \wedge w$ (e) $c \rightarrow w$ (f) $u \rightarrow p$
  (g) $u \rightarrow p$ (h) $m \rightarrow f$ (i) $p \rightarrow m$
  (j) $p \rightarrow m$ (k) $m \leftrightarrow p$

3 (a) Fido has fleas or else he scratches.
  (b) If Fido has fleas, then he scratches.
  (c) It is false that Fido has fleas and also scratches.
  (d) Fido doesn't have fleas, and he doesn't scratch.
  (e) It is false that Fido has fleas or Fido scratches.
  (f) Fido doesn't have fleas or else he doesn't scratch.
  (g) It is false that if Fido has fleas, then he scratches.
  (h) If Fido doesn't scratch, then he doesn't have fleas.
  (i) Either Fido has fleas and scratches or he doesn't have fleas and doesn't scratch.
  (j) If Fido has fleas, then he scratches or else he doesn't have fleas and doesn't scratch.

5 None of these remarks is clearly true or clearly false.

Exercise Set 6.2

1 (a) true   (b) true   (c) true   (d) true
  (e) false   (f) false   (g) true

3 (a) TF   (b) TFFF   (c) TFFF   (d) FFTF
  (e) TF   (f) TTTF   (g) TFFFFFFF
  (h) TFFFFFFF   (i) FFFFFFTF   (j) FTTTTTTT

5 only if all are true

Exercise Set 6.3

1 (a) true　　(b) true　　(c) false　　(d) true
　(e) true　　(f) true　　(g) false　　(h) true
　(i) true　　(j) true

3 (a) because each of the two components can have each of two truth tables
　(b) eight　　　　　　　　　(c) sixteen

5 (a) TFTFTFFF　　　　　　(b) conjunction

7 (a) FTTT　　(b) TTFT　　(c) TTTT　　(d) TTTT
　(e) FFTT

9 $H \rightarrow C$ has truth table TFTT while $C \rightarrow H$ has truth table TTFT

11 Both have TF for a truth table.

13 (a) TF　　　　(b) TF　　　　(c) FT　　　　(d) TTTF
　(e) TFFF　　　(f) TFFFFFFF　　　　　(g) TTTTTTTF
　(h) TTTFFFFF　(i) TTTTTFFF　　　　　(j) TFTT

Exercise Set 6.4

1 (a) $\sim b \rightarrow \sim a$; $a \rightarrow b$; $b \rightarrow a$
　(b) $c \rightarrow (a \lor b)$; $\sim (a \lor b) \rightarrow \sim c$; $\sim c \rightarrow \sim (a \lor b)$
　(c) $\sim q \rightarrow p$; $\sim p \rightarrow q$; $q \rightarrow \sim p$
　(d) If the flowers bloom, it is warm.
　　　If it isn't warm, then the flowers won't bloom.
　　　If the flowers don't bloom, then it is not warm.
　(e) If I am there by noon, then I did not waste time.
　　　If I waste time, then I won't be there by noon.
　　　If I am not there by noon, then I wasted time.
　(f) If it doesn't gather moss, then it rolls.
　　　If it doesn't roll, then it gathers moss.
　　　If it gathers moss, then it doesn't roll.

3 (a) $q \rightarrow p$　　　　(b) $\sim p \rightarrow \sim q$　　(c) true

5 Both have the truth table TTTF.

7 All are TTTT.

9 (a) If set A equals set B, then A is a subset of B and B is a subset of A.
　(b) It could apply to "cat" also.

Exercise Set 6.5

1 (a)

p	q	$\sim p \lor (\sim p \rightarrow q)$
T	T	F　T　F　T　T
T	F	F　T　F　T　F
F	T	T　T　T　T　T
F	F	T　T　T　F　F
		1　5　2　4　3 , and so on.

3 (a)

p	$\sim (p \rightarrow p)$
T	F　T
F	F　T
	2　1 , and so on.

5 (a) Show that $[(p \rightarrow q) \wedge (\sim q)] \rightarrow \sim p$ is a tautology.
  (b) Show that $[(p \vee q) \wedge (\sim p)] \rightarrow q$ is a tautology.
  (c) Show that $(p \wedge q) \rightarrow p$ is a tautology.
  (d) Show that $p \wedge q \rightarrow (p \wedge q)$ is a tautology.
7 (a) $[(p \rightarrow q) \wedge q] \rightarrow p$ has truth table TTFT.
  (b) The truth table is TTTF.
  (c) TTFT   (d) TTFT

CHAPTER 7   NUMBER THEORY

Exercise Set 7.1

1 (a) 77   (b) 9471   (c) 35805; 76846; 7931; 55825   (d) 465; 998; 103; 725; 13
  divides numbers of the form *abcabc* because 1001 divides them, and 13 divides
  1001. Examine the "long" multiplication of 1001 by 465.
  (e) 35805035.769 because 1001 does not divide 465,465,465.
3 (a) no; consider 21 (b) If a number is divisible by 9, then it is divisible by 3.
  (c) yes   (d) 15; 59; 121; 3; 13717
  (e) If the sum of the digits is divisible by 9, the number is divisible by 9.
5 (a) 55   (b) (i) 1275 (ii) 5050 (iii) 500500
7 (a) 225   (b) 3025
9 3, 13, 23, 33, 43, 53
11 (a) $5.12   (b) $5,368,709.12
  (c) $10,737,418.24
13 (a) 1/2, 1/20, 1/200, 1/2000, 1/20000, 1/200000
  (b) 0.5, .05, .005, .0005, .00005, .000005
  (c) 0.5555555 . . . = 5/9
15 (a)   4   9   16   25   36   49
   (b)   1   4   9   16   25   36

  (c)   3   5   7   9   11   13
  (d) arithmetic   (e) $3 + (n-1)\,2 = 1 + 2n$
  (f) $(n+1)^2 - n^2 = n^2 + 2n + 1 - n^2 = 2n + 1$
  (g) true

Exercise Set 7.2

1 2, 3, 5, 7, 11, 13, 17, 19, 23, 29, 31, 37, 41, 43, 47, 53, 59, 61, 67, 71, 73, 79, 83, 89, 97
3 because it has divisors other than 16 and 1
5 (a) composite   (b) composite
  (c) composite   (d) prime   (e) prime
7 (a) 3, 5, 7   (b) 3, 5, 7, 11, 13
  (c) 3, 5, 7, 11, 13, 17, 19, 23, 29, 31, 37 (d) 3, 5, 7, 11, 13, 17, 19
9 (a) 8,446,691,644,909,551,617   (b) 6,700,417
11 It is possible to find a gap larger than any number you have in mind, but the
  gaps do not increase consistently.
13 (a) because of the square root method for testing primeness
  (b) primes up to 13; primes up to 31

Exercise Set 7.3

1 (a) $\{2, 3, 3, 5, 7\}$ (b) $2(3)(3)(5)(7)$
3 (a) $2(3)(3)(3)(3)$ (b) $2(2)(2)(2)(2)(3)(3)$
   (c) $2(2)(2)(5)(5)(5)$ (d) $3(3)(3)(3)(3)$
   (e) 2^{10} (f) none exists   (g) $2^6 \cdot 5^6$
   (h) $3(11)(11)$
5 true
7 (a) It is prime.   (b) It can be factored uniquely into primes.
   (c) nothing   (d) It is prime.
9 (a) $1/7$ (b) $19/110$ (c) $89/123$
   (d) $1/18$ (e) $11/15$ (f) $27/37$
11 (a) 30   (b) 16   (c) 4   (d) 1
   (e) 23   (f) 17   (g) 36   (h) 56
13 (a) no   (b) yes   (c) yes   (d) no
   (e) no   (f) no
15 (a) 1, 3, 5, 15   (b) 1, 2, 3, 6, 9, 18
   (c) 1, 2, 4, 8, 16, 32
   (d) 1, 2, 3, 4, 6, 9, 8, 12, 18, 16, 24, 36, 48, 72, 144
   (e) 1, 2, 5, 4, 10, 25, 20, 50, 100
   (f) 1, 2, 3, 4, 6, 9, 12, 18, 27, 36, 54, 81, 108
17 (a) 126   (b) 63   (c) 36   (d) 420
   (e) 120   (f) 42

Exercise Set 7.4

1 (a) 59 and 61, 79 and 83   (b) no   (c) yes
3 (a) Neither; it is a false statement.
   (c) $133 - 2^1 = 131$, so $133 = 131 + 2^1$
        $135 - 2^1 = 133$;   133 is composite
        $135 - 2^2 = 131$; $135 = 131 + 2^2$
        $137 - 2^1 = 135$
        $137 - 2^2 = 133$
        $137 - 2^3 = 129$
        $137 - 2^4 = 121$
        $137 - 2^5 = 105$
        $137 - 2^6 = 73$, so $137 = 73 + 2^6$
5 (a) $15 = 5 + 5 + 5$ (b) No, $15 = 2 + 2 + 11$.
7 (a) $1001! + 2$; $1001! + 3$; $1001! + 4$; ... ; $1001! + 1001$
   (b) 7   (c) yes
9 that the set is not infinite
11 (a) 1, 2, 31, 4, 62, 8, 124, 16, 248, 496
   (b) $1 + 2 + 31 + 4 + 62 + 8 + 124 + 16 + 248 = 496$
13 33,550,336

CHAPTER 8   AN INTRODUCTION TO ALGEBRA

Exercise Set 8.1

1 (a) Two more than three times a number is eleven.
  (b) A number divided by seven is two.
  (c) Three more than twice a number is the same as four more than the number.
  (d) If a number is subtracted from seven, the result is four.
  (e) If seven is subtracted from a number, the result is four.
3 (a) $S = \{3\}$ (b) $S = \{14\}$
  (c) The solution is 1.   (d) $S = \{3\}; S = \{11\}$
5 all numbers
7 $37 = 2x + 3; S = \{17\}$.
9 90

Exercise Set 8.2

1 (a) $S = \{3\}$ (b) $S = \{\ \}$ (c) $S = \{\ \}$
  (d) $S = \{\ \}$ (e) $S = \{3\}$
3 (a) $\{(0,4), (6,0), (3,2), (9,-2), (1/2,11/3), (-6,8), \ldots\}$
  (b) no
5 $S = \{-4,+4\}$
7 (a) $x - y = 3$ (b) N
  (c) $(8,5), (9,6), (10,7)$ (d) no
9 as ordered triples $(1,2,9), (5,6,1)$ for example

Exercise Set 8.3

1 (a) $S = \{5\}$ (b) $S = \{0\}$ (c) $S = \{1/2\}$
  (d) $S = \{\pi\}$ (e) $S = \{-4\}$ (f) $S = \{1/3\}$
  (g) $S = \{7/2\}$
3 (a) All reals. The graph is conceived of as the entire number line.
  (b) The graph is conceived of as the entire number line.
  (c) no

Exercise Set 8.4

3 (a) quadrant IV   (b) quadrant III   (c) quadrant II
  (d) on the Y axis   (e) on the X axis
4 (a)   (b)   (c)

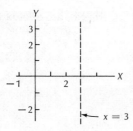   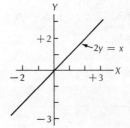   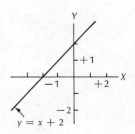

(d)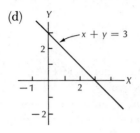

(e)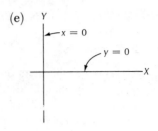

5 $x = 3; y = x; y = x + 2; x + y = 3; y = 0; x = 0$

7 (a)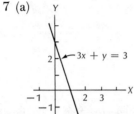

(b)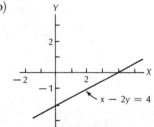

(c)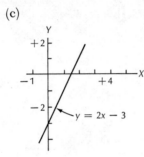

(d)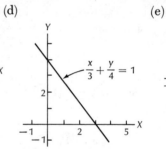

(e)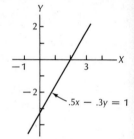

(f) same as (e)

(g)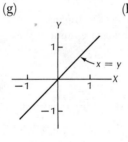

(h), (i)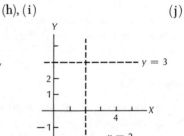

(j)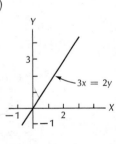

9 (a) $S = \{ (1,1), (2,1/2), (3,0) \}$ (b) No; it is a set of three points.

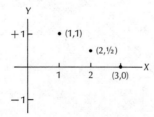

11 (a)

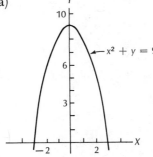

$x^2 + y = 9$

(b)

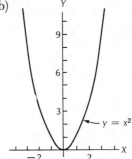

$y = x^2$

(c)

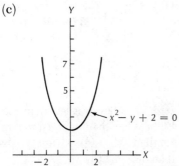

$x^2 - y + 2 = 0$

(d) graph is a circle
with center at
origin and radius 2

(e) graph is a circle, center at (0,0), radius 5

(f) graph is a parabola containing the points
(0,1), (2,0), and (4,1)

(g) graph is a parabola containing the points
(1,3), (1,−3), and (−8,0)

(h) graph is a parabola containing the points
(0,4), (0,−4), and (2,0)

(i)

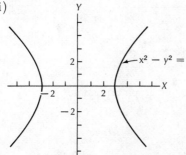

$x^2 - y^2 = 9$

This graph is called a hyperbola.
It contains points (3,0), (−3,0),
(5,4), (5,−4), (−5,4), (−5,−4),
among others.

Exercise Set 8.5

1 $\{(3,5)\}$

3 (a) $x + y = 10$ (b) $\{(6,4)\}$
       $x - y = 2$

5 (a) $S = \{(-1,4)\}$ (b) $S = \{(2,4)\}$
  (c) $S = \{(1,1)\}$ (d) $S = \{(4,0)\}$
  (e) $S = \{(1/2,1/2)\}$ (f) $S = \{(-1,\ -1/3)\}$

7 The graphs are parallel lines. There is no intersection, so the solution set is empty.

9 (a) $\begin{cases} x + y = 25 \\ 25x + 5y = 225 \end{cases}$ (5 quarters and 20 nickels)

(b) $\begin{cases} x + y = 5000 \\ .04x + .05y = 230 \end{cases}$ ($2000 at 4% and $3000 at 5%)

(c) $70 + 90 + x = y$ (third test grade is 95)

$\dfrac{70 + 90 + x}{3} = 85$

(d) 88   (e) 108; 88

(f) $\begin{cases} x + y = 100 \\ 90x + 100y = 9800 \end{cases}$ (20 lb. of 90¢ coffee and 80 lb. of $1.00 coffee)

(g) $\begin{cases} x + y = 30 \\ 2x = 3y \end{cases}$ (18 inches and 12 inches)

Exercise Set 8.6

1 (a) Two is greater than zero.   (b) Seven is less than x.   (c) Nine is greater than or equal to y.   (d) Eight is less than or equal to ten.   (e) x is not equal to four.   (f) x is not greater than two.   (g) y is not less than fifteen.   (h) The sum of x and y is greater than or equal to two.

3 (a) $x > 0$ (b) $x < 0$ (c) $x \le 0$ (d) $x \ge 0$

5 (a) true   (b) true   (c) true   (d) false
  (e) true   (f) true   (g) true   (h) true

7 (a) 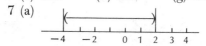   (b)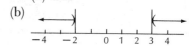

(c) 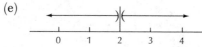   (d)

(e)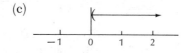

(f) same as (e)   (g) no graph is possible
9 yes

Exercise Set 8.7

1 (a) Rule 2   (b) Rule 1, Rule 2   (c) Rule 1, Rule 2   (d) Rule 1, Rule 2, Rule 4   (e) Rule 1, Rule 2, Rule 5   (f) same as given inequality, Rule 1, Rule 1, Rule 2, Rule 5   (g) Rule 1, Rule 1, Rule 3   (h) Rule 1, Rule 1, Rule 3, the square of no real number is less than −3

3 We get $3 < x < -3$, which is impossible.

5 (a) no   (b) a is less than b

7 (a) no   (b) Kathy is 17 or older.   (c) yes

Exercise Set 8.8

1 (a)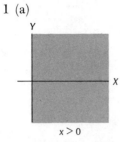

(b)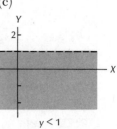

(c)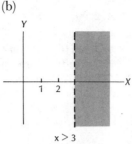

(d)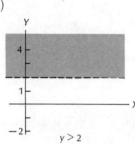

(e)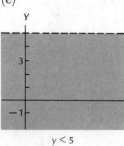

(f)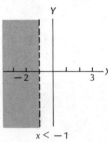

3 (a) true   (b) false   (c) false

5 (a)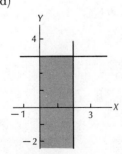

(b)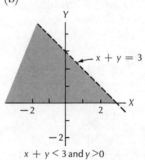

(c)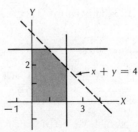

(d)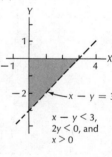

7 (a) $x < 0$ and $y < 0$
  (b) $x > 0$ and $y < 0$
  (c) $0 \leq x \leq 2$ and $0 \leq y \leq 2$
  (d) $-1 \leq x \leq 4$ and $0 \leq y \leq 3$
  (e) $0 < x < 2$ and $0 < y < 2$
  (f) $-1 < x < 4$ and $0 < y < 3$

CHAPTER 9   PROBABILITY

Exercise Set 9.1

1 (a) $\binom{52}{1} = 52$ (b) $\binom{52}{2} = 1326$

  (c) $\binom{52}{48} = \binom{52}{4} = 270{,}725$

  (d) $\binom{52}{13} = 635{,}013{,}559{,}600$

3 Choosing 2 cards from 5 is equivalent to leaving behind 3 cards.   (b) and (e);
  (a) and (f)

5 (a) 1   (b) 3   (c) 3   (d) 1
7 (a) 7   (b) 9   (c) 11   (d) n
9 (a) $9! = 9 \cdot 8(5040) = 362{,}880$
  (b) 15   (c) 105   (d) 42
11 true
13 (a) 12   (b) 21

15 $\binom{88}{3} = 109{,}736$

Exercise Set 9.2

1 (a) $_5C_2 = 10$; $_5P_2 = 20$ (b) 2
3 no
5 (a) 2,598,960   (b) 311,875,200
7 120
9 (a) 12   (b) 120   (c) 34650
11 123453 because $\frac{6!}{2}$ is greater than 5!

Exercise Set 9.3

1 (a) AA, AB, AC, BA, BB, BC, CA, CB, CC
  (b) 1/3   (c) 4/9   (d) 5/9   (e) 1/3
3 1/64
5 (a) 3/4   (b) 75
7 (a) 1/4   (b) 1/2   (c) 1/2   (d) 1/9
9 (b) 7/8   (c) 42   (d) you should

Exercise Set 9.4

1 (a) 0   (b) 1   (c) The number of red marbles is not known.
3 (a) 1/13   (b) 2/13   (c) 3/13   (d) 4/13
   (e) 4/13   (f) 1/2   (g) 7/13
5 7/8
7 (a) 13/20, 7/20   (b) If the marbles are drawn simultaneously, the answers
   are 21/190 (two white) and 78/190 (two red).   (c) 91/190
9 (a) 1/6   (b) 1/36   (c) 1/216
11 (a) 2/9   (b) 1/3   (c) 1/9

Exercise Set 9.5

1 12:1
3 5:3
5 3/14; 11/14
9 (a) 25/18¢   (b) the dealer
   (c) player loses $1.72
11 (a) 1.8   (b) 180   (c) about 2778
13 yes, provided there is no cost to play

CHAPTER 10   A LOOK AT STATISTICS

Exercise Set 10.1

1
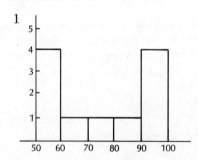

3 (a)

Interval	Frequency	Cumulative Frequency
10–39	10	10
40–69	9	19
70–99	11	30

(b)
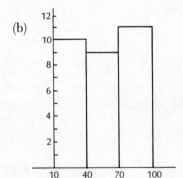

   (c) 70–99   (d) no; no   (e) no
5 mean is 73.8; median is about 75.
7 −15 degrees

Exercise Set 10.2

1 $\bar{x} = 3.5$, mean deviation $= .14$

3 1.2, $s = 1.4$

5 (a) $S = 2\sqrt{2}$; it is twice the standard deviation for the data 1, 2, 3, 4, 5.
  (b) $20\sqrt{2}$

7 $S = 12.09$

9 (a) 5.8   (b) median is 6   (c) mode is 6
  (d) range is 10
  (e)

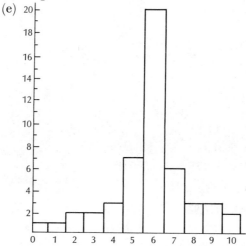

Comment: We constructed the histogram by taking each score at the mid-point of an interval. While there is no possible score of 1.5 or 2.5, this histogram does show central tendency.
  (f) mean deviation is 1.4   (g) $s = 1.9$

11 (a) greater; the range is greater
  (b) $s = 15.8$

13 $s = 0$

15 (a)

Interval	Frequency
1–5	/////// = 7
6–10	/////// = 7
11–15	// = 2
16–20	//// = 4
21–25	///// = 5
26–30	///// = 5

(b)

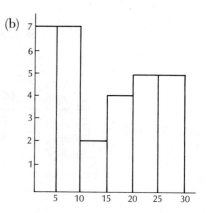

(c) 29   (d) 1–5 and 6–10
(e) 14   (f) 15
(g) 91.8   (h) 9.6
(i) 18

Exercise Set 10.3

1 (a) $\bar{x} = 80$ (b) 13   (c) 54 to 106 (d) 100%
  (e) 63%   (f) not quite (but almost)
3 (a) Between 160 and 194 there are 43 data points.   (b) 45.54%
  (c) 50 + 45.54 = 95% ought to weigh less than 194; actually 93% weigh less than 195.
5 (a) 1024; 1432; 1228   (b) 34; 34   (c) 1479   (d) 496
7 78.81% got lower and 21.19% got higher.
9 (a)

Interval	Frequency		Interval	Frequency
45—49	//		70—74	///////
50—54	/		75—79	////
55—59	////		80—84	//
60—64	//////		85—89	//
65—69	//////		90—94	//

(b) (c)

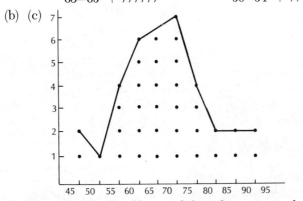

(d) Yes, especially if we disregard the 1 data point in the 50—54 interval.

CHAPTER 11   INFINITY

Exercise Set 11.1

1 (a) 32768   (b) 27,371,520,000   (c) 896,909,967,360,000   (d) no
3 (a) no   (b) $10^{(10^{100})} + 1$
5 (a) Let $S = \{1,3,5,7, \dots\}$ and choose the proper subset $T = \{3,5,7,9, \dots\}$.
  Then $1 \leftrightarrow 3$, $3 \leftrightarrow 5$, $5 \leftrightarrow 7$, and so on is a one-to-one correspondence between S and T. Thus, S is infinite.
  (b) $1 \leftrightarrow 1$, $3 \leftrightarrow 2$, $5 \leftrightarrow 3$, $7 \leftrightarrow 4$, and so on is a one-to-one correspondence between S and N. So S is equivalent to N.
7

$$\begin{array}{cccccccc} 1 & 2 & 3 & 4 & 5 & \dots & n & \dots \\ \updownarrow & \updownarrow & \updownarrow & \updownarrow & \updownarrow & & \updownarrow & \\ 1 & 8 & 27 & 64 & 125 & \dots & n^3 & \dots \end{array}$$

We show that one-to-one correspondence above. Clearly, the given set is equivalent to N, so it is infinite.

9 Let the square $ABCD$ be the base of a tetrahedron.

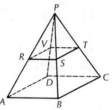

Then it is possible to pass rays through the vertex P intersecting any point of $RSTV$ (essentially a proper subset of the square). The rays also intersect the square. The two points of intersection are in one-to-one correspondence.

11 Rewrite I as $I = \{ 0, +1, -1, +2, -2, +3, -3, \dots \}$ and then establish the correspondence $0 \leftrightarrow 1$, $+1 \leftrightarrow 2$, $-1 \leftrightarrow 3$, $+2 \leftrightarrow 4$, and so on.

13

15 no

17 3×10^{74}

Exercise Set 11.2

1 (a) $n(A) = 5$; $n(B) = 3$; $n(C) = 3$
   (b) $n(A \cup B) = 7$; $n(A \cup C) = 8$
   (c) false   (d) true

3 $\{ 6,7,8,9,10,11, \dots \}$; $\{ -1,-3,-5,-7, \dots \}$; $\{ 7,14,21,28, \dots \}$

5 $T = \{ 1,2,3, \dots, 15 \}$ $S = \{ a,b,c,d,e \}$
   Pick the subset $\{ 10,11,12,13,14 \}$ from T and observe the correspondence $a \leftrightarrow 10$, $b \leftrightarrow 11$, $c \leftrightarrow 12$, $d \leftrightarrow 13$, $e \leftrightarrow 14$. So S is equivalent to a proper subset of T, but we cannot show that T is equivalent to a subset of S. Therefore the cardinality of T is greater than the cardinality of S.

7 The cardinality of N is $\aleph_0$; the cardinality of $\{ 0,-1 \}$ is 2. Form $N \cup \{ 0,-1 \}$. This is easily shown to have cardinality $\aleph_0$. So $\{ 0,-1 \} \cup \{ 1,2,3,4, \dots \}$ $= \{ 0,-1,1,2,3,4, \dots \}$ implies $2 + \aleph_0 = \aleph_0$.

9 (a) $n(A) = \aleph_0$
   (b) $n(B) = \aleph_0$
   (c) $A \cup B = \{ 1,2,3,4,5, \dots \}$
   (d) $n(A \cup B) = \aleph_0$

11 N ought to have $2^{\aleph_0}$ subsets.

13 Only sets which can be counted have cardinality $\aleph_0$.

Exercise Set 11.3

1 If space is curved, any straight line (geodesic) would be curved also.

3 no

5 It would go north toward the pole and then south away from the pole.

7 two

CHAPTER 12   THE COMPUTER AGE

Exercise Set 12.2

1 (a) prints 9   (b) prints 1,2,3,4, . . . , 49, 50.
   (c) adds the first 50 counting numbers and stores them in A
   (d) averages scores and prints result

3 (a)

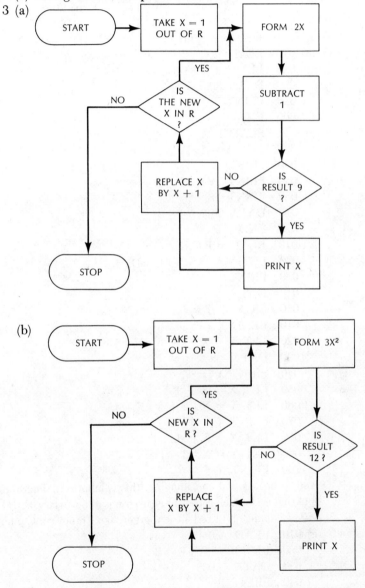

All other flowcharts are similar. The idea is to take the first number, evaluate the left side of the equation; if the value is the same as the number on the right side of the equation, print the number. Then take the next number (if it is in the set) and repeat the procedure. Notice that the charts test each of the seven numbers.

Exercise Set 12.3

1 (a) 3(4)   (b) 2(4 + 5)   (c) 3^2
   (d) $\sqrt{3}$ (e) $3^{1/2}$ (f) $3^{.5}$ (g) 7 − 2(3)

   (h) (7 − 2)3   (i) 3 + 9/12 (j) $\dfrac{3+9}{12}$

   (k) $z + 5(3 + 6)$ (l) $(z + 5)(3 + 6)$
3 (a) 18   (b) 36   (c) 14   (d) 20   (e) 32
   (f) 7   (g) 15   (h) 19   (i) 23   (j) 35
5 1.25, 40, and 50 are printed
7 (a) 010   READ L,W
       020   LET A = L∘W
       030   DATA 30.5, 5.4
       040   PRINT A
       050   END
   (b) 010   READ L,W
       020   LET A = L∘W
       030   PRINT A
       040   PRINT W
       050   PRINT L
       060   DATA 30.5, 5.4
       070   END
   (c) 010   READ A,B,C,D
       020   LET Y = (A+B+C+D)/4
       030   PRINT Y
       040   DATA _____, _____, _____, _____
       050   END
   (d) 010   READ X
       020   LET Y = X↑2 + 3∘X − 18
       030   DATA − 8
       040   PRINT Y
       050   LET X = X + 1
       060   LET Y = X↑2 + 3∘X − 18
       070   PRINT Y
       080   LET X = X + 1
       090   LET Y = X↑2 + 3∘X − 18
       100   PRINT Y
(And so on. As you can imagine, this is laborious. There are further instructions that make such repetitious programs unnecessary. See *Computing with the Basic Language* by Fred Gruenberger, San Francisco: Canfield Press, 1972.)
   (e) 010   READ A,B
       020   LET R = (A+B)↑2
       030   DATA 1,.002
       040   PRINT R
       050   END

(f) 020   READ A,R

040   LET S = A/(1−R)

060   DATA 1, .2

080   PRINT S

081   END

(g) 010   READ A,N,D

020   LET L = A + (N−1) ∘ D

030   DATA 3, 7, 4

9 010   READ A,B,C

020   READ D,E,F

030   LET X = (C∘E − B∘F)/(A∘E − B∘D)

040   DATA 1, 2, 1, −1

050   DATA 3, 4

060   PRINT X

070   END

Comment: Lines 040 and 050 could be rewritten as

055   DATA 1, 2, 1, −1, 3, 4

ADDITIONAL READINGS

Adler, Alfred. "Mathematics and Creativity." *The New Yorker Magazine*, February 19, 1972, pp. 39–45.

Beckmann, Petr. *A History of π (pi)*. Boulder, Colorado: The Golem Press, 1971, pp. 179–185. Discusses how π has been calculated to 500,000 decimal places, using computers.

Beiler, Albert H. *Recreations in the Theory of Numbers: The Queen of Mathematics Entertains*. New York: Dover Publications, 1964, pp. 239–247. About the invention of a remarkable factoring machine.

Bell, E. T. *Men of Mathematics*. New York: Simon and Schuster, 1937.

Benacerraf, P. and Putnam, H. (eds.). *Philosophy of Mathematics; Selected Readings*. Englewood Cliffs, N. J.: Prentice-Hall, 1964.

Bierce, Ambrose. "Moxon's Master," in *The Collected Writings of Ambrose Bierce*. New York: The Citadel Press, 1946, pp. 429–437. A bizarre short story about a man and a chess-playing machine.

Bishop, Morris. "Pascal." *Horizon Magazine*, vol. V, no. 3 (January 1963), pp. 94–104.

Courant, Richard and Robbins, Herbert. *What Is Mathematics?* New York: Oxford University Press, 1941.

Dantzig, Tobias. *Number: The Language of Science*. New York: Macmillan, 1949. Treats all aspects of numbers and explores "the anatomy of the infinite."

De Morgan, Augustus. *On the Study and Difficulties of Mathematics*. La Salle, Ill.: Open Court Publishing Company, 1943. A reprint of an original treatise bearing the date 1831.

Einstein, Albert. *Sidelights on Relativity*. London: Methuen, 1922. A collection of addresses given by Einstein that is remarkably easy to comprehend.

Eves, Howard. *An Introduction to the History of Mathematics*. New York: Holt, Rinehart and Winston, 1962.

Halmos, Paul R. "Innovation in Mathematics." *Scientific American*, vol. 199, no. 3 (September 1958), pp. 66–73.

Hansen, Morris H. "How to Count Better: Using Statistics to Improve the Census," in *Statistics: A Guide to the Unknown*, Judith Tanur *et al*, eds., San Francisco: Holden-Day, 1972, pp. 276–284.

Hardy, G. H. *A Mathematician's Apology*. London: Cambridge University Press, 1967, chapter 23. A section of a delightful little book dealing with the contrast between pure and applied mathematics.

Havemann, Ernest. "Wonderful Wizard of Odds." *Life*, vol. 51, no. 14 (October 6, 1961), pp. 30ff. An article about Ziv Mayer, a famous race track bettor.

Helm, E. Eugene. "The Vibrating String of the Pythagoreans." *Scientific American*, vol. 217, no. 6 (December 1967), pp. 93–103.

Hiller, Lejaren A. "Computer Music," in *Computers and Computation* (Readings from *Scientific American*). San Francisco: W. H. Freeman, 1971, pp. 113–122.

Kemeny, John G. "Machines as Extensions of Human Brains," in *Random Essays on Mathematics, Education and Computers*. Englewood Cliffs, N. J.: Prentice-Hall, 1964, pp. 123–127.

Kemeny, John G. *A Philosopher Looks at Science*. Princeton, N. J.: D. Van Nostrand, 1959.

Kline, Morris. *Mathematical Thought from Ancient to Modern Times*. New York: Oxford University Press, 1972.

Lieber, Lillian R. *Mits, Wits, and Logic*, third edition. New York: W. W. Norton, 1960.

Morse, Marston. "Mathematics and the Arts." *The Yale Review*, vol. XL, no. 4 (June 1951), pp. 604–612.

Mueller, Robert E. "Idols of Computer Art." *Art in America*, May-June 1972, pp. 68–73.

O'Brien, Katharine. "Hair." *Mathematics Magazine*, vol. 47, no. 3 (May 1974), p. 149. An amusing poem about the hair styles of famous mathematicians.

Poincare, Henri. "Mathematical Discovery," in *Science and Method* (translated by Francis Maitland). New York: Dover Publications, pp. 46–63.

Russell, Bertrand. "The Study of Mathematics," in *Mysticism and Logic and other Essays*. London: Longmans, Green, 1918 (New York: Barnes & Noble).

Salomon, Louis B. "Univac to Univac (sotto voce)." *Harper's Magazine*, vol. 216, no. 1294 (March 1958) pp. 37–38. A tongue-in-cheek poem, a conversation between two computers.

Synge, J. L. *Science: Sense and Nonsense*. New York: W. W. Norton, 1950, pp. 26–30. An imaginary conversation between Euclid and a twelve year old boy.

Whitehead, Alfred North. "Mathematics as an Element in the History of Thought," in *Science and the Modern World*. New York: Macmillan, 1941, chapter 2. An essay from Whitehead's Lowell Lectures of 1925.

INDEX